Fundamental Structural Analysis

W. J. Spencer

Fundamental Structural Analysis

Springer Science+Business Media, LLC

Originally published by Springer-Verlag New York Inc. in 1998
Softcover reprint of the hardcover 1st edition 1998

First published 1988

Published by
MACMILLAN EDUCATION LTD
Houndmills, Basingstoke, Hampshire RG21 2XS and London
Companies and representatives throughout the world

Sole distributors in the USA and Canada

ISBN 978-1-4757-2008-2 ISBN 978-1-4757-2006-8 (eBook)
DOI 10.1007/978-1-4757-2006-8

Contents

List of Examples

Preface

Significant changes have occurred in the approach to structural analysis over the last twenty years. These changes have been brought about by a more general understanding of the nature of the problem and the development of the digital computer. Almost all structural engineering offices throughout the world would now have access to some form of digital computer, ranging from hand-held programmable calculators through to the largest machines available. Powerful microcomputers are also widely available and many engineers and students have personal computers as a general aid to their work. Problems in structural analysis have now been formulated in such a way that the solution is available through the use of the computer, largely by what is known as matrix methods of structural analysis. It is interesting to note that such methods do not put forward new theories in structural analysis, rather they are a restatement of classical theory in a manner that can be directly related to the computer.

This book begins with the premise that most structural analysis will be done on a computer. This is not to say that a fundamental understanding of structural behaviour is not presented or that only computer-based techniques are given. Indeed, the reverse is true. Understanding structural behaviour is an underlying theme and many solution techniques suitable for hand computation, such as moment distribution, are retained. The most widely used method of computer-based structural analysis is the matrix stiffness method. For this reason, all of the fundamental concepts of structures and structural behaviour are presented against the background of the matrix stiffness method. The result is that the student is naturally introduced to the use of the computer in structural analysis, and neither matrix methods nor the computer are treated as an addendum.

Matrix algebra is now well taught in undergraduate mathematics courses and it is assumed that the reader is well acquainted with the subject.

In many instances the solution techniques require the manipulation of matrices and the solution of systems of simultaneous linear equations. These are the operations that the digital computer can most readily handle and they are operations which are built into computer application programs in structural engineering. For the student, however, it is important that these operations are understood, so that it is desirable to have a form of matrix manipulation computer program available. Many programmable pocket calculators currently provide for such operations and there is no doubt that the capacity and speed with which these machines can carry out these tasks will increase with further developments. Some computer languages, notably some versions of BASIC, provide for general matrix manipulation, and scientific library subroutines for handling matrices are provided with other languages such as Fortran. A third possibility is to provide a computer program in the form of a problem-orientated language, with a command structure directly aimed at facilitating the manipulation of matrices. Such a computer program, known as MATOP and developed by the author, is presented as an appendix to the text and used with illustrative examples throughout. The program is not unique and other such programs have been widely available for a number of years.

The text is seen as a first course in structural mechanics or the theory of structures, although it is assumed that students will have done a first course in the more general field of applied mechanics including simple beam theory and stress analysis. The material is probably more than can be covered in two semesters, and indeed it has been delivered over three semesters. The first two chapters outline the fundamental principles and introduce students to the nature of structures and the structural analysis problem. A detailed study of equilibrium and statical and kinematic determinacy is presented in chapter 2. In chapter 3, the foundations of the matrix stiffness method are presented and the ideas of element and structure stiffness matrices are developed. The classical slope–deflection equations are developed from simple beam theory in this chapter, and presented in matrix notation to give the general beam element stiffness matrix.

The matrix stiffness method is further developed in chapter 4, where it is applied to continuous beams and rectangular frames. At this stage coordinate transformation is not introduced and axial deformation of the element is ignored. The approach leads to some powerful applications where the analysis results can be obtained quite rapidly, particularly with the use of the direct stiffness method. It is shown in many instances that the solution is reduced to one of handling matrices of a size that can be adequately dealt with on a pocket calculator.

The moment distribution method has been retained as a useful hand method of analysis and this is detailed in chapter 5, with applications to beams and rectangular frames. The work is closely related to that of chapter 4 and the moment distribution method is shown as a logical variation of the matrix stiffness method. Chapter 6 returns to the matrix stiffness method to introduce the general stiffness method and coordinate transformation. A wider range of structures is now considered, including composite structures where elements of different types are introduced into the one structure.

A fundamental study of structural analysis must include a reference to the principle of virtual work which is presented in chapter 7. Both the principle of virtual displacements and the principle of virtual forces are considered. The principle of virtual forces, particularly with regard to expressions for the deflection of structures, leads logically into the flexibility method of analysis presented in chapter 8. This provides an alternative approach to the stiffness method and gives a balance to the overall study.

The author is convinced that the general use of computer programs for structural analysis makes demands for greater, rather than less, awareness and understanding of structural behaviour on the part of users. Structural computations must still be checked, results must still be interpreted and engineering judgement must still be exercised. To facilitate this, a chapter on approximate methods of analysis is included (chapter 9). It is presented at this stage since it is felt that approximate methods can only be introduced against a background of general knowledge of structural behaviour.

In a final chapter, some general guidance to computer application programs in structural analysis is presented. Some aspects of modelling of structures are also discussed. Numerous examples are given throughout the text and a common thread is achieved through the use of the program MATOP, details of which are given in an appendix with a program listing in Fortran 77. Much of the data presented throughout the text is collected together in another appendix as a 'Structural Mechanics Students' Handbook'. The significant data here is a collective statement of the element stiffness matrices for various element types.

W. J. Spencer

Chapter 1
Introduction to Structural Engineering

The analysis of structures has long been a subject of enquiry by intellectual man. Attempts to determine the nature of forces within structures has kept pace with man's determination to build. Serious study commenced in the 16th and 17th Centuries with scholars such as Leonardo Da Vinci and Galileo Galilei. Rapid progress was made during the 18th and 19th Centuries, and particularly in the Industrial Revolution, when many classical theories of structural behaviour were first put forward. More recently, the advent of the digital computer has led to a re-appraisal of the theory of structures. It is now expected that computations associated with the analysis of structures will be carried out by computer.

Engineers conceive a structure in the form in which it will be built, however the analysis must be based on a mathematical model which approximates to the behaviour of the structure. As will be seen, part of the art of structural engineering is to model the structure accurately in this mathematical sense. It is more convenient though, in an educational sense, to start with an understanding of mathematical models representing certain structures. Simple line diagrams can be used to represent beams, columns and ties, and these are at once mathematical models.

1.1 THE NATURE OF STRUCTURES

A structure may be regarded as a number of components, referred to as elements, connected together to provide for the transmission of forces. The forces arise from loads on the structure and the elements are designed to transmit these forces to the foundations. In addition, a structure will have a particular form to enable it to perform a useful function such as providing an enclosed space. Broadly speaking, this is the function of the structure of a building.

Figure 1.1 is a line diagram of a structure made up of pin-connected elements. In response to the loads applied, the elements develop axial forces which are transmitted along the elements to the support points at A and B, where the reactions develop. At the reactions, the structure exerts forces on the foundations while the foundations exert balancing forces back on the structure.

The fundamental objective of structural analysis is to determine the response of the structure to the application of loads. As such, this involves consideration of the loads, materials and the geometry and the form of the structure. The response of the structure may be measured in many ways. Clearly one of these is the resulting deflections of the structure which can be measured or calculated at discrete points. However, unless the structure is a mechanism, the deflections are simply the aggregate effect of internal deformations of the elements which cause strain and stress within the element. The conditions of stress, strain and deflection are all inter-related according to appropriate laws of mechanics. The response, or behaviour, of the structure must meet certain minimum requirements for the structure to be considered satisfactory. Deflections must be confined to reasonable limits otherwise the structure may not be able to perform its intended function. The stresses must be similarly limited to values which will not cause failure within the elements or connections, perhaps leading to the collapse of the structure.

Methods of analysis are frequently based on numerical techniques which require the structure to be modelled as a series of discrete elements that are connected together at node points. Since the behaviour of the structure is the aggregated effect of the behaviour of the element, a study of the behaviour of the element is particularly important. Figure 1.2(a) shows a typical portal frame commonly used in industrial buildings. The action or behaviour of the structure is considered to apply in the plane of the frame so that the problem is a two-dimensional one. Further, the

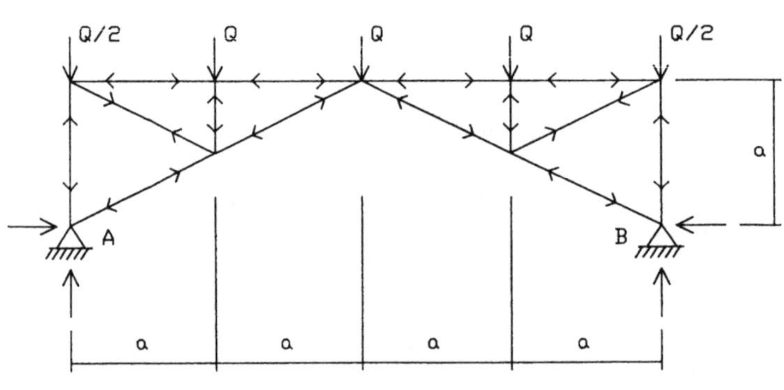

Figure 1.1 *Force transmission in a structure.*

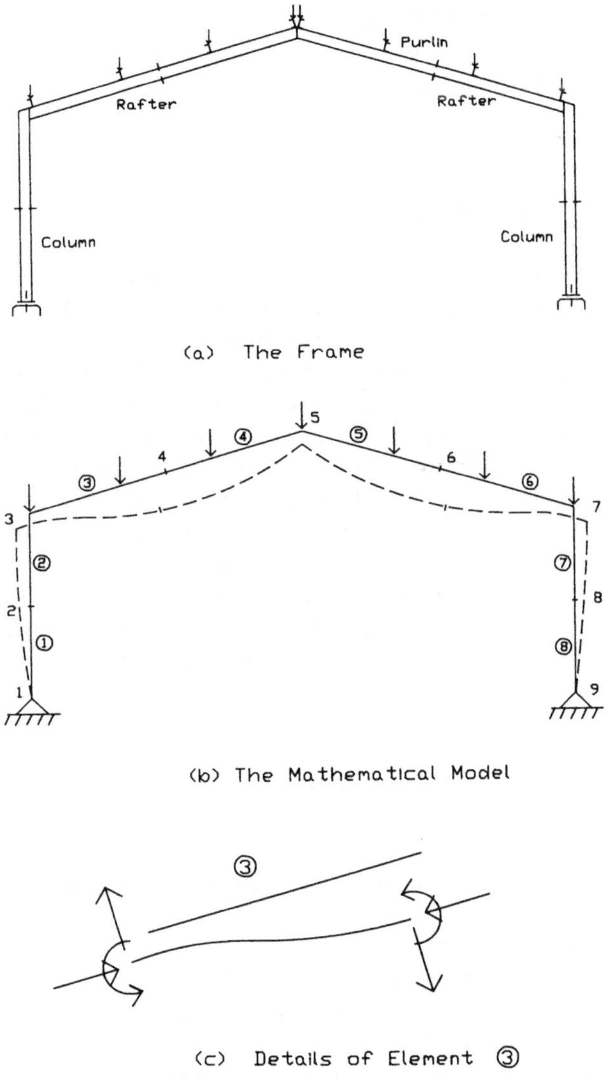

Figure 1.2 Modelling of a portal frame.

mathematical model can be presented as a series of one-dimensional elements, or line elements, since the actions on such an element can later be used to determine the stresses over the depth of the section.

In response to the loads applied, the structure may deform after the pattern shown in figure 1.2(b), where the structure is also shown as having nine nodes and eight elements, although more or fewer elements and nodes could be used in the model. The deformation causes internal stresses, the resultants of which can be shown on a free body diagram once the element is isolated, as is the case for element three as shown in figure 1.2(c). In many instances, a study of structural behaviour can begin with a study of the behaviour of such an element under the action of the stress resultants shown.

1.2 EQUILIBRIUM AND COMPATIBILITY

In a satisfactory response of a structure to the loads applied to it, the structure develops reactive forces which are in equilibrium with the loads. The conditions of equilibrium must be satisfied for the structure as a whole and for each element considered as a free body. It may be noted from figure 1.2 that each element must deform in such a way that, following deformation, the assembly of elements must conform to the continuous nature of the structure. This is, in effect, a description of compatibility. Compatibility can be described in a number of ways but the essential feature is that the structure, in satisfying compatibility, remains in one piece after deformation.

It is possible then to set down certain equations of equilibrium and equations of compatibility to be used in the analysis of structures. This will be presented in more detail in subsequent chapters. It is sufficient to note at this stage that the structure must, in general, satisfy the conditions of equilibrium and compatibility.

1.3 STRESS–STRAIN RELATIONSHIPS AND BOUNDARY CONDITIONS

It has already been suggested that the deformation of the structure sets up strains and related internal stresses within the elements. Stress is related to strain through a stress–strain law which is a function of the type of material and the nature of the strain. The best known stress–strain law is that which defines linear elastic behaviour. In this case, stress is proportional to strain and the constant of proportionality is Young's Modulus, E. There are other stress–strain laws defining a wide range of behaviour but it should be appreciated that all stress–strain laws are approximations.

At the boundaries of the element, where the elements are interconnected through the nodes, the internal stresses may be summed to give the resultant forces such as those shown in figure 1.2(c). Such forces act on the node through all the elements connected at the node and keep it in equilibrium with any external loads applied there. If the node happens to be a boundary node, then the external actions there are the reactions. Boundary nodes are important in the overall solution to the problem since they provide certain conditions which must be met. For example, there must be sufficient restraint on the structure as a whole to meet the conditions of equilibrium, and the displacements at these restraints must be zero, or at least some prescribed value.

In summary, it can be seen that the solution to a problem in structural analysis, where the full behaviour of the structure is investigated, involves a consideration of equilibrium, compatibility, stress-strain relationships and the boundary conditions.

1.4 STRUCTURAL ANALYSIS—AN ILLUSTRATIVE EXAMPLE

Figure 1.3(a) shows two rods of a linear elastic material suspended from two supports and connected together to provide support for a vertical load. In this case the structure consists of two elements and three nodes, and node 2 can displace, because of elongation of the bars, with translations in the x and y directions shown. However, because of the symmetry of the structure and the load, the displacement at node 2 will be restricted to y translation only.

The analysis can commence with a consideration of the behaviour of element ① isolated from the structure and shown as a free body in figure 1.3(b). The deformation of the element can be considered in two parts. Namely, rigid body rotation followed by an axial extension e_1, so that end 2 moves to meet the requirement that it should descend vertically by the

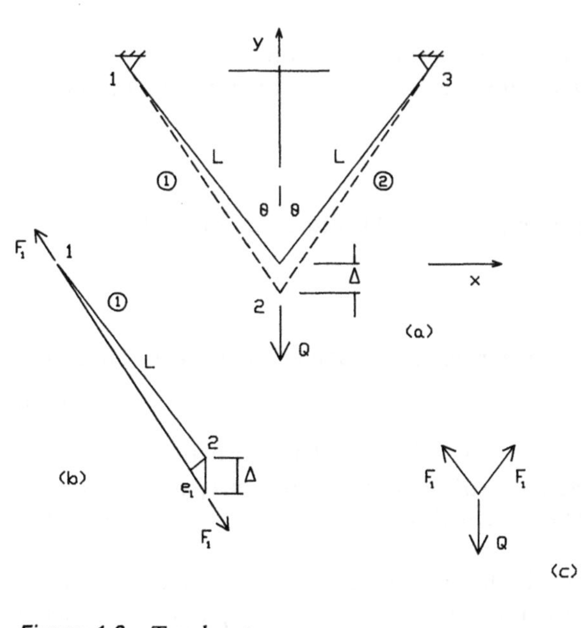

Figure 1.3 *Two bar truss.*

amount Δ as shown in figure 1.3(a). It is assumed that the displacements are small so that the element still has the same inclination to the vertical. This enables a simple geometric relationship to be found between e_1 and Δ, so that

$$e_1 = (\cos\theta)\Delta \tag{1.1}$$

Equation (1.1) is an expression of compatibility since it ensures that the element will deform to meet the overall displacement requirements of the structure. It is also clear that element ② will undergo similar deformation.

The elongation of element ① results in internal stress which gives rise to the stress resultant F_1. The linear-elastic stress–strain law then gives

$$F_1 = \frac{EA}{L} e_1 \tag{1.2}$$

where E is Young's Modulus for the material and A is the cross-sectional area of the element. The same force, F_1, obviously exists in element ② as well.

A free body diagram of node 2 is shown in figure 1.3(c). The element forces react on the node so that equilibrium of the node is maintained. Recalling that the displacements are small, the geometry of the forces on the node is taken as the geometry of the undeformed structure. On this basis the following equilibrium equation can be written:

$$2F_1 \cos\theta = Q \tag{1.3}$$

While the solution to equation (1.3) is immediately obvious for given values of Q and θ, it is instructive to substitute for F_1 in equation (1.3) to give

$$2\frac{EA}{L} e_1 \cos\theta = Q \tag{1.4}$$

Then, substituting for e_1 from equation (1.1) into equation (1.4) gives

$$2\frac{EA}{L} \cos^2\theta\Delta = Q \tag{1.5}$$

The term $2(EA/L)\cos^2\theta$ may be written as k, so that equation (1.5) becomes

$$k\Delta = Q \tag{1.6}$$

where k is now an expression of the stiffness of the structure. (It should be noted that it is not the general stiffness of the structure, since it is valid only when the structure carries vertical load.) This result is summarised in the graphs of figure 1.4.

A feature of the analysis is that it has been assumed that the change in geometry of the system was not significant. This is a qualitative concept that is expanded on in section 1.5 and it must be treated with some care,

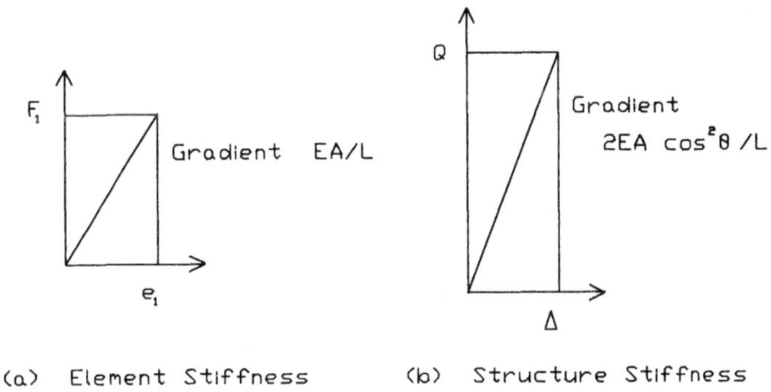

Figure 1.4 Structure and element stiffness.

although in the majority of cases it is clear that the displacements do not cause a significant change in geometry. From equation (1.5), it is seen that as θ approaches 90, the stiffness of the system approaches zero, suggesting that two horizontal bars pin-connected together cannot carry a vertical load. Rigid body statics, where no deformations or change in geometry are admitted, supports this view but in reality the system will deflect, allowing the elements to carry the load. The analysis under these circumstances requires the change in geometry to be considered and the problem becomes non-linear because of this effect.

1.5 LINEARITY, STABILITY AND LOADING

Structural systems are frequently assumed to be linear elastic systems. However, structures may behave both in a non-linear manner and inelastically. The characteristics of structural behaviour can be summarised by the series of load–deflection curves of figure 1.5. As shown in the graphs of figure 1.5, a structure is said to behave elastically when the loading path is retraced during unloading. The graphs of figures 1.5(e) and 1.5(f) can be considered as special cases of linear inelastic behaviour. These are important since they represent the fundamental basis of the plastic theory of structures which concentrates on the behaviour of structures once the elastic limit is reached. The characteristics of figure 1.5 may apply to either an element of a structure or the structure as a whole. Non-linear behaviour may be due to the material properties (material non-linearity), or to the geometry of the system (geometric non-linearity).

For structures that behave in a linear elastic manner, the principle of superposition can be applied. This principle is frequently used to advantage in the analysis of structures, and it simply refers to the fact that a superposition of effects from several load systems is equivalent to the effects

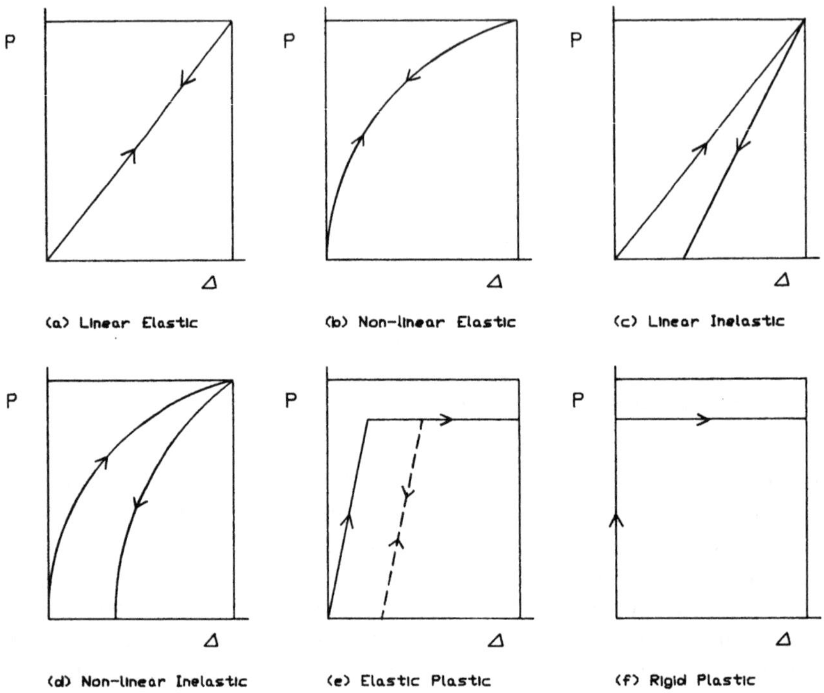

Figure 1.5 *Chararacteristics of structural behaviour.*

caused by all the load systems acting simultaneously. The effects refer to any condition at a given point such as a reaction, an internal action or a deflection.

1.5.1 Geometric Non-linearity

In general, the changes in geometry due to the loads applied to a structure are not significant and, provided the material is linear elastic, this leads to the linear elastic behaviour of the structure. The neglect of changes of geometry is not so difficult to accept when it is appreciated that in most structures the deflections are very small when compared with the dimensions of the structures. However the assumption is not always appropriate and this can be illustrated by two classical cases, one involving a tension element, the other involving a compression element.

The cable stays of a guyed mast, for instance, are clearly tension elements with some initial geometry. Under wind load, the mast will sway and the cable geometry will change. The resisting force developed by the cables due to the cable tension can be shown to be a non-linear function of the mast displacement. In a relative sense, the change in geometry of such a structure is significant and it cannot be ignored.

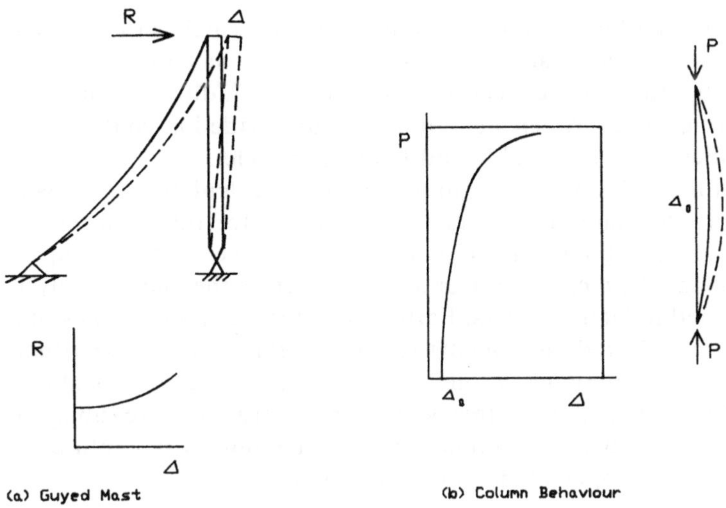

Figure 1.6 Examples of geometric non-linearity.

On the other hand, a column with some initial lateral deflection at mid height will exhibit non-linear behaviour as the column load is increased. Usually, the non-linearity is not significant at normal design loads, but it becomes highly pronounced as the load approaches the critical elastic load for the column. In spite of this, structures which include compression elements may still be considered to be linear elastic systems. This is due to the fact that it is assumed that the compression elements will be designed in such a way that the element load is confined to that part of the load-deflection curve which is essentially linear. These two non-linear effects are illustrated in figures 1.6(a) and 1.6(b) where it can be noted that the effective stiffness of the cable actually increases with load, while the stiffness of the column decreases.

1.5.2 The Stability of Structures

A simple tension element can be loaded essentially up to the yield stress of the material without any undue effect on the structure. On the other hand, compression elements are limited not only by the yield stress, but also by their tendency to buckle elastically. This introduces the question of the stability of columns and of structures generally. The Euler buckling load of a column under various end conditions is widely known and it is well treated in a number of texts on applied mechanics.

Mathematically, the Euler load, or critical elastic load, represents a bifurcation point in the load–displacement curve. That is to say, there are two equilibrium states beyond the critical load, one of which is the buckled condition. Just as a column acting as a single element has a critical load, so too does a structure, as a collection of elements including compression ones, have a critical load. An element of a structure loaded in compression beyond the critical load is unstable and liable to buckle.

A study of the stability of structures is aimed at calculating the elastic critical load and deducing appropriate design loads for the compression elements, to ensure that buckling does not occur. This is generally a complex procedure although the techniques can be built up from the matrix analysis methods presented in later chapters. Fortunately, the stability analysis of a structure can be considered subsequent to the linear elastic analysis. Further, in many cases Codes of Practice offer sufficient guidance for a stability analysis not to be necessary. Nevertheless, important structures are subjected to stability analysis and the computational effort required is continually being reduced by developments in computer applications.

1.5.3 Loads on Structures

Structural analysis is simply one part of the design process for a structural engineering project. Throughout this text, the analytical starting point is the mathematical model of the structure with nominated loads acting on it. The design of structures is much broader than that and includes such things as the assessment of loads and proposals for the form that the structure should take. Loads arise from the materials of which the structure is built and from the function and use of the structure. Many different types of load are readily identified and they include dead load, live load, and wind load as well as loads due to other natural effects such as temperature, earthquakes, snow and ice. A full discussion on loads is beyond the scope of this text although it must be treated in the overall design context. In many cases, loads can be assumed to be static; that is, they are simply applied to the structure and do not vary with time. Even the effect of wind forces acting on a structure is generally approximated by a set of static loads. For this reason, routine structural analysis is often referred to as static analysis of a structure.

Of course loads do vary with time and if the time variation is significant compared with certain dynamic characteristics of the structure, then the loads should be regarded as dynamic. Broadly speaking, dynamic loads cause a variation in the structures behaviour that oscillates about a mean position given by the static analysis. Such behaviour can introduce vibration and fatigue problems and it introduces the topic of the dynamic analysis of structures. As with the techniques for the stability analysis of structures, the procedures for the dynamic analysis of structures can be developed from the matrix methods of static analysis.

1.6 SCOPE OF THE BOOK—FUNDAMENTAL ASSUMPTIONS

Some reference to the scope of this text has already been made in the preface. However it is pertinent to make some further reference here and, in particular, to point out some underlying assumptions which provide limits to the material subsequently presented.

The text has been written as a first course in structural mechanics for students undertaking civil engineering or structural engineering degree courses. It is assumed that students will have done an introductory course in the more general field of applied mechanics, including an introduction to simple beam theory and beam stress analysis. Some familiarity with drawing bending moment and shear force diagrams for beams is also assumed, along with fundamental notions of equilibrium.

There are some underlying assumptions of structural analysis which apply to all of the material presented in the text. On occasions, particularly when first introduced, these assumptions are restated. However it is expected that the assumptions as stated here will be understood to apply throughout. The assumptions are:

(a) that the structures behave in a linear elastic manner; and
(b) that small deflection theory applies to the structures under analysis.

In presenting a fundamental text on structural analysis, it is appropriate to restrict the material to that which is applicable to linear elastic structures. Most of the common types of structure under loads within their serviceability limits act in a manner that is approximately linear elastic. In any event, a fundamental understanding of linear elastic behaviour is essential before other types of behaviour are studied.

1.6.1 Sign Conventions

In introducing students to beam behaviour and bending moment diagrams, it is usual to introduce a sign convention based on a rigorous mathematical approach using the first quadrant right-hand set of cartesian axes. This leads logically to the notion of positive bending moment being associated with positive curvature, or more simply, a sagging beam. Such moments are then often plotted with positive ordinates above a datum line in the conventional manner of any graph, which after all is what a bending moment diagram is.

However it is a widely held convention that the ordinates of the bending moment diagram should be plotted off the tension face of a line diagram of the structure. Where bending moment diagrams are shown in this text, that is the convention that has been followed. Beyond that it is not necessary to indicate whether the bending moment value is positive or negative. Although the concept of positive and negative bending can still be applied to continuous beams, it becomes rather meaningless for frames. Bending moments may equally well be plotted off the compression face of an element; the important point is simply that the convention be stated and adhered to.

The sign conventions for other quantities such as displacements and internal actions and loads are defined as they arise. Further comment on the question of sign convention is also made in the introductions given to both parts of the 'Structural Mechanics Students' Handbook' presented as appendix B.

FURTHER READING

Norris, C. H. and Wilbur, J. S., *Elementary Structural Analysis*, 2nd edn, McGraw-Hill, New York, 1960.

Parnell, J. P. M., *An Illustrated History of Civil Engineering*, Thames and Hudson, London, 1964.

Straub, H., *A History of Civil Engineering*, Basel, 1949.

White, R. N., Gergely, P. and Sexsmith, R. G., *Structural Engineering*, Combined edn, Vols 1 and 2, Wiley, New York, 1976.

Chapter 2
Equilibrium Analysis and Determinacy of Structures

The equilibrium of forces throughout a structure at rest represents an important basis for the analysis of those forces. Provided the structure stays at rest as the loads are applied, the structure can be described as being in a state of static equilibrium. There are conditions of dynamic equilibrium relating to bodies in motion, including structures, but this text is concerned only with statics—that is, the interaction of bodies at rest.

The early Greek mathematicians developed the important principles of statics which underlie the behaviour of all structures. A significant contribution was made to the theory of structures in the 18th Century by the French engineer C. A. Coulomb, when he clearly stated for the first time the conditions of equilibrium of forces acting on a beam—conditions which are the basis of shear force and bending moment diagrams.

2.1 THE EQUATIONS OF EQUILIBRIUM

Although most structures are elastic bodies, the deformations under load are considered to be small so that changes in geometry are generally ignored. For this reason, when considering equilibrium, the structure is regarded as a rigid body and the operation is often described as one of applying rigid body statics. Any system of forces and moments acting on a rigid body can be resolved into components acting along a set of Cartesian axes. This is illustrated in figure 2.1 where an arbitrary set of actions on a rigid body in three-dimensional space is shown. A typical force F_i has components as shown in figure 2.1(a), which gives rise to the equivalent actions at the axes as shown in figure 2.1(b). The resultant forces acting on the body of figure 2.1(c) are shown with respect to the coordinate system, the origin and orientation of which is quite arbitrary. For the body to be in equilibrium

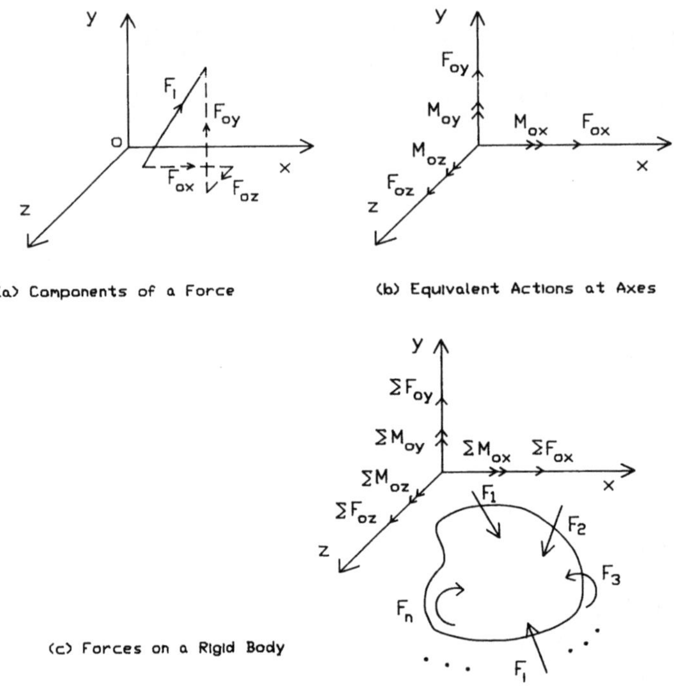

Figure 2.1 *Generalised forces in 3-D space.*

there must be no resultant force acting, and the equations of equilibrium for a rigid body in three-dimensional space are then given as

$$\begin{aligned} &\sum F_x = 0; \quad \sum F_y = 0; \quad \sum F_z = 0 \\ &\sum M_{ox} = 0; \quad \sum M_{oy} = 0; \quad \sum M_{oz} = 0 \end{aligned} \tag{2.1}$$

2.1.1 Moment of a Force about an Axis

The resolution of any force F into components parallel to a set of Cartesian axes is well known as a function of its direction cosines. Not so well known, however, is the moment of any arbitrary force F about an axis. This may be defined as the product of the resolute of the force on a plane normal to the axis and the perpendicular distance between the line of action of the resolute and the intersection of the plane and the axis. It follows that a force will have no moment about an axis if the force is parallel to the axis or intersects it. It is necessary to understand the moment of a force with respect to an axis to apply the moment equations of equation (2.1) successfully to three-dimensional structures, as will be seen.

Frequently it is sufficient to work in two-dimensional space, nominating a suitable origin and x and y coordinates. Under these circumstances, the equations of equilibrium become

$$\sum F_x = 0; \qquad \sum F_y = 0; \quad \text{and} \quad \sum M_{oz} = 0 \tag{2.2}$$

The last of these equations is still strictly the sum of the moments about an axis oz taken normal to the x–y plane from the origin. However, since this axis is not usually shown, the equation is often written simply as $\sum M = 0$. The moment of a force about a point in two-dimensional statics, which may be defined as the product of the force and its perpendicular distance from the point, should now be seen as a special case of the moment of a force about an axis.

2.1.2 Sufficiency of Restraint

For a structure acting as a rigid body under load to be in static equilibrium, a sufficient number of restraining forces or reactions must develop. A three-dimensional system requires a minimum of six independent reaction components, since any arbitrary applied load may result in a force component in any of three directions or a moment about any of the three axes. For a structure in two-dimensional space, the necessary minimum number of restraints is three. These are necessary, but not sufficient, conditions however, since certain conditions also apply to the arrangement of the reactions, as will be shown.

Figure 2.2(a) shows a two-dimensional rigid body and serves to introduce the notation for support conditions. At A, the body is attached to the foundation by a roller support, capable of developing a reaction, x_a, only in the direction normal to the direction in which the roller is free to move. At B, the connection to the foundation is through a pin connection, capable of developing a reaction in any direction, which gives rise to two independent components expressed as X_b and Y_b as shown in the free body diagram of figure 2.2(b). A further type of restraint, which is not illustrated, can be defined as a fixed or clamped support where the body is built into the foundation so that it cannot rotate at that point. The fixed support can develop three independent reaction components since a moment restraint can develop in addition to the reactive forces in the x and y directions.

Applying the equations of equilibrium as expressed by equation (2.2) to figure 2.2(b) gives

$$\sum F_x = 0: \qquad X_a + X_b \qquad = 0 \tag{i}$$

$$\sum F_y = 0: \qquad -Q + Y_b \qquad = 0 \tag{ii}$$

$$\sum M = 0: \quad -X_a \frac{L}{2} - Q\frac{L}{2} + Y_b L = 0 \tag{iii}$$

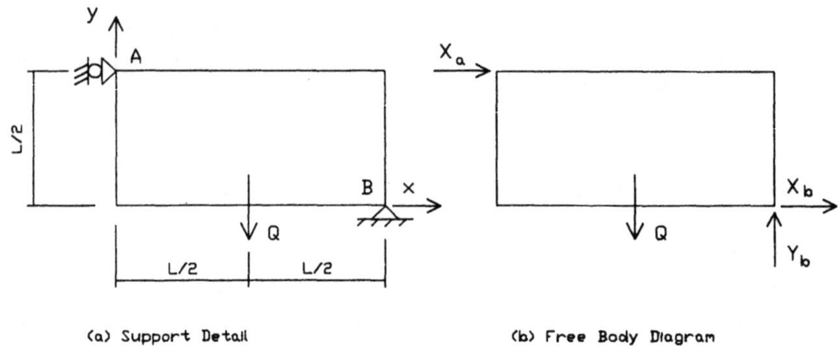

Figure 2.2 *Reactions on a 2-D body.*

From equation (ii):

$$Y_b = Q$$

Substituting into equation (iii):

$$-X_a\frac{L}{2} = -Q\frac{L}{2}$$

$$X_a = Q$$

and using equation (i):

$$X_b = Q$$

It should be evident that a consistent sign convention has been used in applying the equations. Moments have been taken as acting positive in an anticlockwise sense about the implied o*z* axis, consistent with the right-hand grip rule often used. Of course, any other convention could equally well have been used.

The more general equations of equilibrium given by equation (2.1) are applicable to the three-dimensional rigid body, in this case a space truss, of figure 2.3(a). The structure is attached to the foundations at A, B and C. The connection at A is capable of developing a reaction in any direction and, as the equivalent to a pin connection in two-dimensional statics, may be described as a spherical pin connection. It is more convenient, though, to represent such a connection by its three independent components shown as rigid pin-ended links as in figure 2.3(a). To satisfy equilibrium with respect to such links alone, the reaction component in the link must act along its direction. On a similar basis, the rigid links have been introduced at B and C to define the reactions there. A similar notation to define reactions can be used with two-dimensional statics.

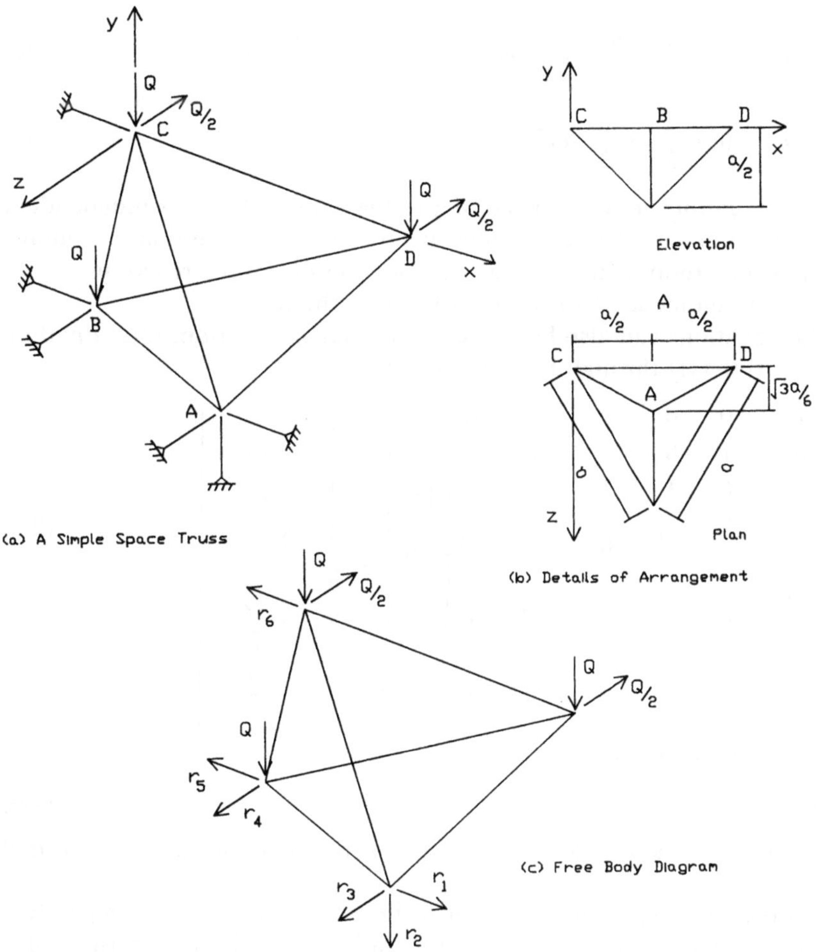

Figure 2.3 Reactions on a 3-D body.

For the equilibrium analysis to proceed it is helpful to assume that all of the reactions are acting as tension positive. This leads to the free body diagram of figure 2.3(c) where all of the forces acting on the structure are shown.

Applying the equations of equilibrium as expressed by equation (2.1) gives

$$\sum F_x = 0: r_1 - r_5 - r_6 = 0 \qquad \text{(i)}$$

$$\sum F_y = 0: -r_2 - 3Q = 0 \qquad \text{(ii)}$$

$$\sum F_z = 0: r_3 + r_4 - Q = 0 \qquad \text{(iii)}$$

$$\sum M_{ox} = 0: r_2 \frac{\sqrt{(3)}a}{6} - r_3 \frac{a}{2} + Q \frac{\sqrt{(3)}a}{2} = 0 \qquad \text{(iv)}$$

$$\sum M_{oy} = 0: r_1 \frac{\surd(3)a}{6} - r_3 \frac{a}{2} - r_4 \frac{a}{2} - r_5 \frac{\surd(3)a}{2} + Q\frac{a}{2} = 0 \qquad \text{(v)}$$

$$\sum M_{oz} = 0: r_1 \frac{a}{2} - r_2 \frac{a}{2} - Q\frac{a}{2} - Qa = 0 \qquad \text{(vi)}$$

and these equations may be solved to give the values of the six independent reaction components. There may have been some computational advantage in putting the origin of the axes at A, since the first three reactions would then have no moments with respect to any of the axes.

The equations can also be expressed in matrix notation; after dividing equations (iv)–(vi) through by a, the result is

$$\begin{bmatrix} 1 & 0 & 0 & 0 & -1 & -1 \\ 0 & -1 & 0 & 0 & 0 & 0 \\ 0 & 0 & 1 & 1 & 0 & 0 \\ 0 & \frac{\surd 3}{6} & -\frac{1}{2} & 0 & 0 & 0 \\ \frac{\surd 3}{6} & 0 & -\frac{1}{2} & -\frac{1}{2} & -\frac{\surd 3}{2} & 0 \\ \frac{1}{2} & -\frac{1}{2} & 0 & 0 & 0 & 0 \end{bmatrix} \begin{Bmatrix} r_1 \\ r_2 \\ r_3 \\ r_4 \\ r_5 \\ r_6 \end{Bmatrix} = \begin{Bmatrix} 0 \\ 3Q \\ Q \\ -\frac{\surd(3)Q}{2} \\ -\frac{Q}{2} \\ \frac{3Q}{2} \end{Bmatrix} \qquad (2.3)$$

which may be written as

$$A \cdot R = P \qquad (2.4)$$

where A is a statics matrix, R is a vector of the reactions and P is a load vector.

It was previously mentioned that while a minimum of six independent reaction components was a necessary condition for the equilibrium of a three-dimensional body, that condition alone was not sufficient. The arrangement of the restraints is also important. This can be illustrated by returning to the structure of figure 2.3(a) but now introducing a spherical pin connection at B while eliminating the restraint at C. This gives the required six reactions and the equilibrium analysis can proceed as before. The statics matrix, A, may now be shown to be

$$A = \begin{bmatrix} 1 & 0 & 0 & -1 & 0 & 0 \\ 0 & -1 & 0 & 0 & 0 & -1 \\ 0 & 0 & 1 & 0 & 1 & 0 \\ 0 & \frac{\surd 3}{6} & -\frac{1}{2} & 0 & 0 & \frac{\surd 3}{2} \\ \frac{\surd 3}{6} & 0 & -\frac{1}{2} & -\frac{\surd 3}{2} & -\frac{1}{2} & 0 \\ \frac{1}{2} & -\frac{1}{2} & 0 & 0 & 0 & -\frac{1}{2} \end{bmatrix} \qquad (2.5)$$

Attempts to find a solution to the equation of the form of equation (2.4) will fail since the matrix A will be found to be singular. A physical interpretation of this is immediately given by considering an origin of the axes at A, with an axis along AB. Since all the reactions now intersect such an axis, no reactive moment can develop about it and the structure is free to rotate about AB.

While it is possible to derive certain rules about the arrangement of the reactions, they will not be pursued here. In most cases the necessary arrangement is fairly obvious and in any event, as was shown, the reactions cannot be calculated for an unsatisfactory arrangement. A similar situation occurs in two-dimensional statics, where the statics matrix A can be written and examined in that case.

Throughout this discussion some emphasis has been placed on the notion of independent reaction components. Engineers find it convenient to define reactions with respect to forces in the x, y and z directions of a suitably nominated set of cartesian axes. These components are said to be independent if the reaction is free to develop in any direction. For the pin connection in two-dimensional statics, there are two unknowns associated with the reaction: namely its magnitude and direction. (Its sense may be simply nominated, to be verified by equilibrium.) Instead of describing the unknowns as say, R_b and θ_B, the reaction is described in terms of its independent components of say, X_b and Y_b. This is shown in figure 2.4(a), which also illustrates a condition when the reaction must develop in a specified direction. The pin connection at B in figure 2.4(b) is only subjected to two forces and for these to be in equilibrium, they must be co-linear. The reaction can still be resolved into two components, but these are not independent components.

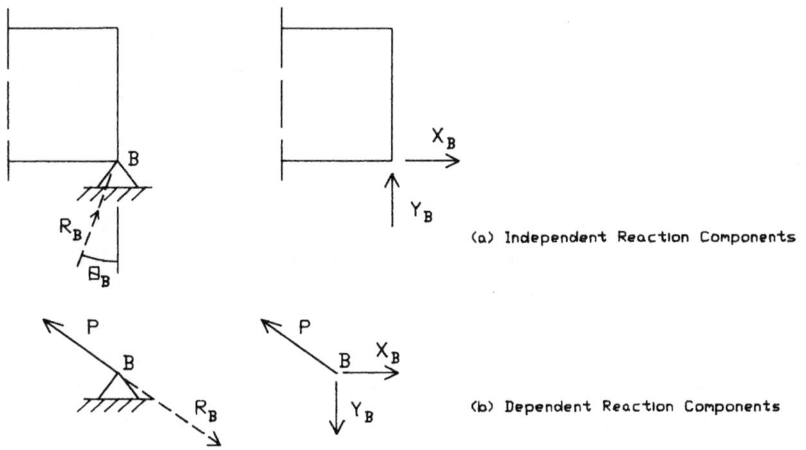

Figure 2.4 The nature of reactions.

2.1.3 Determinacy on the Basis of Equilibrium

If more than the required number of reactions is provided for a rigid body, then the equilibrium equations will not be sufficient to define the reactions. In such a case the statics matrix, A, of equation (2.4) will not be square and the equation cannot be solved for R. The problem is then said to be statically indeterminate; in other words, the reactions cannot be determined by statics alone. This issue will be developed further in section 2.4.

On the other hand, if an insufficient number of restraints were introduced then the body would be unstable. For example, a beam resting on two roller supports actually represents an unstable system, since it has no resistance to lateral force. If the loads are confined to vertical loads though, the system has a solution. In this case the first equation of equation (2.2), that is $\sum F_x = 0$, is being ignored and the loads are not general. It is convenient to use this approach in the case of continuous beam analysis since the transverse loads represent one loading condition, while the longitudinal loads represent another and the system can be analysed separately for this condition.

2.2 A CLASSIFICATION OF STRUCTURES

Part of the definition of a structure given in chapter 1 was that a structure may be defined as a series of elements connected together in a certain manner. The elements may be physically recognised, for example, as beams, columns, ties, slabs and plates. In each case, for the analysis of the structure to proceed, the element must be represented by a suitable idealisation.

Elements which can be represented by a single line are described as one-dimensional elements, characterised by the fact that one dimension is very much greater than the other two. Structures which are built up from one-dimensional elements are known as skeletal structures or skeletal frames. In general, beams, columns, ties and struts are all one-dimensional elements and they are certainly considered as such in skeletal frame analysis.

A two-dimensional element is one where two dimensions, of about the same magnitude, are very much greater than the third. This definition introduces slabs, plates and shells as two-dimensional elements. In a three-dimensional element, all dimensions are recognized and the structure is seen as being built up from three-dimensional blocks. The emphasis in this text is on skeletal frame analysis and the elements are all necessarily considered as one dimensional. Within that constraint though, the resulting analysis may be of a two-dimensional or a three-dimensional structure.

2.2.1 Skeletal Frames

Clearly a three-dimensional form can be built up from one-dimensional elements as is evidenced by the skeletal form of any city building. Frequently, because of the orthogonal arrangement, such structures are analysed as two-dimensional systems in different planes. When the structure is considered in three-dimensional space it is described as a space frame. Further, if the connections of the elements are modelled as pin connections, then the structure is a space truss. The more general space frame is a structure with moment transfer possible at the connections which are considered as rigid jointed.

A plane frame or planar structure, as a class of skeletal frame, is defined as one where all of the elements of the frame, and the loads acting on it, lie in the one plane. This leads to the two-dimensional skeletal frames which are dominant in the examples of this text. As a special case of the plane frame, when all the connections are modelled as pin connections, the structure is defined as a plane truss. Otherwise the plane frame may have connections that are modelled as rigid joints, allowing moment transfer.

Two other classes of skeletal frame remain. The first of these is the plane grid or grillage. In this system all of the elements of the structure lie in the one plane with the loads confined to act normal to that plane. The deck of a bridge structure may be modelled as a plane grid. The final classification is perhaps the simplest form of a skeletal frame and that is simply a beam, either acting with a single span or continuous over several spans. The classification is summarized in table 2.1 of section 2.5, where additional characteristics of the structures are also noted.

2.3 INTERNAL ACTIONS IN STRUCTURES

A one-dimensional element in a space frame will develop internal stresses due to the strains of its deformation under load. On any nominated section taken through the element, the stresses can be defined by a stress resultant. Since the stress resultant can be resolved into three components with respect to a selected set of cartesian axes taken at that section, six internal actions can be defined, as shown in figure 2.5(a). The six internal actions reduce to three for a plane frame as shown in figure 2.5(b); for a beam subjected to transverse loads alone, only shear and moment apply, while for a plane truss, axial force is the only internal action.

A knowledge of the type of internal actions in a given class of structure enables free body diagrams to be drawn for parts of the structure. Provided the structure is statically determinate, an analysis is then possible by considering equilibrium. The structure itself must have a stable form so that,

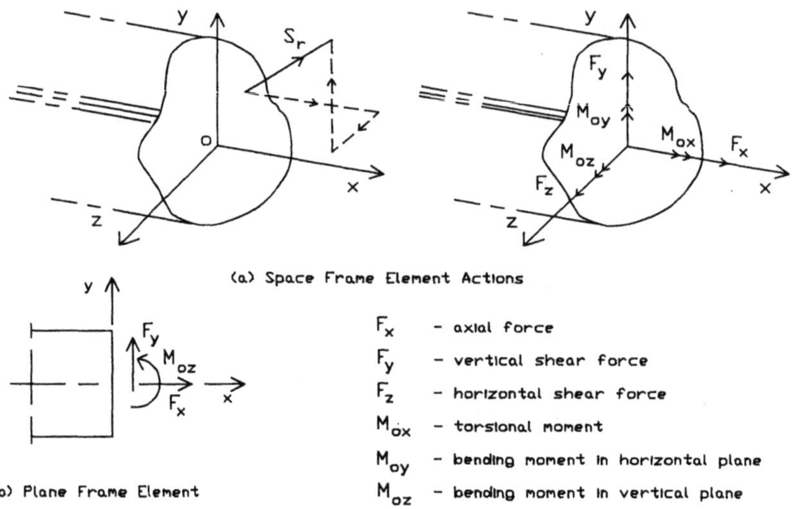

Figure 2.5 Internal actions in one-dimensional elements.

for the purposes of statics, it is a rigid body. A simple beam will meet this requirement as will a plane truss built up from a series of basic triangles or a space truss built up from a series of basic tetrahedrons.

2.4 STATICAL DETERMINACY

A structure is said to be statically determinate when the reactions and internal actions can be determined from a consideration of statics only. That is, the solution is possible by considering equilibrium alone. It has been seen that there are certain minimum requirements with regard to the restraints on a structure in order that the equations of equilibrium can be satisfied overall for the structure. If the minimum requirements are exceeded, the structure would have redundant reactions and be statically indeterminate.

However, the question of statical determinacy does not rest with the reactions. The arrangement and the number of elements also affect the determinacy with regard to the internal actions. It is possible then for a structure to be statically determinate with respect to the reactions, but indeterminate with respect to the internal actions, and such a structure must be classified as statically indeterminate.

The series of plane trusses shown in figure 2.6 illustrates the point. A truss has a sufficient and satisfactory arrangement of elements if the form can be built up from a basic triangle with two additional elements being added for each additional node, such as is shown in figure 2.6(a). The three

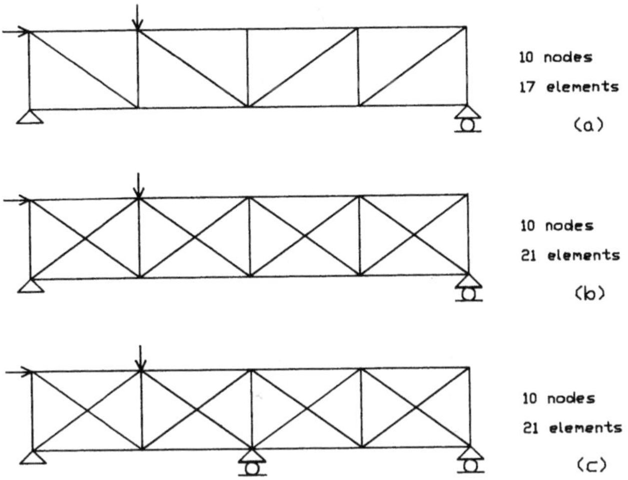

Figure 2.6 *Statical determinacy of a truss.*

equations of equilibrium for a planar structure require a minimum of three independent reaction components to develop at the restraints, in order that the equations can be satisfied, and again a satisfactory arrangement is shown in figure 2.6(a). The truss of that figure can be described as statically determinate.

In figure 2.6(b), additional elements have been introduced into each of the panels so that the total number of unknowns exceeds the number of equations of equilibrium that are available for the solution. This can be seen by considering each node in a free body diagram, representing a system of co-planar concurrent forces for which two equations of equilibrium can be written. For n nodes there are $2n$ such equations available. The unknowns are represented by the total number of independent reaction components, r, and the number of elements, b, since each element has the unknown axial force as the internal action. Thus it can be seen that if $2n$ equals $b+r$, the truss is statically determinate; while if $2n$ is less than $b+r$, the truss is statically indeterminate. The structure would have an unstable form if $2n$ were greater than the number of unknowns. While the truss of figure 2.6(b) is statically indeterminate, it may be noted that the reactions may still be calculated from statics. The truss of figure 2.6(c) has both the additional elements and an additional reaction component, and this is therefore statically indeterminate.

A further example is given by the series of plane frames shown in figure 2.7. The rigid jointed plane frame of figure 2.7(a) has a sufficient and satisfactory arrangement of restraints and both the reactions and the internal actions can be determined from a consideration of equilibrium. The flexural

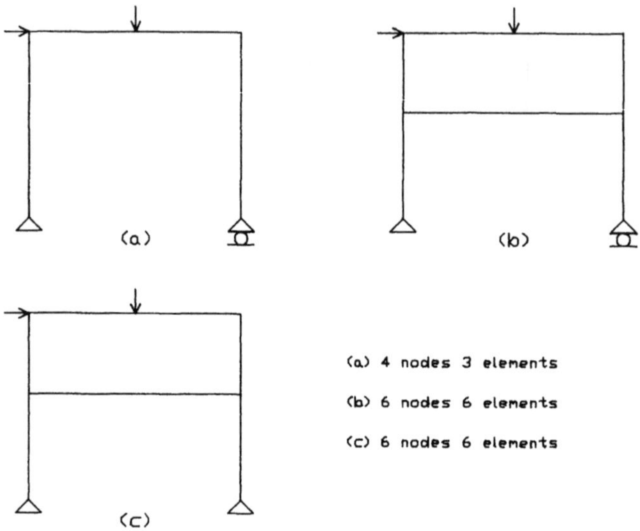

Figure 2.7 *Statical determinacy of a frame.*

element of a plane frame carries axial force, shear force and moment as the unknown internal actions. For the frame of figure 2.7(a), these can be determined at any point in the frame by applying the method of sections and considering the resulting free body diagram. Clearly such a frame is statically determinate. The additional element introduced in figure 2.7(b) renders that structure indeterminate, even though the reactions can still be calculated. The frame of figure 2.7(c) is redundant both with regard to the reactions and the internal actions.

It is possible to generalize the question of statical determinacy of a frame in a similar manner to that given for a truss. Since each node, as a free body diagram, represents a system of non-concurrent co-planar forces, there are three equations of equilibrium available per node. The total number of unknowns is the sum of the independent reaction components, r, and the total number of internal actions, which is three times the number of elements, b. Thus if $3n$ is equal to $3b+r$, the frame is statically determinate, while if $3n$ is less than $3b+r$, then the frame is statically indeterminate. An unstable form would result if $3n$ was greater than $3b+r$.

2.4.1 Equations of Condition

Frequently the internal actions at some point in a structure are prescribed by a condition introduced through the connection of the elements. For instance, if two beam elements are pin-connected together then the prescribed condition is that there can be no bending moment at the pin, since

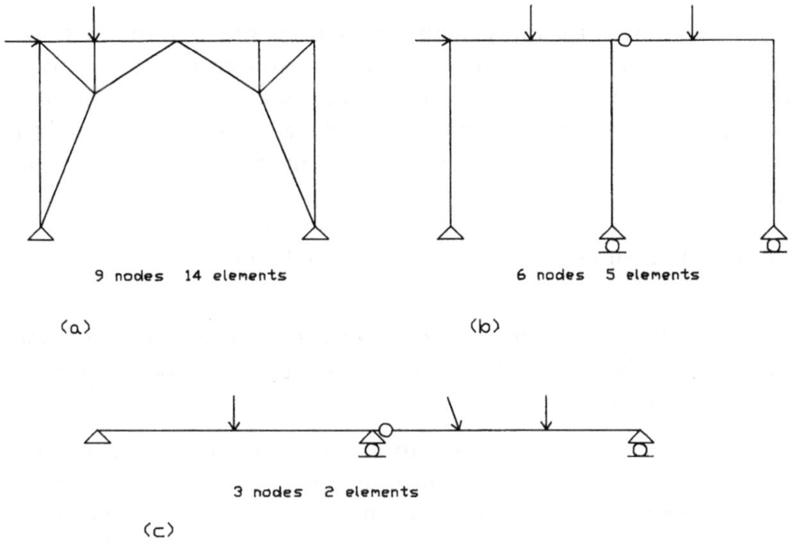

Figure 2.8 *Statically determinate forms.*

the pin cannot resist or transmit moment. Such a condition effectively provides an additional equation of equilibrium, usually described as an equation of condition. In all cases then, the number of equilibrium equations available becomes the sum of the nodal equilibrium equations and the equations of condition.

Equations of condition effectively amount to the release of an internal action that would otherwise be present in the structure. However, in terms of understanding statical determinacy, the number of equations is simply increased by the number of equations of condition present in the structure. On this basis, all of the structures shown in figure 2.8 are statically determinate and this may be verified by applying the rules previously given. The portal truss of figure 2.8(a) is particularly interesting. The structure acts as a three pinned arch, and the fact that there are four independent reaction components is compensated for by the equation of condition at the crown where there is no bending moment.

It should be noted that while the moment release is one of the most common forms of equation of condition, it is by no means the only one. Any or all of the internal actions of an element may be released by the nature of a particular connection.

2.4.2 Degree of Statical Indeterminacy

The extent to which a structure is statically indeterminate may be described by the degree of statical indeterminacy. This is simply the amount by which the number of unknowns for the structure (reactions plus internal actions) exceeds the number of equilibrium equations available for the solution (including equations of condition). The information is of some value in comparing structural forms and ultimately it is of significance in the analysis of such structures. For the present, it is simply instructive to understand the notion.

An alternative definition which leads to a preferred method of understanding statical determinacy, and hence structural behaviour, may be given as follows:

> *The degree of statical indeterminacy of a structure is the number of releases that must be introduced into the structure in order to give a statically determinate primary form*

A release may be specified with regard to an internal action or a reaction component. It can be seen that the introduction of a release effectively introduces an equation of condition. The nature of the release is also a function of the element type. For example, since a truss element only carries axial force as the internal action, the only internal release possible is achieved by cutting the element. On the other hand, completely severing a flexural element of a frame amounts to three releases, corresponding to the internal actions of axial force, shear force and moment.

This approach leads to the notion of cutting back the structure and a useful concept with regard to frames is to introduce the idea of a tree structure. The unique form of a tree has the main trunk fixed at the base with branches from the trunk that are not directly connected to each other. Any frame which takes this form will be found to be statically determinate. This can be readily confirmed since the static analysis can start at the end of any cantilever branch and proceed throughout the structure. The concept is illustrated in figure 2.9 with the original frame of figure 2.9(a) released

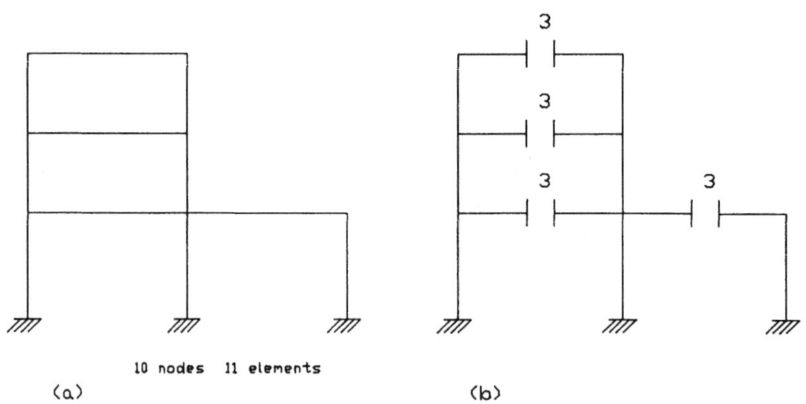

Figure 2.9 Tree structure concept.

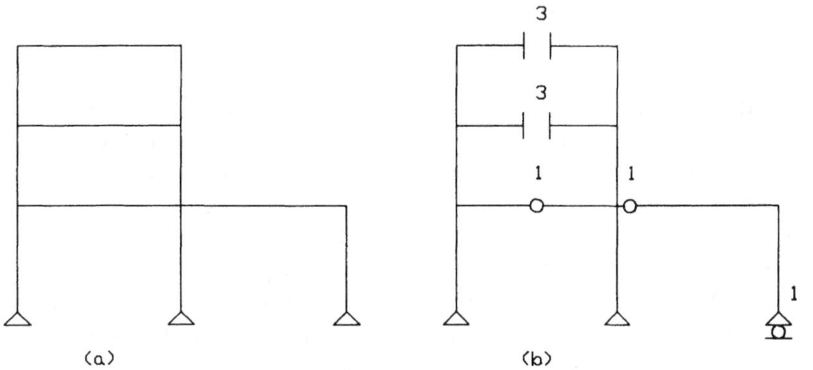

Figure 2.10 Frame releases.

by the cuts shown in figure 2.9(b) to produce three tree structures. The frame has a degree of statical indeterminacy of 12, which can be seen as both the total number of releases and as $3b+r-3n$.

Of course not all frames have fixed bases, so that introducing three releases into each beam element is not always appropriate and may in fact result in an unstable form. A suitable arrangement of releases for the frame of figure 2.10(a) is shown in figure 2.10(b), where the true tree form is restricted to the upper levels and the lower level has a statically determinate form similar to that of the structure of figure 2.8(b).

2.5 KINEMATIC DETERMINACY

Kinematic determinacy is given little attention in the general literature in structural engineering, but it is particularly relevant to the general stiffness method of structural analysis. As the term suggests, kinematic determinacy is related to determining the displacements of the structure. However, since a linear elastic structure has an infinite number of displacements throughout its displaced shape, the displacements in question need to be confined to the nodes. A structure can be said to be kinematically determinate when the displacements at the nodes are restrained to be zero. The restraints may come from the supports as specified or they may be provided by additional restraints imposed on the structure at the nodes, which would otherwise be free to move. The simplest form of a kinematically determinate structure is a beam built in at both ends which is often described as an encastré beam. Such a beam has a degree of statical indeterminacy of three, while it is kinematically determinate. Kinematic determinacy is closely linked to the concept of degrees of freedom of a structure and it is necessary to examine this aspect before proceeding further.

2.5.1 Degrees of Freedom

The number of degrees of freedom of any structure can be defined as the number of displacement terms necessary to completely define the displaced shape of the structure. For a linear elastic structure, the deflected shape will be continuous throughout the entire structure. This implies that there is an infinite number of degrees of freedom in a structure, as is indeed the case. However, the definition can be refined somewhat by selecting the displacement terms at the nodes only. All other displacements follow as a consequence. This is because the nodal displacements lead to the end actions on the elements, and the elastic curve of the element is a function of loads, geometry and element properties. Thus, the number of degrees of freedom of a structure is a function of the number of nodes selected to define the structure and the number of displacement terms associated with each node—that is, the nodal degrees of freedom.

2.5.2 Nodal Degrees of Freedom

For a three-dimensional structure, a node may have up to six degrees of freedom. This corresponds to the three translations and three rotations that may occur with respect to a set of three mutually orthogonal coordinate axes taken through the node. For a two-dimensional structure, a node may have three degrees of freedom corresponding to the two translations and one rotation with respect to the coordinate axes taken through the node. The general nodal degrees of freedom for a range of skeletal structures are shown in table 2.1. It should be noted that with pin-jointed trusses, rotations are not included since the behaviour of a truss is completely defined by the translations of each node.

Table 2.1 Nodal Degrees of Freedom for Various Structures

Structure type	Nodal degrees of freedom	Translation			Rotation		
		x	y	z	θ_x	θ_y	θ_z
3-D Space							
Frame	6	√	√	√	√	√	√
Truss	3	√	√	√	—	—	—
2-D x–y plane							
Frame	3	√	√	—	—	—	√
Truss	2	√	√	—	—	—	—
Beam	2	—	√	—	—	—	√
2-D x–z plane							
Grid	3	—	√	—	√	—	√

2.5.3 Structure Degrees of Freedom

The number of degrees of freedom for a structure then follows from a summation of the nodal degrees of freedom, excluding those degrees of freedom that are restrained by the boundary conditions. The application of this principle is shown in figure 2.11 where a series of structures are shown along with the degrees of freedom of the structure.

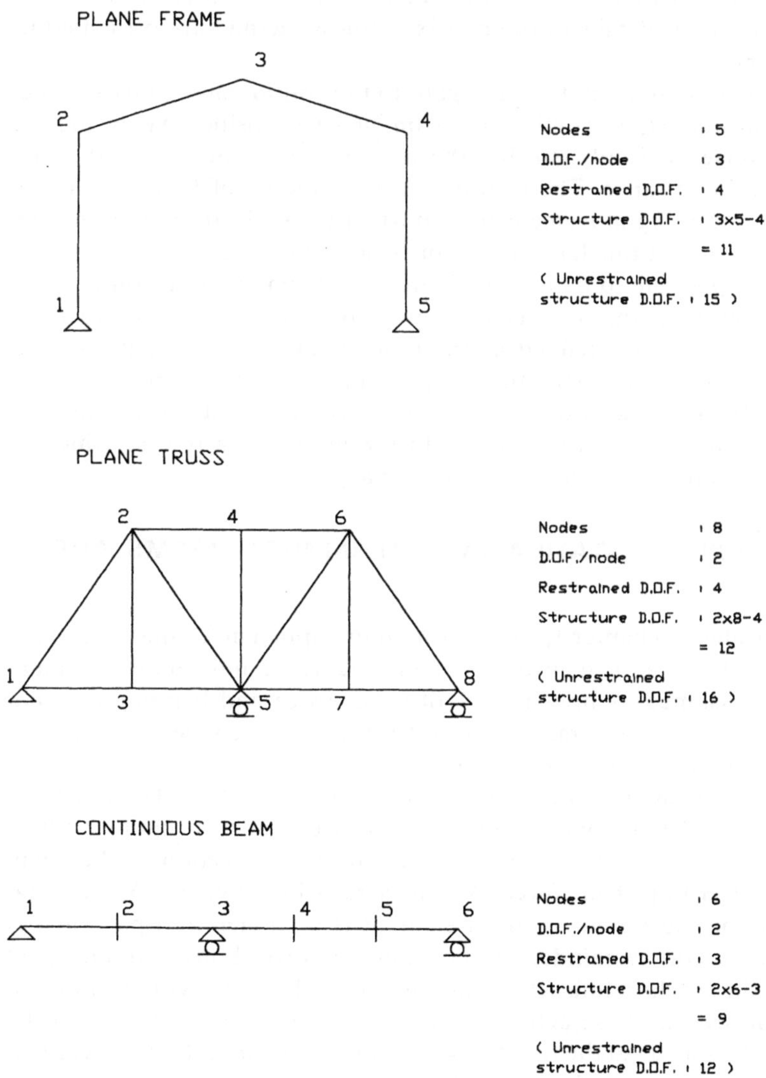

Figure 2.11 Structure degrees of freedom.

There are two aspects of figure 2.11 that warrant particular attention. Firstly it may be noted that for each structure, the unrestrained structure degrees of freedom are quoted. The reason for this is that in the general application of the matrix stiffness method of analysis, the structure stiffness matrix may be formed without regard to the boundary conditions initially. The number of degrees of freedom for the unrestrained structure represents the number of equations initially set up and represented in the structure stiffness matrix, so that this information is of general interest. This will become apparent as details of the matrix stiffness method are presented in later chapters.

The second point is made with regard to the continuous beam example. In the present context, the continuous beam must be considered to be loaded under transverse load only, so that the matter of reactions along the line of the beam do not arise. The restraint type at node 1 of figure 2.11(c) is capable of restraining movement both normal to the beam and along the line of the beam, but this latter restraint is not considered.

It is now possible to define the degree of kinematic indeterminacy of a structure simply as the number of degrees of freedom of that structure. Alternatively, it may be defined as the number of restraints that must be placed on the nodes of a structure to prevent movement of the structure completely. It can be seen that the notion of the restraint, in the form of props and clamps, is exactly opposite to the notion of a release which is central to the concept of statical indeterminacy.

2.6 ANALYSIS OF STATICALLY DETERMINATE BEAMS AND FRAMES

As was indicated in chapter 1, a background in applied mechanics involving simple beam theory and an introduction to shear force and bending moment diagrams is assumed. A brief review of shear force and bending moment will be presented here, mainly to introduce sign conventions before proceeding to more substantive work.

Consider the simply supported beam of figure 2.12(a). The reactions are readily calculated from equilibrium and are shown in figure 2.12(b). For transverse loads only, the internal actions at any section of the beam consist of a resisting shear force, V, and a resisting moment, M. Internal actions may be illustrated when an element is sectioned and part of it is removed so that the remainder can be shown in a free body diagram. That is to say all the equilibrating actions are shown on the body, with the internal actions representing those actions that the removed portion exerts on the remainder. This leads to the free body diagrams of figure 2.12(c) where two cases are illustrated with the internal actions shown in a positive sense; in case 1, the section is taken before the load Q, while in case 2 the section is beyond the load.

Since all the free body diagrams of a stable structure must be in equilibrium, the equilibrium equations can be applied as follows, taking an origin of coordinates at A:

For case 1: $0 < x < a$

$$\sum F_x = 0: X_a = 0 \qquad \text{(i)}$$

$$\sum F_y = 0: Y_a - V = 0 \qquad \text{(ii)}$$

$$\sum M = 0: -Vx + M = 0 \qquad \text{(iii)}$$

From (ii)

$$V = Y_a = Q\frac{(L-a)}{L}$$

From (iii)

$$M = Vx = Q\frac{(L-a)x}{L}$$

For case 2: $a < x < L$

$$\sum F_x = 0: X_a = 0 \qquad \text{(i)}$$

$$\sum F_y = 0: Y_a - Q - V = 0 \qquad \text{(ii)}$$

$$\sum M = 0: -Vx - Qa + M = 0 \qquad \text{(iii)}$$

From (ii)

$$V = Y_a - Q = -Q\frac{a}{L}$$

From (iii)

$$M = -Q\frac{a}{L}x + Qa = Q\frac{(L-x)a}{L}$$

It is assumed that the beam is capable of developing the necessary internal actions without failure of the material, so that the shear force and bending moment due to the applied loads are equal to the resisting shear force, V, and the resisting moment, M, respectively. The expressions for shear force and bending moment, which can be seen as functions of x, can be plotted to give the shear force diagram and the bending moment diagram respectively, as shown in figure 2.12(d). In this case, as with all the bending moment diagrams in this text, the bending moment values have been plotted off the tension face of the beam.

From this simple illustration it is possible to define shear and bending moment at any point in a flexural element as follows;

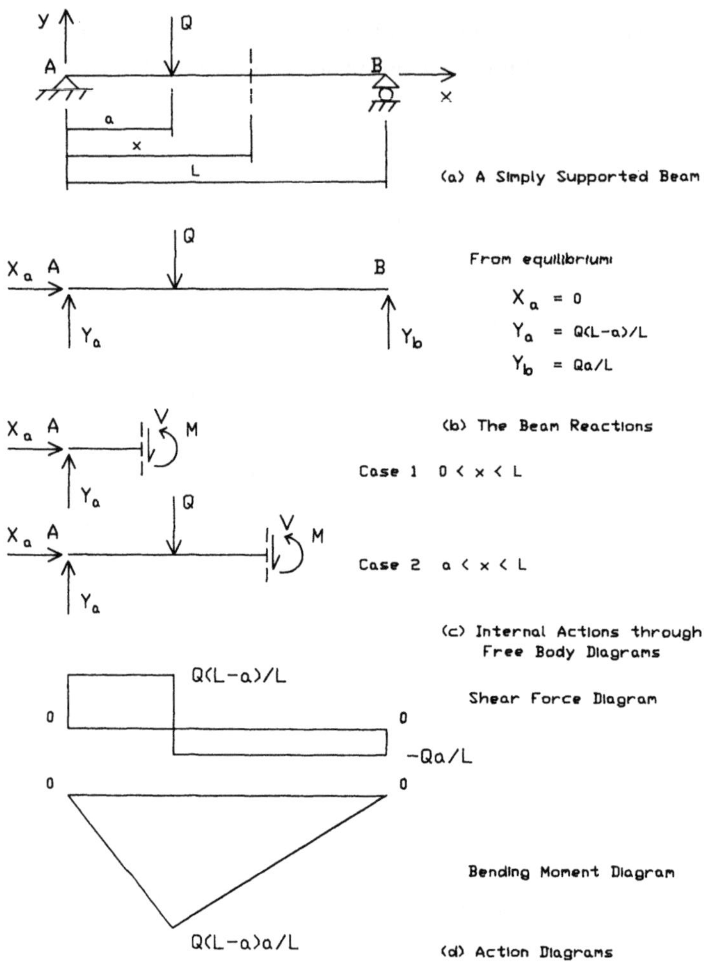

Figure 2.12 *Equilibrium analysis of a simple beam.*

> *The shear force, V, acting transversely across a section is equal to the sum of all the forces acting parallel to the section and taken on one side of the section*

> *The bending moment, M, acting at a section is equal to the sum of the moments of all the forces and moments about the section and taken on one side of the section*

Both definitions are of course an expression of equilibrium based on rigid body statics. In dealing with the elements of a frame, a liberal interpretation must be placed on the notion of 'one side of the section'. The expression really refers to the continuation of the entire structure beyond either side of the section in question.

For the element to perform satisfactorily it must be capable of developing the internal stresses. For example, a beam, through its section and material properties, must develop a moment of resistance equal to the bending moment imposed on it by the applied loads. This is a matter for stress analysis and element design which are considered after the structure has been analysed to determine the shear forces and bending moments acting throughout the structure. Other internal actions, such as axial force and torsional moment, can be considered in a similar manner for structures where such actions occur.

Rigid jointed plane frames represent an important class of structure to the structural engineer. Such structures maintain stability through the capacity of the rigid connection between the beam and column to transfer bending moment, and the flexural action is dominant in their behaviour. Simple cases are often referred to as bents or portal frames when a beam and two columns are involved. Example 2.1 demonstrates the equilibrium analysis of a frame and serves to introduce some important concepts in drawing frame bending moment diagrams. The frame of example 2.1 can be regarded as simply a bent beam although it must be stressed that while the rest of the structure is free to bend or flex, the angular relationship between the elements at node 2 must remain the same; this is a characteristic of a rigid joint.

In example 2.1, the internal action diagram shows how typical actions may be calculated along the element 1–2. A similar diagram can be drawn for a section along the element 2–3. For frames, it is convenient to plot the action diagrams using a line diagram of the structure as the base line; the ordinates are simply plotted normal to the beam section at which they apply. On this basis, the shear force and bending moment diagrams of example 2.1 result. At node 2, the shear in the element 1–2 is balanced by the applied load, while the shear in element 2–3 is balanced by the axial force in element 1–2.

On the other hand, since there is no applied moment at node 2, the moment at end 2 of element 1–2 must be consistent with the moment at end 2 of element 2–3, which is indeed the case. Since the frame turns through 90 degrees, the same ordinate information is displayed twice. Again the bending moment diagram is plotted from the tension face of the elements, as may be confirmed by the deflected shape of the frame as shown. The axial deformation of the column may be ignored and the frame sways in response to the lateral load since the roller support is free to move. Although node 2 sways and rotates, the joint angle remains at 90 degrees.

A more elaborate frame is analysed to find the reactions and bending moments in example 2.2. In this case, an equation of condition must be used, based on the pin connection of the beam element 4–6 to the column

Example 2.1: Equilibrium Analysis of a Simple Bent

Given data:

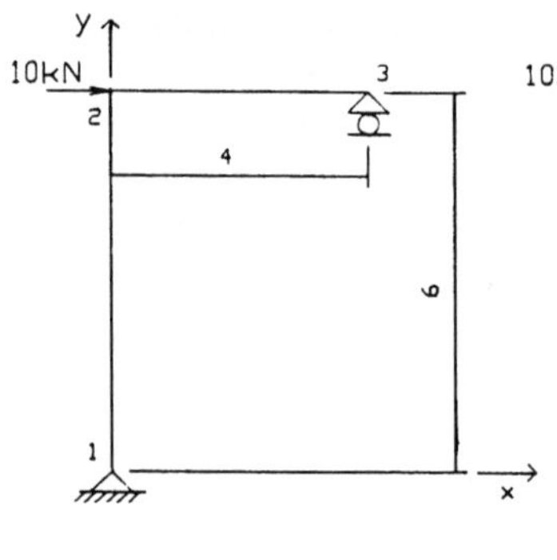

(a) The Frame

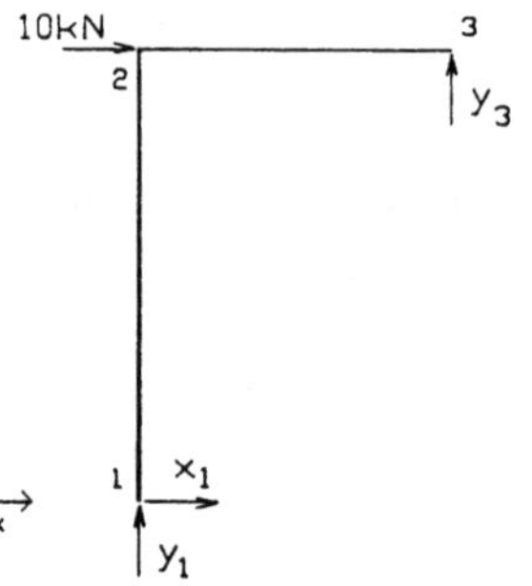

(b) Frame Reactions

Reactions from equilibrium:

$\sum F_x = 0$: $\quad X_1 + 10 = 0$ (i)

$\sum F_y = 0$: $\quad Y_1 + Y_3 = 0$ (ii)

$\sum M = 0$: $\quad -10(6) + Y_3(4) = 0$ (iii)

From (i): $X_1 = -10$ kN
From (iii): $Y_3 = 15$ kN
From (ii): $Y_1 = -15$ kN

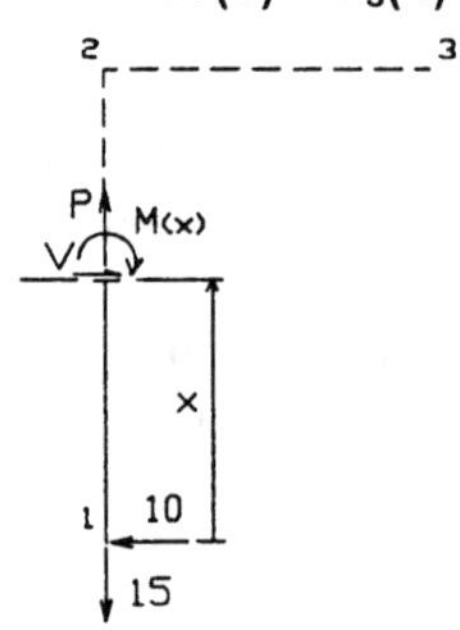

$0 \leqslant x \leqslant 6$

$V = 10$

$M(x) = 10(x)$

$P = 15$

(c) Internal Actions

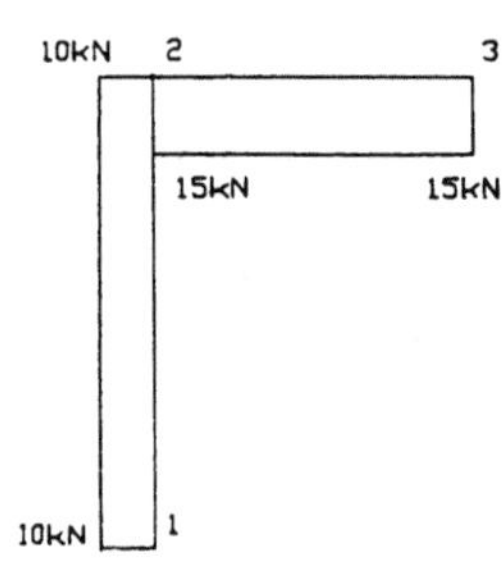

(d) Shear Force Diagram

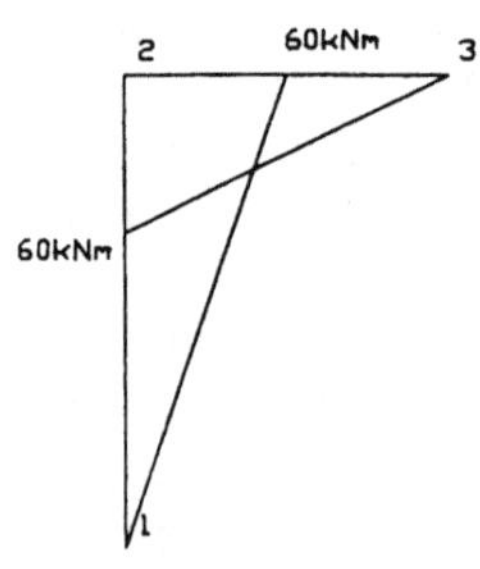

(e) Bending Moment Diagram

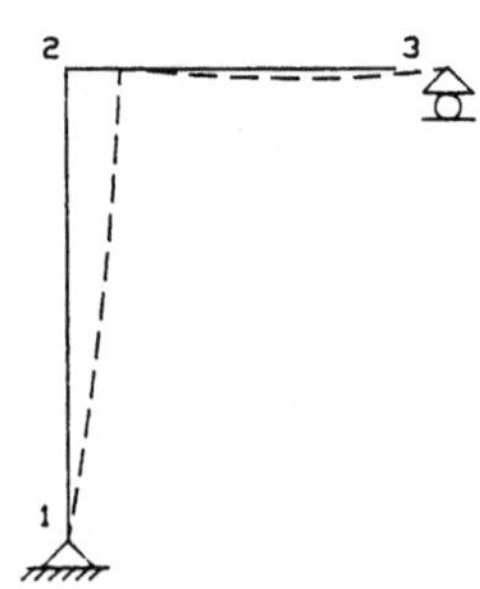

(f) Deflected Shape

at node 4. Four independent reaction components are identified from the support conditions but, with the three equations of equilibrium and one equation of condition, a solution for the reactions is readily found. With the reactions known, the bending moment diagram can be drawn by progressively considering sections along the elements. There is no bending in element 1-2 or 5-6, since there is no horizontal reaction at either node 1 or 5. Continuity of moment is again preserved at nodes 3 and 6. The same frame is used in chapter 7, where the frame deflections are calculated and a sketch of the deflected shape is given in example 7.3. The frame is used again in chapter 8 where this time a statically indeterminate version is analysed.

Example 2.2: Equilibrium Analysis of a Building Frame

Given data:

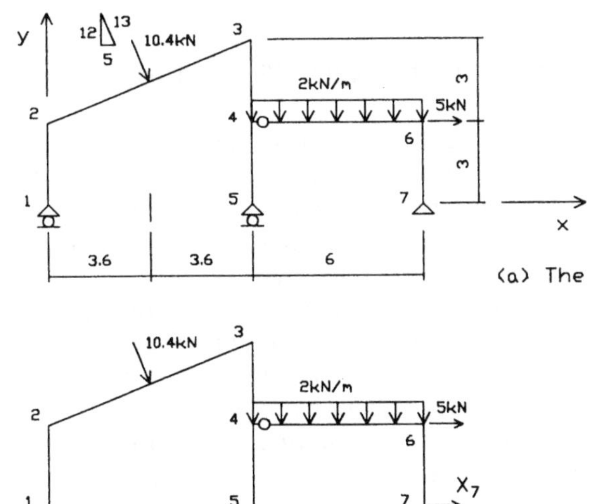

(a) The Frame

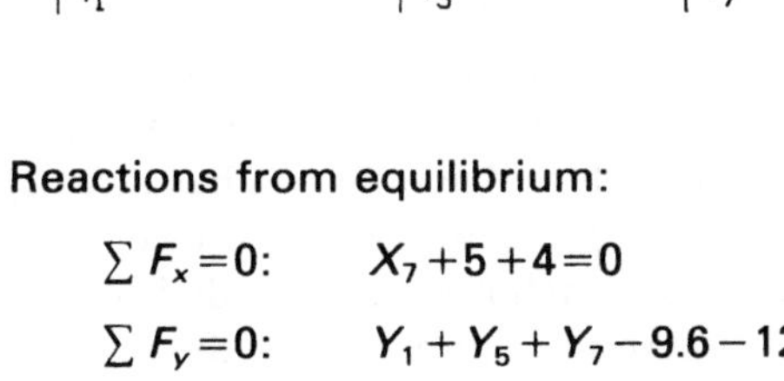

(b) Frame Reactions

Reactions from equilibrium:

$$\sum F_x = 0: \qquad X_7 + 5 + 4 = 0 \qquad \text{(i)}$$

$$\sum F_y = 0: \qquad Y_1 + Y_5 + Y_7 - 9.6 - 12 = 0 \qquad \text{(ii)}$$

$$\sum M = 0: \qquad Y_5(7.2) + Y_7(13.2) - 5(3) - 12(10.2) - 4(4.5) - 9.6(3.6) = 0 \qquad \text{(iii)}$$

Equation of condition: (bending moment at pin is zero)

$$Y_7(6)+X_7(3)-12(3)=0 \tag{iv}$$

From (i): $X_7=-9$ kN
From (iv): $Y_7=10.5$ kN
From (iii): $Y_5=7.133$ kN
From (ii): $Y_1=3.966$ kN

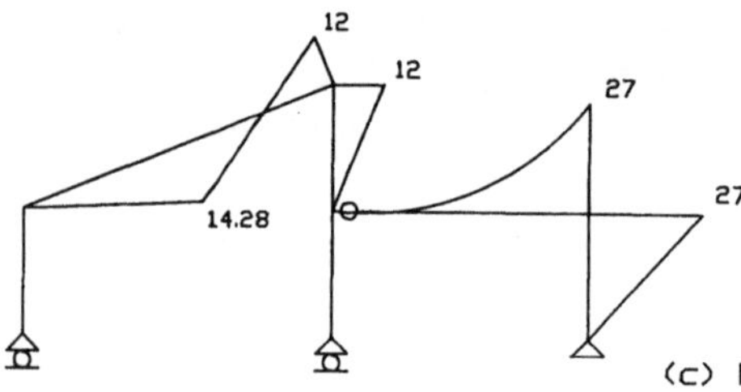

(c) Frame Bending Moment Diagram

2.7 ANALYSIS OF STATICALLY DETERMINATE TRUSSES

Trusses are an efficient form of structure in which the elements are arranged in a triangular pattern, with the connection of the elements at the node assumed to act as a pin. This assumption, coupled with the additional requirement that the loads are only applied at the nodes, ensures that the elements carry axial force as the only internal action.

If a node is examined as a free body under the action of the applied loads and the internal forces acting on it, then the resulting forces represent a system of concurrent forces. Concurrent force systems necessarily satisfy the moment equations of the equilibrium equations of equation (2.1). For a space truss, this leaves three equations of equilibrium available for the solution for the internal forces at any node, while for a plane truss, two such equations are available.

The equilibrium analysis of a truss may proceed on the same basis as that already presented for beams and frames. That is, the reactions can be calculated considering the structure as a rigid body in an overall sense, and the solution can then go on to calculate the internal actions. As has already been intimated, the solution for internal actions is based on free body diagrams. Traditionally, the technique where each node is regarded in turn as a free body has been described as the method of joints, while the method in which the truss is sectioned to provide free body diagrams is known as the method of sections. It is sometimes convenient to use a combination of both techniques and the procedures are applicable to both space trusses and plane trusses, the only difference being the number of equilibrium equations used as previously indicated.

The analysis of a truss requires a systematic approach which is certainly emphasised in computer-based analysis. A certain degree of rigour is applied in the following examples, in so far as all of the unknown internal actions are initially assumed to be in tension, and the reactions are assumed to act in the positive directions of the coordinate axes. It should be recognised, though, that intuitive guesses as to the sense and magnitude of certain forces is an aid to understanding and may reduce the computational effort. The latter point is certainly true when elements that have zero force are identified intuitively, even though the basis of this must be equilibrium.

In example 2.3, a plane truss is analysed by the method of joints, after first calculating the reactions. Since there are only two equilibrium equations applicable at each node, the nodes are taken in an order which introduces a maximum of two unknowns.

A statically indeterminate form of the plane truss of example 2.3 is studied in chapter 3, and it is useful to consider the equilibrium equations in more general terms as a guide to understanding subsequent work. If each of the nodes is considered in turn without regard to the nature of the internal actions or the reactions, the following set of equations is found:

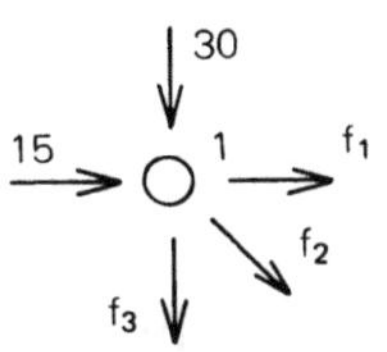

$$\sum F_x = 0: \quad 15 + 1.0f_1 + 0.6f_2 = 0$$
$$\sum F_y = 0: \quad -30 - 0.8f_2 - 1.0f_3 = 0$$

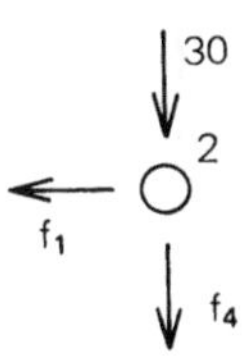

$$\sum F_x = 0: \quad 0 - 1.0f_1 = 0$$
$$\sum F_y = 0: \quad -30 - 1.0f_4 = 0$$

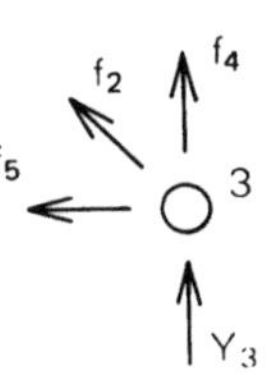

$$\sum F_x = 0: \quad 0 - 0.6f_2 - 1.0f_5 = 0$$
$$\sum F_y = 0: \quad Y_3 + 0.8f_2 + 1.0f_4 = 0$$

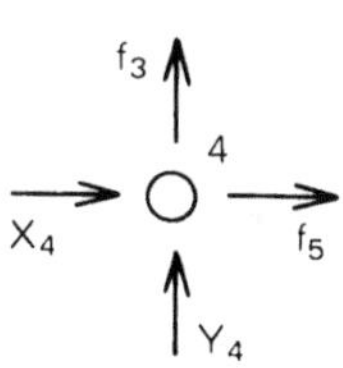

$$\sum F_x = 0: \quad X_4 + 1.0f_5 = 0$$
$$\sum F_y = 0: \quad Y_4 + 1.0f_3 = 0$$

Example 2.3: Equilibrium Analysis of a Truss

Given data:

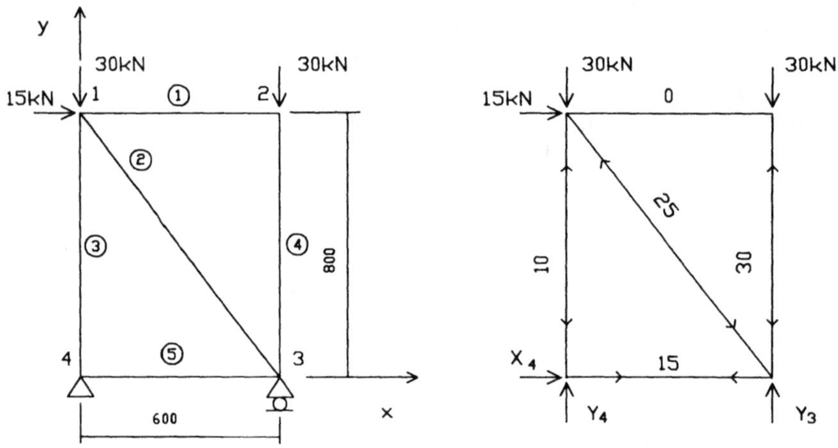

(a) The Truss (b) Truss Reactions and Forces

Reactions from equilibrium:

$$\sum F_x = 0: \qquad X_4 + 15 = 0 \tag{i}$$

$$\sum F_y = 0: \qquad Y_4 + Y_3 - 30 - 30 = 0 \tag{ii}$$

$$\sum M = 0: \qquad Y_3(600) - 15(800) - 30(600) = 0 \tag{iii}$$

From (i): $X_4 = -15$ kN
From (iii): $Y_3 = 50$ kN
From (ii): $Y_4 = 10$ kN

Equilibrium equations applied to each node: (local axes)

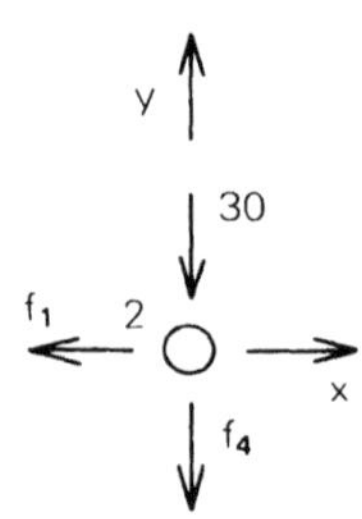

Select node 2

$$\sum F_x = 0: \qquad 0 - f_1 = 0$$

$$\sum F_y = 0: \qquad -30 - f_4 = 0$$

hence $f_1 = 0$; $f_4 = -30$ kN

Select node 4

$\sum F_x = 0$: $\quad -15 + f_5 = 0$

$\sum F_y = 0$: $\quad 10 + f_3 = 0$

hence $f_5 = 15$ kN; $f_3 = -10$ kN

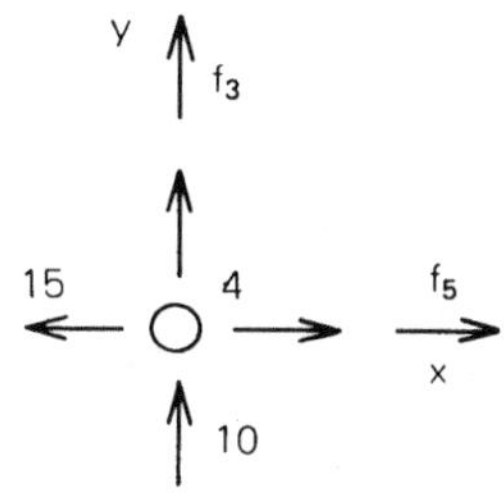

Select node 1

$\sum F_x = 0$: $\quad 15 + f_1 + 0.6 f_2 = 0$

$\sum F_y = 0$: $\quad -30 - 0.8 f_2 - f_3 = 0$ (check)

hence $f_2 = -25$ kN

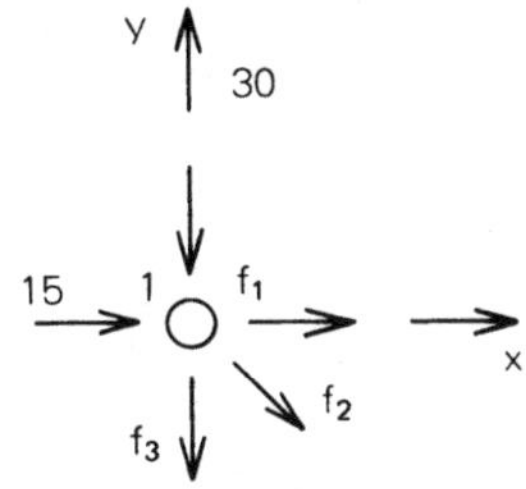

These equations may be written in matrix form as:

$$\begin{Bmatrix} 15 \\ -30 \\ 0 \\ -30 \\ 0 \\ Y_3 \\ X_4 \\ Y_4 \end{Bmatrix} = \begin{bmatrix} -1.0 & -0.6 & 0.0 & 0.0 & 0.0 \\ 0.0 & 0.8 & 1.0 & 0.0 & 0.0 \\ 1.0 & 0.0 & 0.0 & 0.0 & 0.0 \\ 0.0 & 0.0 & 0.0 & 1.0 & 0.0 \\ 0.0 & 0.6 & 0.0 & 0.0 & 1.0 \\ 0.0 & -0.8 & 0.0 & -1.0 & 0.0 \\ 0.0 & 0.0 & 0.0 & 0.0 & -1.0 \\ 0.0 & 0.0 & -1.0 & 0.0 & 0.0 \end{bmatrix} \begin{Bmatrix} f_1 \\ f_2 \\ f_3 \\ f_4 \\ f_5 \end{Bmatrix} \tag{2.6}$$

and in the matrix notation of

$$P = A \cdot f \tag{2.7}$$

where P is a load vector, A is a statics matrix and f is the vector of unknown internal forces. However, since the equilibrium equations of equation (2.6) include the unknown reactions as part of the load vector, a solution to equation (2.7) is not immediately possible. The load vector and the statics matrix may both be partitioned to distinguish between the known load terms, denoted collectively as P_F, and the unknown reactions, denoted by P_R. Equation (2.7) can then be written as

$$\begin{Bmatrix} P_F \\ \hline P_R \end{Bmatrix} = \begin{bmatrix} A_F \\ \hline A_R \end{bmatrix} \{f\} \tag{2.8}$$

Expanding equation (2.8) gives

$$P_F = A_F \cdot f \tag{2.8a}$$

and

$$P_R = A_R \cdot f \tag{2.8b}$$

Equation (2.8a) can be solved for f and the reactions can be recovered by the use of equation (2.8b). More generally, a solution can be found for the internal forces and the reactions, simultaneously, if the equations are augmented as follows.

Equation (2.8a) is also $P_F = A_F \cdot f + 0 \cdot P_R$ where 0 is a null matrix, and equation (2.8b) is also $0 = A_R \cdot f - I \cdot P_R$ where I is the identity, or unit matrix. These two equations can be written in the form

$$\left\{\begin{array}{c} P_F \\ \hline 0 \end{array}\right\} = \left[\begin{array}{c|c} A_F & 0 \\ \hline A_R & I \end{array}\right] \left\{\begin{array}{c} f \\ \hline P_R \end{array}\right\} \tag{2.9}$$

and equation (2.9) can be solved directly. Taken node by node, the equations of equilibrium will not always appear in the ordered fashion of equation (2.6), but they can always be re-arranged to that form.

Equation (2.9) has been used as the basis for analysing the truss of example 2.3 in conjunction with the matrix operations program, MATOP, which is presented in appendix A. The following output file from MATOP indicates the nature of the operations and presents the solution.

```
REMARK. Analysis of the Truss of Example 2.3 using MATOP
LOAD.AT.8.8
PRINT.AT  The Augmented Statics Matrix
-.100000E+01 -.600000E+00 0.000000E+00 0.000000E+00 0.000000E+00
0.000000E+00 0.000000E+00 0.000000E+00
0.000000E+00 0.800000E+00 0.100000E+01 0.000000E+00 0.000000E+00
0.000000E+00 0.000000E+00 0.000000E+00
0.100000E+01 0.000000E+00 0.000000E+00 0.000000E+00 0.000000E+00
0.000000E+00 0.000000E+00 0.000000E+00
0.000000E+00 0.000000E+00 0.000000E+00 0.100000E+01 0.000000E+00
0.000000E+00 0.000000E+00 0.000000E+00
0.000000E+00 0.600000E+00 0.000000E+00 0.000000E+00 0.100000E+01
0.000000E+00 0.000000E+00 0.000000E+00
0.000000E+00 -.800000E+00 0.000000E+00 -.100000E+01 0.000000E+00
0.100000E+01 0.000000E+00 0.000000E+00
0.000000E+00 0.000000E+00 0.000000E+00 0.000000E+00 -.100000E+01
0.000000E+00 0.100000E+01 0.000000E+00
0.000000E+00 0.000000E+00 -.100000E+01 0.000000E+00 0.000000E+00
0.000000E+00 0.000000E+00 0.100000E+01
LOAD.P.8.1
PRINT.P  The Modified Load Vector
0.150000E+02
-.300000E+02
0.000000E+00
-.300000E+02
0.000000E+00
0.000000E+00
0.000000E+00
0.000000E+00
SOLVE.AT.P
SELECT.F.P.5.1.1.1
PRINT.F  The Element Forces
-.947159E-15
-.250000E+02
-.100000E+02
-.300000E+02
0.150000E+02
SELECT.R.P.3.1.6.1
PRINT.R  The Reactions
-.500000E+02
0.150000E+02
-.100000E+02
End of File
```

A computer program for the analysis of statically determinate plane trusses could be developed using the matrix operations outlined. There is not much point in this though, since more general procedures that will analyse trusses, irrespective of the nature of their determinacy, will be outlined in later chapters.

2.8 PROBLEMS FOR SOLUTION

2.1 For each of the beams shown in figure P2.1, determine the reactions and sketch the shear force and bending moment diagrams.

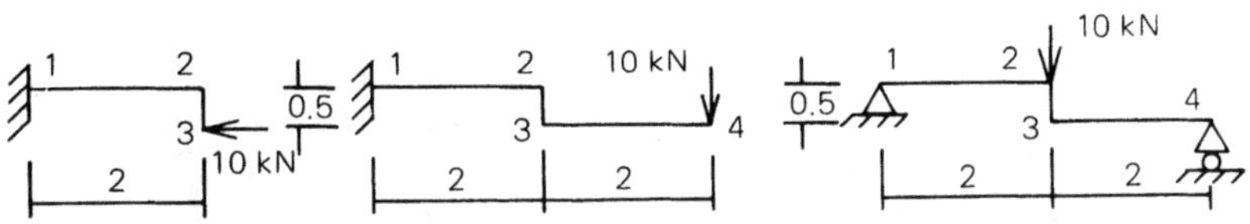

Figure P2.1.

2.2 For each of the plane frames shown in figure P2.2, determine the reactions and sketch the bending moment diagram.

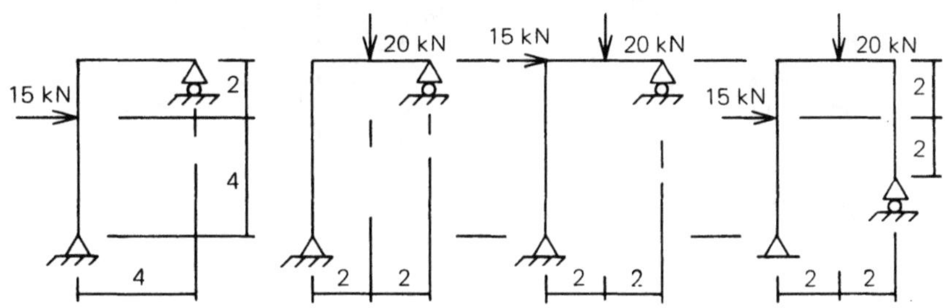

Figure P2.2.

2.3 For each of the plane grid structures shown in figure P2.3, calculate the reactive forces and sketch the shear force, and bending and torsional moment diagrams for the element 1–2.

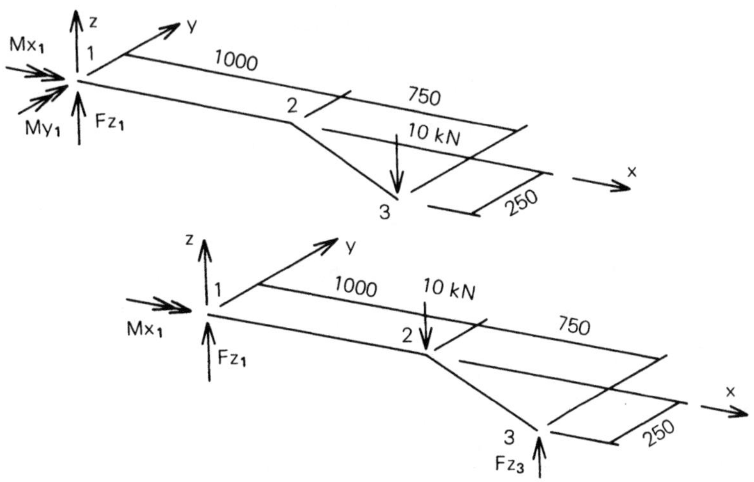

Figure P2.3.

2.4 For each of the space frames shown in figure P2.4, calculate the reactive forces and sketch the axial thrust, shear force, bending moment and torsional moment diagrams for the element 1–2.

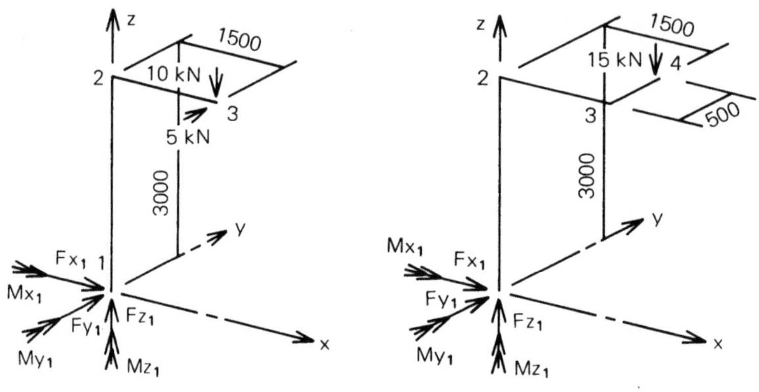

Figure P2.4.

2.5 Determine the reactions and all element forces for both of the plane trusses shown in figure P2.5.

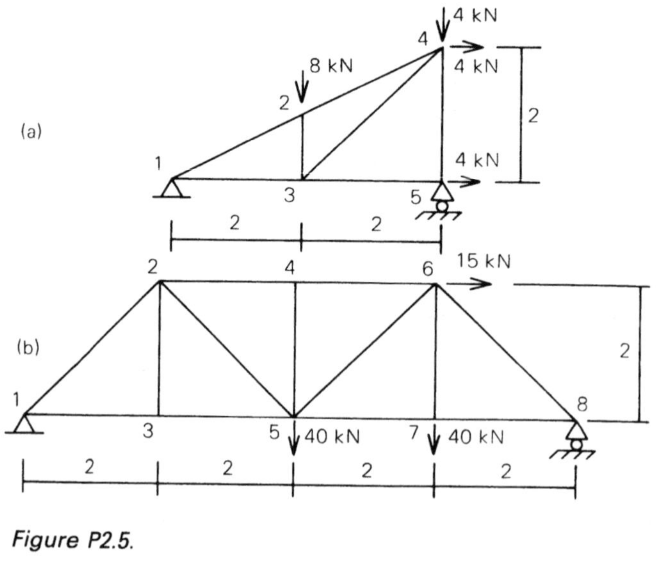

Figure P2.5.

2.6 The boom of a tower crane is pin-connected to the tower at B, and held in place by a tie rod, AC, connected to the top of the tower as shown in figure P2.6. Calculate the forces acting on the tower face due to the load shown on the boom. For this analysis, the boom may be regarded as a plane truss.

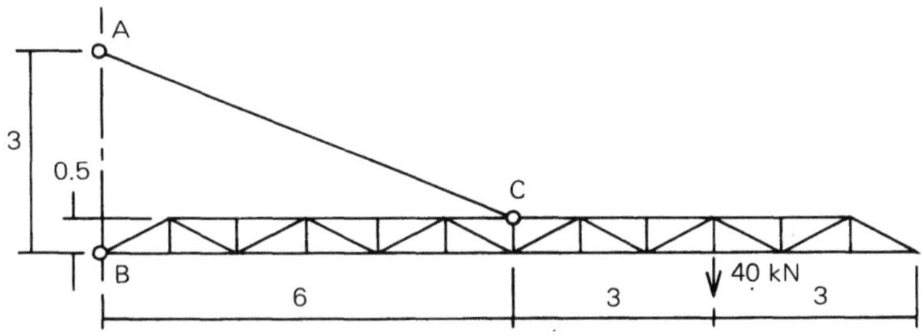

Figure P2.6.

2.7 The boom of the tower crane of problem **2.6** is in fact a space truss with the general arrangement of elements as shown in figure P2.7, which only shows a portion of the boom from the cantilever end. Determine the forces in all the elements shown due to the loads indicated.

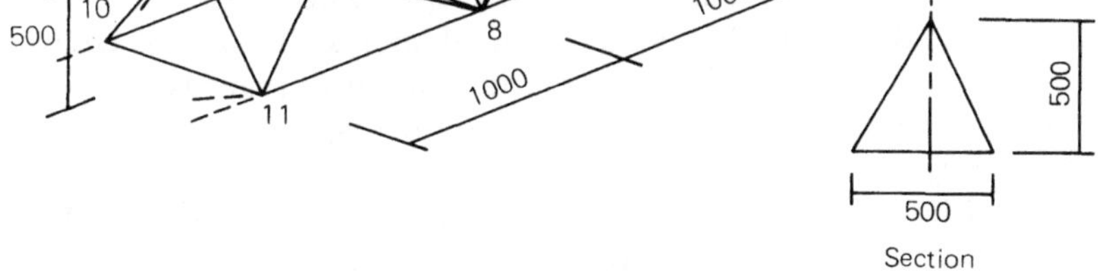

Figure P2.7.

FURTHER READING

Norris, C. H. and Wilbur, J. B., *Elementary Structural Analysis*, 2nd edn, McGraw-Hill, New York, 1960.

Chapter 3
Basic Concepts of the Stiffness Method

The underlying philosophy of the stiffness method was informally introduced in chapter 1 and it is now appropriate to examine the fundamental principles which form the basis of what is known as the stiffness or displacement method of structural analysis. The method is applicable to all classes of structure and while particular details may vary in different applications, the basic concepts remain the same. The method begins by regarding the structure in the manner in which it has already been defined, that is as a collection of elements connected together at node points. In response to certain loads, generally at the nodes but not necessarily so, the structure will displace and the displaced shape can be defined by the collective terms describing the displacements at the nodes.

The stiffness method regards the displacements as the fundamental unknowns and it is for this reason that the method is sometimes referred to as the displacement method. With the displacements known, it is possible to determine the internal actions on the elements. By way of contrast, the flexibility method, which is presented in chapter 8, regards any redundant actions as the fundamental unknowns and the displacements are calculated subsequently.

Consider the behaviour of a single linear elastic spring, which is a one degree of freedom system as shown in figure 3.1(a). In response to the force, F, the spring will extend by the displacement, x, which will be a function of the spring constant, k. The force–extension relationship can be described by the graph of figure 3.1(b), and may be written as

$$F = k \cdot x \tag{3.1}$$

The spring of figure 3.1(a) represents a simple form of structure and the notion of the force–extension relationship can be extended to a multi-degree of freedom system. However, the external forces acting on the structure,

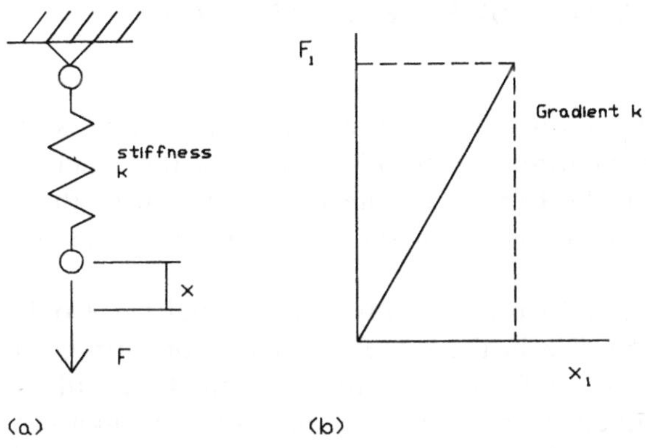

Figure 3.1 *Behaviour of a linear elastic spring.*

and the resulting displacements, must be described in terms of vectors which are known as load and displacement vectors respectively. The relationship which is the equivalent of equation (3.1) is then given in a matrix equation, which may be written as

$$P = K \cdot d \tag{3.2}$$

where P and d are both n by 1 vectors, and K is an n by n matrix known as the structure stiffness matrix. The size of the arrays is therefore defined by n, which is directly related to the degrees of freedom of the structure.

3.1 ELEMENT AND STRUCTURE STIFFNESS

In general, a structure is built up from a series of elements, each of which has its own characteristics with regard to action and response. These characteristics are described in the element stiffness which must be expressed in matrix form where the nodes of the element have more than one degree of freedom. Collectively, the elements contribute to the stiffness of the structure as expressed in the structure stiffness matrix and it remains for a convenient way of determining the terms of this matrix to be found. There are in fact several ways of achieving this and the details vary depending on the precise nature of the structure and the selected technique. In the following section, a general technique for analysing pin-jointed trusses is described. Although alternative techniques are available and the method has some limitations, it is a useful method for illustrating the basic concepts of the stiffness method of structural analysis.

3.2 FORMING THE STRUCTURE STIFFNESS MATRIX BY DIRECT MULTIPLICATION

The analysis of a pin-jointed truss will be examined in some detail as an illustration of the principles of the stiffness method. In this case, the structure stiffness matrix will be seen to be formed by the multiplication of three matrices in a technique that the author suggests should be known as the global stiffness method.

The elements of a pin-jointed truss are subjected to axial force only and the individual elements behave in the same manner as the spring of figure 3.1(a). Consider the truss of figure 3.2. Initially, attention will be focused only on the unrestrained degrees of freedom, with the unknown displacements being described in the vector with terms from d_1 to d_5 and the corresponding possible loads being defined in the vector with terms from P_1 to P_5. It should be noted that the positive directions of the loads and displacements are defined with respect to the global axes of reference as shown in figure 3.2. In this case the structure stiffness matrix is a 5 by 5 matrix and it can be developed through a consideration of equilibrium, compatibility and the stress–strain law governing element behaviour.

Each node is in equilibrium through the action of the external loads acting there and the resulting internal actions which can be described in an internal force vector, f, the terms of which are f_1 to f_6. Defining the internal forces as tension positive, the nodal equations of equilibrium are

$$P_1 = -1.0f_1 - 0.6f_2 + 0.0f_3 + 0.0f_4 + 0.0f_5 + 0.0f_6$$

$$P_2 = 0.0f_1 + 0.8f_2 + 1.0f_3 + 0.0f_4 + 0.0f_5 + 0.0f_6$$

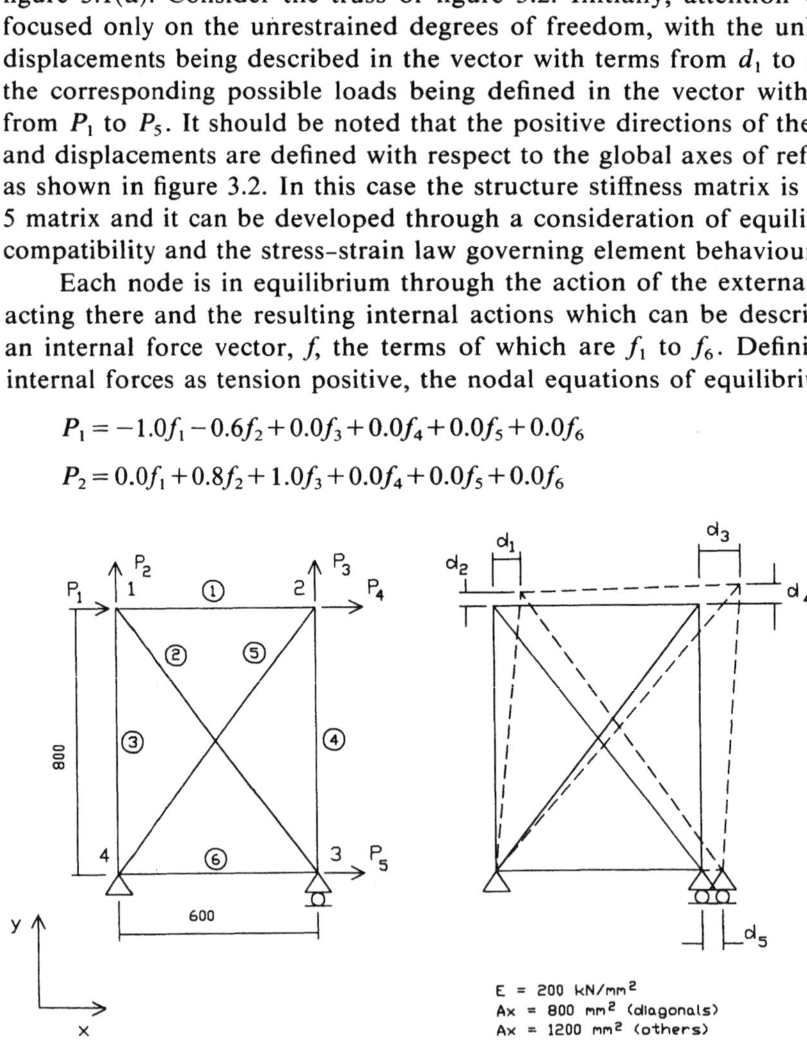

Figure 3.2 Behaviour of a truss.

$$P_3 = 1.0f_1 + 0.0f_2 + 0.0f_3 + 0.0f_4 + 0.6f_5 + 0.0f_6$$

$$P_4 = 0.0f_1 + 0.0f_2 + 0.0f_3 + 1.0f_4 + 0.8f_5 + 0.0f_6$$

$$P_5 = 0.0f_1 + 0.6f_2 + 0.0f_3 + 0.0f_4 + 0.0f_5 + 1.0f_6$$

These equations may be expressed in matrix form as

$$\begin{Bmatrix} P_1 \\ P_2 \\ P_3 \\ P_4 \\ P_5 \end{Bmatrix} = \begin{bmatrix} -1.0 & -0.6 & 0.0 & 0.0 & 0.0 & 0.0 \\ 0.0 & 0.8 & 1.0 & 0.0 & 0.0 & 0.0 \\ 1.0 & 0.0 & 0.0 & 0.0 & 0.6 & 0.0 \\ 0.0 & 0.0 & 0.0 & 1.0 & 0.8 & 0.0 \\ 0.0 & 0.6 & 0.0 & 0.0 & 0.0 & 1.0 \end{bmatrix} \begin{Bmatrix} f_1 \\ f_2 \\ f_3 \\ f_4 \\ f_5 \\ f_6 \end{Bmatrix}$$

and written in the notation

$$P = A \cdot f \tag{3.3}$$

where A is known as a statics matrix, and P and f are the vectors previously defined.

Under the action of the internal forces, each element will extend. The extensions may be described in an element extension vector with terms from e_1 to e_6. For a linear elastic structure, the force–extension relationship is given by the equation

$$f_i = \frac{E_i A_i}{L_i} e_i$$

where E_i is Young's Modulus, A_i is the section area and L_i is the length of the element. It may be noted that the equation is an expression of element stiffness and in this case it can be expressed in a scalar algebraic equation. However, the full set of such equations may be written in matrix form as

$$\begin{Bmatrix} f_1 \\ f_2 \\ f_3 \\ f_4 \\ f_5 \\ f_6 \end{Bmatrix} = \begin{bmatrix} \frac{E_1A_1}{L_1} & & & & & \\ & \frac{E_2A_2}{L_2} & & & & \\ & & \frac{E_3A_3}{L_3} & & & \\ & & & \frac{E_4A_4}{L_4} & & \\ & & & & \frac{E_5A_5}{L_5} & \\ & & & & & \frac{E_6A_6}{L_6} \end{bmatrix} \begin{Bmatrix} e_1 \\ e_2 \\ e_3 \\ e_4 \\ e_5 \\ e_6 \end{Bmatrix}$$

and written in the notation:

$$f = S \cdot e \tag{3.4}$$

where the matrix S is a matrix of element stiffnesses. Clearly equation (3.4) is an expression of the stress–strain law governing the material behaviour.

The element extensions are directly related to the displacements through the compatibility equations, the nature of which may be determined by considering the displacements at the end of each element in turn. For example, consider element ⑤ as shown in figure 3.3. End 4 of element ⑤ is held against translation by the restraints, while end 2 moves with components of displacement d_3 and d_4, each of which contributes to the element extension. Considering each component in turn and assuming small displacement theory, then

$$e_5 = 0.6d_3 + 0.8d_4$$

as is shown by figure 3.3(a) and (b), and it is unrelated to the remaining displacements. Compatibility is expressed in the fact that the nodal displacements are common to all elements that terminate at that node.

Repeating the process for all elements results in the equations:

$$e_1 = -1.0d_1 + 0.0d_2 + 1.0d_3 + 0.0d_4 + 0.0d_5$$

$$e_2 = -0.6d_1 + 0.8d_2 + 0.0d_3 + 0.0d_4 + 0.6d_5$$

$$e_3 = 0.0d_1 + 1.0d_2 + 0.0d_3 + 0.0d_4 + 0.0d_5$$

$$e_4 = 0.0d_1 + 0.0d_2 + 0.0d_3 + 1.0d_4 + 0.0d_5$$

$$e_5 = 0.0d_1 + 0.0d_2 + 0.6d_3 + 0.8d_4 + 0.0d_5$$

$$e_6 = 0.0d_1 + 0.0d_2 + 0.0d_3 + 0.0d_4 + 1.0d_5$$

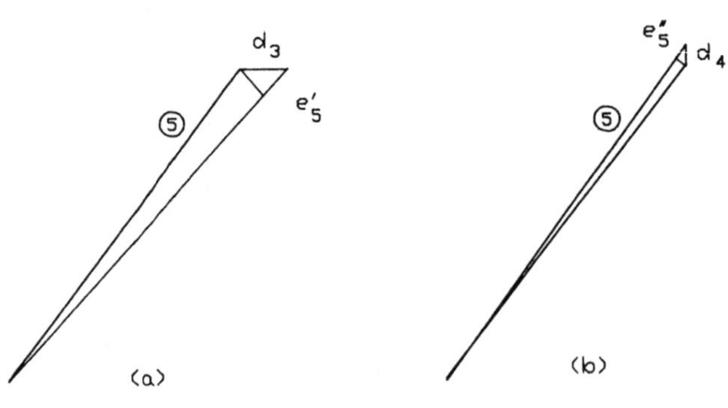

Figure 3.3 *Details of element extension.*

which can be written in matrix form as

$$\begin{Bmatrix} e_1 \\ e_2 \\ e_3 \\ e_4 \\ e_5 \\ e_6 \end{Bmatrix} = \begin{bmatrix} -1.0 & 0.0 & 1.0 & 0.0 & 0.0 \\ -0.6 & 0.8 & 0.0 & 0.0 & 0.6 \\ 0.0 & 1.0 & 0.0 & 0.0 & 0.0 \\ 0.0 & 0.0 & 0.0 & 1.0 & 0.0 \\ 0.0 & 0.0 & 0.6 & 0.8 & 0.0 \\ 0.0 & 0.0 & 0.0 & 0.0 & 1.0 \end{bmatrix} \begin{Bmatrix} d_1 \\ d_2 \\ d_3 \\ d_4 \\ d_5 \\ d_6 \end{Bmatrix}$$

and expressed in matrix notation as

$$e = B \cdot d \tag{3.5}$$

where the matrix B is known as a displacement transformation matrix or a kinematics matrix. It may be seen that the matrix B is the transpose of the matrix A, and this is a necessary condition as may be seen from the proof presented in chapter 7.

In summary, the equations may be written as

$$P = A \cdot f \tag{i}$$

$$f = S \cdot e \tag{ii}$$

and

$$e = B \cdot d \tag{iii}$$

Substituting equation (iii) into equation (ii) gives

$$f = S \cdot B \cdot d \tag{iv}$$

and substituting equation (iv) into (i) gives

$$P = A \cdot S \cdot B \cdot d \tag{v}$$

which can be written as

$$P = K \cdot d \tag{3.6}$$

where K is the structure stiffness matrix now given by the multiplication of three defined matrices as $K = A \cdot S \cdot B$.

Equation (3.6) can be solved for a given set of loads to yield the displacements. With the displacements known, the internal forces can be calculated from the equation (iv), in a back-substitution process. Although the technique has been presented in terms of a particular problem, it is readily seen that it is a general one applicable to both statically determinate and statically indeterminate trusses.

The method of forming the structure stiffness matrix by matrix multiplication can be applied to other classes of structure such as continuous beams. However it has the disadvantage of involving the multiplication of large matrices for structures with a high degree of freedom, and other techniques are more frequently used. The principles expressed in the technique are fundamental to the matrix stiffness method and it is worth while to review briefly the process before proceeding further. The initial objective of the stiffness method is to develop a relationship between the external loads applied to the structure and the resulting displacements. This relationship is given by the structure stiffness matrix. The external loads are related to the internal actions on the elements through the nodal equilibrium equations, while the internal actions are related to the element displacements (strains) by the element stiffness expressed through a stress–strain law. The element displacements are then linked to the structure displacements by the requirements of compatibility. Finally, the boundary conditions are applied either by inference or by direct consideration, as will be seen in the following section. The solution for the displacements follows from the formation of the stiffness matrix and the internal actions are recovered by back-substitution.

The boundary conditions have been implicitly taken into account through the nomination of zero displacements at the restraints, as shown in figure 3.2. However it is possible to take a more general approach than that outlined so far and include both the reactions and the restraint displacements. This is done through augmenting the statics matrix and the kinematics matrix and by incorporating the reactions and the restrained displacements in general load and displacement vectors respectively.

Applying the equations of equilibrium to the restrained nodes with respect to the restrained degrees of freedom gives the following additional equations:

$$P_6 = 0.0f_1 - 0.8f_2 + 0.0f_3 - 1.0f_4 + 0.0f_5 + 0.0f_6$$

$$P_7 = 0.0f_1 + 0.0f_2 + 0.0f_3 + 0.0f_4 - 0.6f_5 - 1.0f_6$$

$$P_8 = 0.0f_1 + 0.0f_2 - 1.0f_3 + 0.0f_4 - 0.8f_5 + 0.0f_6$$

which may be written as

$$\begin{Bmatrix} P_6 \\ P_7 \\ P_8 \end{Bmatrix} = \begin{bmatrix} 0.0 & -0.8 & 0.0 & -1.0 & 0.0 & 0.0 \\ 0.0 & 0.0 & 0.0 & 0.0 & -0.6 & -1.0 \\ 0.0 & 0.0 & -1.0 & 0.0 & -0.8 & 0.0 \end{bmatrix} \begin{Bmatrix} f_1 \\ f_2 \\ f_3 \\ f_4 \\ f_5 \\ f_6 \end{Bmatrix}$$

and expressed in the notation

$$P_R = A_R \cdot f$$

The terms P_6 to P_8 describe the reactions and follow the ordered set of terms P_1 to P_5. In more general terms, equation (3.3) may be written as

$$\left\{\begin{array}{c} P_F \\ \hline P_R \end{array}\right\} = \left[\begin{array}{c} A_F \\ \hline A_R \end{array}\right]\{f\} \tag{3.7}$$

where the general load vector P has now been partitioned to incorporate the load terms, P_F, and the reactions, P_R, so that P and A of equation (3.3) now become P_F and A_F respectively.

In a similar manner it can be seen that in general terms equation (3.5) can be written as

$$\{e\} = [B_F \mid B_R]\left\{\begin{array}{c} d_F \\ \hline d_R \end{array}\right\} \tag{3.8}$$

where the compatibility equations have been extended to include prescribed displacements at the restraints.

The general displacement vector has now been partitioned to incorporate the unrestrained displacements, d_F, and the restrained displacements, d_R. The terms of d_R for the example under study are d_6 to d_8 and they follow the ordered set of terms d_1 to d_5. The terms B and d of equation (3.5) now become B_F and d_F respectively. The result is that the more general expression of the structure stiffness is given in the equation

$$\left\{\begin{array}{c} P_F \\ \hline P_R \end{array}\right\} = \left[\begin{array}{c} A_F \\ \hline A_R \end{array}\right][S][B_F \mid B_R]\left\{\begin{array}{c} d_F \\ \hline d_R \end{array}\right\} \tag{3.9}$$

where the matrix $[B_F \mid B_R]$ is still given as $\left[\begin{array}{c} A_F \\ \hline A_R \end{array}\right]^T$ and the result may be written as

$$\left\{\begin{array}{c} P_F \\ \hline P_R \end{array}\right\} = \left[\begin{array}{c|c} K_F & C \\ \hline C^T & K_R \end{array}\right]\left\{\begin{array}{c} d_F \\ \hline d_R \end{array}\right\} \tag{3.10}$$

where K_F is stiffness matrix relating to the unrestrained displacements,
K_R is the reaction stiffness matrix,
and C is a connection stiffness matrix with the transpose C^T.

Equation (3.10) is the general form of equation (3.6) which should now be written in the particular form of $P_F = K_F \cdot d_F$.

3.3 SOLUTION TO OBTAIN DISPLACEMENTS

While equation (3.10) has been developed from a consideration of the analysis of a truss, it will be seen that it is in fact a general form of presentation of structure stiffness in a load-displacement relationship. As such it is appropriate to consider how the solution can proceed from such a relationship.

Equation (3.10) can be expanded by matrix multiplication to give the matrix equations

$$P_F = K_F \cdot d_F + C \cdot d_R \tag{3.10a}$$

$$P_R = C^T \cdot d_F + K_R \cdot d_R \tag{3.10b}$$

A solution follows from equation (3.10a) since

$$d_F = K_F^{-1} \cdot P_F - K_F^{-1} \cdot C \cdot d_R$$

and d_R is known as a prescribed set of values (often zero). With d_F known, the reactions can be determined from equation (3.10b) and the internal actions can be recovered from the action–displacement relationships.

The solution technique implied is not a preferred method although it is convenient to describe the solution in this manner at this stage. One of the problems is that the equations may not have been ordered in such a way that the load and displacement vectors, and subsequently the stiffness matrix, can be conveniently partitioned as required by equation (3.10). Of course the equations can be re-arranged since it should always be remembered that the relationship simply presents a set of linear simultaneous equations. However this is a tedious process and alternative procedures are available and will be presented in chapter 6. A further disadvantage is that the solution requires the inverse of the matrix K_F, particularly when d_R is specified as non-zero. While the solution of linear equations is readily given through the inverse of a matrix, this is not a computationally efficient method. Particular solution techniques, such as Gaussian Elimination or Choleski Decomposition, are more efficient and should be used wherever possible. Details will not be presented here, although it is appropriate to point out that the SOLVE routine of the matrix manipulation program MATOP is based on Gaussian Elimination.

In spite of the adverse comments, the solution presented is a concise way of introducing the topic and it does offer ready understanding. In many instances d_R is zero so that solution immediately follows from

$$P_F = K_F \cdot d_F$$

This is the standard form of the solution of a set of linear simultaneous equations, often expressed in mathematics texts as the problem

$$Ax = B$$

where x is the vector of unknowns, A is the matrix of coefficients and B is the vector of constants or right-hand side terms.

3.4 NATURE OF THE STRUCTURE STIFFNESS MATRIX

There are several characteristics of a structure stiffness matrix that are worthy of comment. The first of these is the rather obvious statement that the matrix is square. This follows from the one-to-one correspondence that exists between the terms of the load and displacement vectors. In addition the matrix is necessarily symmetrical, although this will not be proved here since the proof is dependent on a subsequent study. For a stable structure the terms on the leading diagonal must all be positive and typically they are dominant. That is to say the diagonal term is greater than the sum of the off-diagonal terms of the same row.

If an attempt is made to analyse a geometrically unstable structure, the resulting stiffness matrix will be deficient and the solution technique should be capable of detecting this. For instance, if a diagonal element were to be omitted from a panel of an otherwise statically determinate truss, the truss would have no shear capacity and it would be unstable. A computer-based solution routine should detect this and probably display an error message advising that the matrix is singular. A similar situation will arise if the restraints are either insufficient in number or incorrectly arranged.

Generally, a structure stiffness matrix is sparse and banded because of the nature of the connectivity between the elements. A sparse matrix is simply one with many zero terms, while a banded matrix has the non-zero terms concentrated about the leading diagonal. This is not apparent in the example treated at this stage, but it will become evident later on. Most of

---

Example 3.1: Analysis of a Truss

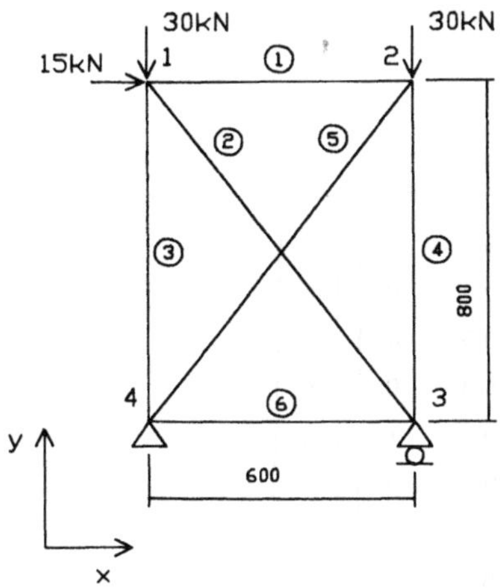

Given data:

E = 200 kN/mm^2 throughout
Ax = 800 mm^2 (diagonals)
Ax = 1200 mm^2 (others)

Procedure: Form complete structure stiffness by matrix multiplication, $K = A \cdot S \cdot B = A \cdot S \cdot A^T$

Statics Matrix:

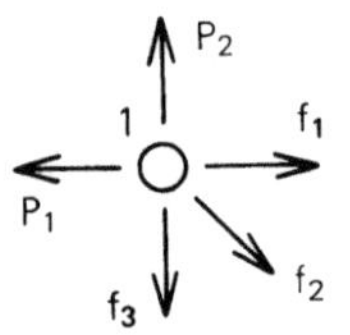

$P_1 = -f_1 - 0.6f_2;$ $\quad P_2 = 0.8f_2 + f_3$

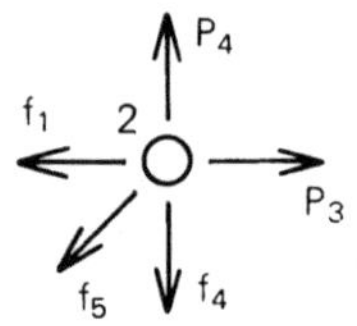

$P_3 = f_1 + 0.6f_5;$ $\quad P_4 = f_4 + 0.8f_5$

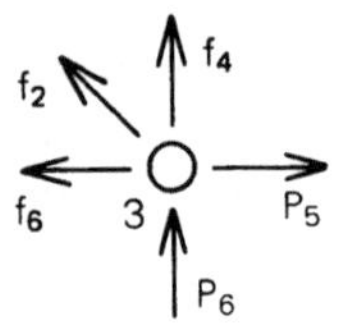

$P_5 = 0.6f_2 + f_6;$ $\quad P_6 = -0.8f_2 - f_4$

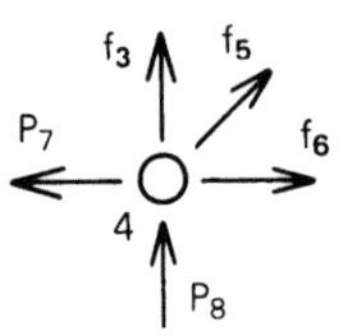

$P_7 = -0.6f_5 - f_6;$ $\quad P_8 = -f_3 - 0.8f_5$

Hence

$$\begin{Bmatrix} P_1 \\ P_2 \\ P_3 \\ P_4 \\ P_5 \\ P_6 \\ P_7 \\ P_8 \end{Bmatrix} = \begin{bmatrix} -1 & -0.6 & 0 & 0 & 0 & 0 \\ 0 & 0.8 & 1 & 0 & 0 & 0 \\ 1 & 0 & 0 & 0 & 0.6 & 0 \\ 0 & 0 & 0 & 1 & 0.8 & 0 \\ 0 & 0.6 & 0 & 0 & 0 & 1 \\ 0 & -0.8 & 0 & -1 & 0 & 0 \\ 0 & 0 & 0 & 0 & -0.6 & -1 \\ 0 & 0 & -1 & 0 & -0.8 & 0 \end{bmatrix} \begin{Bmatrix} f_1 \\ f_2 \\ f_3 \\ f_4 \\ f_5 \\ f_6 \end{Bmatrix}$$

Matrix of Element Stiffnesses:

$$f_i = \frac{E_i A_i}{l_i} e_i$$

$$\begin{Bmatrix} f_1 \\ f_2 \\ f_3 \\ f_4 \\ f_5 \\ f_6 \end{Bmatrix} = \begin{bmatrix} 400 & 0 & 0 & 0 & 0 & 0 \\ 0 & 160 & 0 & 0 & 0 & 0 \\ 0 & 0 & 300 & 0 & 0 & 0 \\ 0 & 0 & 0 & 300 & 0 & 0 \\ 0 & 0 & 0 & 0 & 160 & 0 \\ 0 & 0 & 0 & 0 & 0 & 400 \end{bmatrix} \begin{Bmatrix} e_1 \\ e_2 \\ e_3 \\ e_4 \\ e_5 \\ e_6 \end{Bmatrix}$$

With the matrices A and S defined, the solution can proceed by matrix operations in any suitable manner, the program MATOP of appendix B is specifically designed for the necessary operations. For this example, the MATOP input file is as follows:

```
REMARK. Example 3.1  Chapter 3
LOAD.A.8.6  The Full Statics Matrix
-1 -0.6 0 0 0 0
0 0.8 1 0 0 0
1 0 0 0 0.6 0
0 0 0 1 0.8 0
0 0.6 0 0 0 1
0 -0.8 0 -1 0 0
0 0 0 0 -0.6 -1
0 0 -1 0 -0.8 0
NULL.S.6.6
MODDG.S  Create the matrix of element stiffnesses
6
1 400
2 160
3 300
4 300
5 160
6 400
TRANS.A.B  The Displacement Transformation Matrix
MULT.A.S.TEMP
MULT.TEMP.B.K
PRINT.K  The Structure Stiffness Matrix
SELECT.KF.K.5.5.1.1
LOAD.PF.5.1  The Load Vector
15
-30
0
-30
0
PRINT.PF  The Load Vector
SOLVE.KF.PF
PRINT.PF  Now the Displacement Vector DF
REMARK.  Back substitute to find Forces
SELECT.BF.B.6.5.1.1
DELETE.TEMP
MULT.S.BF.TEMP
MULT.TEMP.PF.FORCES
PRINT.FORCES  The Element Actions
SELECT.CT.K.3.5.6.1
MULT.CT.PF.REACT
PRINT.REACT  The Reactions
QUIT
```

The operations can be understood by referring to the manual of MATOP in appendix A. In the first phase of the operations, the structure stiffness matrix is formed and

$$P = K \cdot d$$

is found to be

$$\begin{Bmatrix} P_1 \\ P_2 \\ P_3 \\ P_4 \\ P_5 \\ P_6 \\ P_7 \\ P_8 \end{Bmatrix} = \begin{bmatrix} 457.6 & -76.8 & -400 & 0 & -57.6 & 76.8 & 0 & 0 \\ -76.8 & 402.4 & 0 & 0 & 76.8 & -102.4 & 0 & -300 \\ -400 & 0 & 457.6 & 76.8 & 0 & 0 & 57.6 & -76.8 \\ 0 & 0 & 76.8 & 402.4 & 0 & -300 & -76.8 & -102.4 \\ -57.6 & 76.8 & 0 & 0 & 457.6 & -76.8 & -400 & 0 \\ 76.8 & -102.4 & 0 & -300 & -76.8 & 402.4 & 0 & 0 \\ 0 & 0 & -57.6 & -76.8 & -400 & 0 & 457.6 & 76.8 \\ 0 & -300 & -76.8 & -102.4 & 0 & 0 & 76.8 & 402.4 \end{bmatrix} \begin{Bmatrix} d_1 \\ d_2 \\ d_3 \\ d_4 \\ d_5 \\ d_6 \\ d_7 \\ d_8 \end{Bmatrix}$$

which can be expressed in the form

$$\begin{Bmatrix} P_F \\ \hline P_R \end{Bmatrix} = \left[\begin{array}{c|c} K_F & C \\ \hline C^T & K_R \end{array}\right] \begin{Bmatrix} d_F \\ \hline d_R \end{Bmatrix}$$

The solution for d_F follows from $P_F = K_F \cdot d_F$ since d_R is zero.

The remaining operations give the displacements, the element forces and the reactions, and the relevant part of the output file is

```
SELECT.KF.K.5.5.1.1
LOAD.PF.5.1  The Load Vector
PRINT.PF  The Load Vector
0.150000E+02
-.300000E+02
0.000000E+00
-.300000E+02
0.000000E+00
SOLVE.KF.PF
PRINT.PF  Now the Displacement Vector DF
0.193403E+00
-.436864E-01
0.187579E+00
-.110353E+00
0.316764E-01
REMARK.  Back substitute to find Forces
SELECT.BF.B.6.5.1.1
DELETE.TEMP
MULT.S.BF.TEMP
MULT.TEMP.PF.FORCES
PRINT.FORCES  The Element Actions
-.232944E+01
-.211176E+02
-.131059E+02
-.331059E+02
0.388240E+01
0.126706E+02
SELECT.CT.K.3.5.6.1
MULT.CT.PF.REACT
PRINT.REACT  The Reactions
0.500000E+02
-.150000E+02
0.100000E+02
End of File
```

Summary of Solution

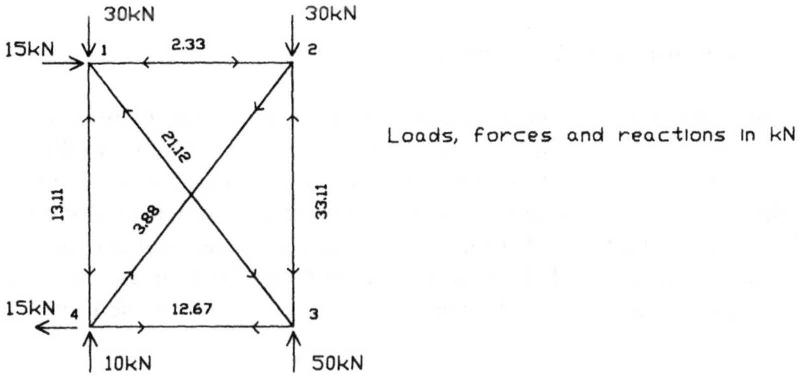

the solution techniques available reduce the computational effort by using the fact that the matrix is symmetrical and banded.

The truss of figure 3.2 is analysed in example 3.1 using the matrix stiffness method with the formation of the structure stiffness matrix based on the global stiffness approach.

3.5 DEVELOPMENT OF THE SLOPE–DEFLECTION EQUATIONS

The beam is an important structural element both in its own right as a single element structure under various end conditions and as part of a more complex structure. As such it is important to understand the behaviour of a beam with regard to its response under load. The beam is a flexural element transmitting load primarily through bending action and its response to load can be measured in terms of rotation and translation at the nodes. Classical studies of the deflection of beams begin with the governing differential equation relating bending moment and curvature in a flexural element. The equation is known as the Bernoulli–Euler equation after the 18th Century mathematicians who first proposed and solved it, and is given by the expression

$$\frac{d^2y}{dx^2}=\frac{M(x)}{EI} \tag{3.11}$$

While it is possible to proceed directly to the solution of the differential equation and obtain general expressions for slope (rotation) and deflection, an important interpretation of the solution is given by the moment-area theorems. These theorems are particularly attractive since they are based on properties of area of the bending moment diagram of the beam which are familiar to both students and engineers.

3.5.1 The Moment–Area Theorems

Consider an element 1–2 taken from any region of a flexural member with second moment of area, I, and modulus of elasticity, E, as shown in figure 3.4(a). The element is subjected to the end actions of moment and shear to maintain equilibrium with any arbitrary transverse load acting along its length. The bending moment, $M(x)$, over the length of the element can be readily evaluated in terms of the end actions and the load, and it may be conveniently plotted as an M/EI diagram as shown superimposed on the element.

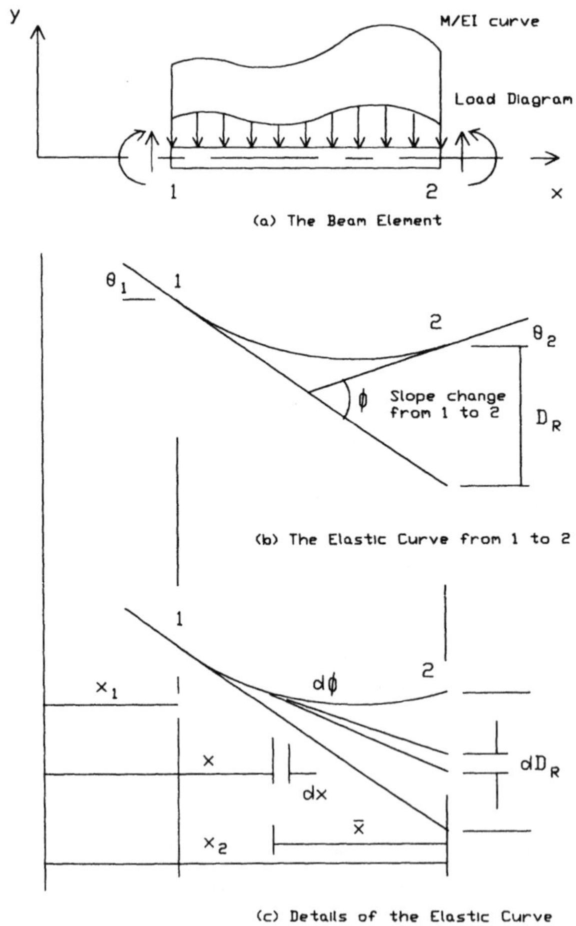

Figure 3.4 Beam element in bending.

The bending moment produces the elastic curve (that is, the deflected shape of the beam centre line) of figure 3.4(b), where the ends of the element have rotated and translated relative to the initial position of the unloaded beam. It is assumed that the flexural displacements are small and that the deformation caused by shear stresses is negligible. The first of the moment-area theorems is concerned with the change of slope over the length of the element. This is given directly by the expression of equation (3.11) since

$$\frac{d^2y}{dx^2} = \frac{d\phi}{dx}$$

where ϕ is the slope of the element at any point x. Thus

$$\frac{d\phi}{dx} = \frac{M}{EI} \tag{3.12}$$

and

$$\phi = \int_{x_1}^{x_2} \frac{M}{EI}\, dx \tag{3.13}$$

The integral of equation (3.13) can be interpreted as the area of the M/EI diagram between the ends of the element. This leads to a formal statement of the first moment-area theorem as

The change in slope between two points on a flexural member is equal to the area of the M/EI diagram between those points

The second of the moment-area theorems is concerned with the deflection of one point with respect to a tangent taken at another point. In figure 3.4(b) the deflection of point 2 with respect to a tangent at point 1 is defined as D_R. From figure 3.4(c) it is apparent that D_R is built up from increments such as δD_R due to the slope change over the incremental length dx. The magnitude of δD_R is given by

$$\delta D_R = \bar{x}\, d\phi$$

where $\bar{x}$ is the distance from point 2 to the increment dx. Integrating over the length of the element gives

$$D_R = \int_{x_1}^{x_2} \bar{x}\frac{M}{EI}\, dx \tag{3.14}$$

The integral of equation (3.14) can be interpreted as the first moment of area of the M/EI diagram between points 1 and 2, taken about point 2. This leads to a formal statement of the second moment-area theorem as

The deflection of a point 2 on a flexural member, taken relative to a tangent from point 1, is equal to the first moment of area of the M/EI diagram between those points, taken about point 2

The two theorems provide a direct method of calculating slopes and deflections in flexural elements. For instance, consider the simple cantilever beam of figure 3.5(a), shown with both the deflected shape and the bending moment diagram imposed on the beam. The deflection at the free end is given directly by the second moment-area theorem, since the tangent at point 1 is coincident with the line of the undeflected beam. Hence

$$d_2 = \frac{1}{EI}(\tfrac{1}{2} \cdot L \cdot QL \cdot \tfrac{2}{3}L)$$

$$= \frac{QL^3}{3EI}$$

The first moment-area theorem gives the slope at end 2 directly, again because the slope at point 1 is zero. Then

$$\theta_2 = \frac{1}{EI}(\tfrac{1}{2} \cdot L \cdot QL)$$

$$= \frac{QL^2}{2EI}$$

The theorems say nothing about the sign associated with either of the terms evaluated, and although a sign convention can be introduced it is not considered necessary. An appropriate sign can be assigned to the results based on a sign convention that will be introduced in section 3.5.2.

A further example is given by considering the simply supported beam of figure 3.5(b). In this case, taking advantage of symmetry, it is appropriate to consider point 1 at mid span on the deflection curve. The slope at point 1 is zero and the tangent is now parallel to the line of the undeflected beam. The displacement of point 2 from the tangent at point 1 is then the mid span deflection of the beam. Hence

$$d_2 = \frac{1}{EI}\left(\frac{1}{2} \cdot \frac{L}{2} \cdot \frac{QL}{4} \cdot \frac{2}{3} \cdot \frac{L}{2}\right)$$

$$= \frac{QL^3}{48EI}$$

Again, since the slope at point 1 is zero, the rotation at either end is given in magnitude by

$$\theta_2 = \frac{1}{EI}\left(\frac{1}{2} \cdot \frac{L}{2} \cdot \frac{QL}{4}\right)$$

$$= \frac{QL^2}{16EI}$$

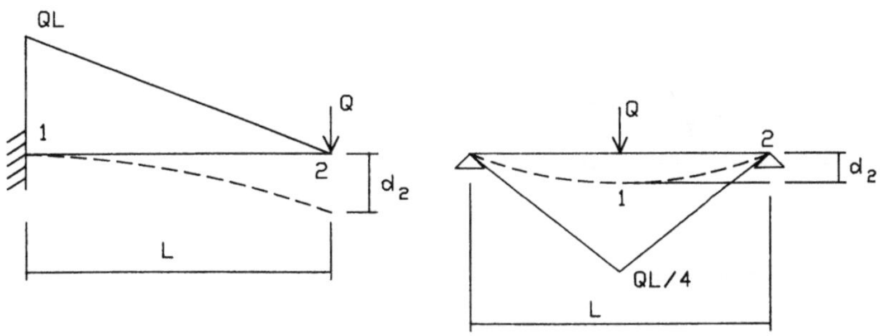

Figure 3.5 Application of the moment-area theorems.

Clearly the use of the theorems is dependent on a knowledge of the properties of area, and some appropriate details for standard moment diagram shapes are given in appendix B.

It is possible to generalise the approach, outlined by the two examples given, to calculate the slopes and deflections at any point in a beam. However this will be left as an exercise for the student, while the moment-area theorems will be used to develop the beam element stiffness matrix and the slope deflection equations.

3.5.2 The Beam Element Stiffness Matrix

Consider now the general behaviour of a flexural element as shown in figure 3.6(a). The beam element is subjected to end moments and shears as a result of which it deforms in the general manner shown relative to the set of axes taken through point 1. The element is of length, L, and has flexural properties defined by EI, where E is Young's Modulus and I is the second moment of area of the beam section. The deformation can be considered as the combination of the deformations shown in figures 3.6(b) and (c), and it is convenient to consider the action-displacement relationship for the element in terms of these two patterns before combining the result. It is important to node that all of the terms shown in figure 3.6(a) are shown in a positive sense and that, as such, they define the sign convention to be used.

The moment-area theorems may be applied to the element of figure 3.6(b), where the slope change from point 1 to point 2 is $\theta_2 - \theta_1$ and where the deflection of point 2 from a tangent at point 1 is equal to $-\theta_1 L$. Hence the first moment-area theorem gives

$$\theta_2 - \theta_1 = \frac{1}{EI}(-\tfrac{1}{2} \cdot m_{12} \cdot L + \tfrac{1}{2} \cdot m_{21} \cdot L) \tag{3.15}$$

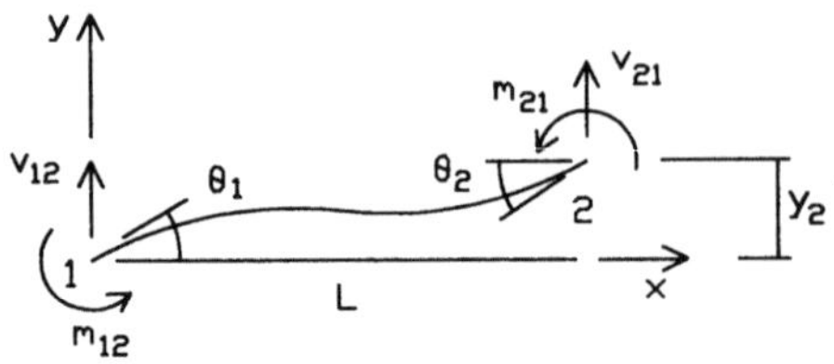

(a) General Deformation

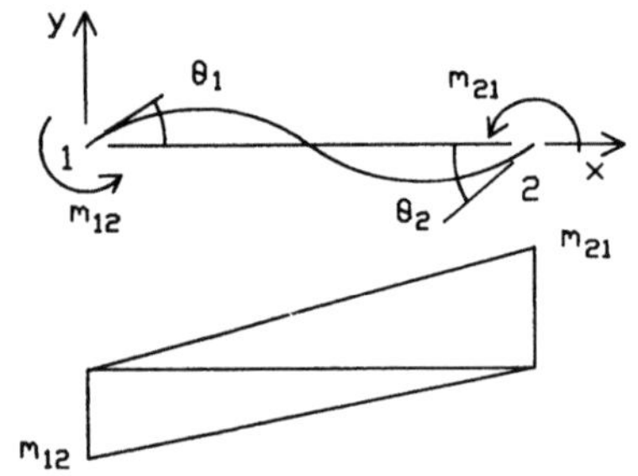

(b) End Rotation Only

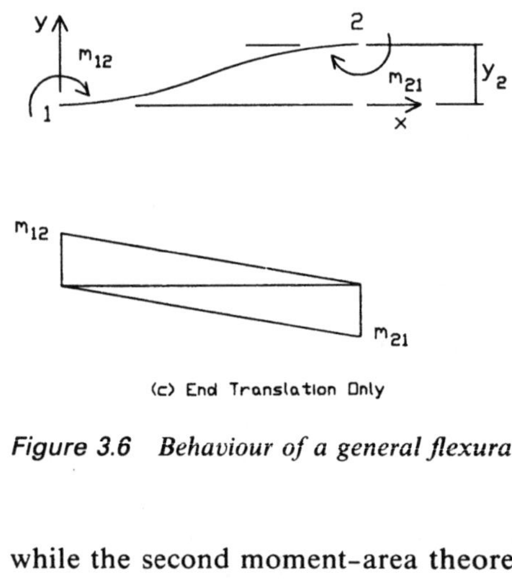

(c) End Translation Only

Figure 3.6 Behaviour of a general flexural element.

while the second moment-area theorem gives

$$-\theta_1 L = \frac{1}{EI}\left(-\tfrac{1}{2}\cdot m_{12}\cdot L\cdot \tfrac{2}{3}L + \tfrac{1}{2}\cdot m_{21}\cdot L\cdot \tfrac{1}{3}L\right) \tag{3.16}$$

Equations (3.15) and (3.16) may be written as

$$\theta_2 - \theta_1 = -\frac{m_{12}L}{2EI} + \frac{m_{21}L}{2EI}$$

$$-\theta_1 L = -\frac{m_{12}L^2}{3EI} + \frac{m_{21}L^2}{6EI}$$

and solving for θ_1 and θ_2 gives

$$\theta_1 = \frac{m_{12}L}{3EI} - \frac{m_{21}L}{6EI}$$

$$\theta_2 = -\frac{m_{12}L}{6EI} + \frac{m_{21}L}{3EI}$$

The results may be written in matrix form as

$$\begin{Bmatrix} \theta_1 \\ \theta_2 \end{Bmatrix} = \begin{bmatrix} \dfrac{L}{3EI} & -\dfrac{L}{6EI} \\ -\dfrac{L}{6EI} & \dfrac{L}{3EI} \end{bmatrix} \begin{Bmatrix} m_{12} \\ m_{21} \end{Bmatrix} \tag{3.17}$$

Solving equation (3.17) for m_{12} and m_{21} results in

$$m_{12} = \frac{4EI}{L}\theta_1 + \frac{2ET}{L}\theta_2$$

$$m_{21} = \frac{2EI}{L}\theta_1 + \frac{4EI}{L}\theta_2$$

which can be written in matrix form as

$$\begin{Bmatrix} m_{12} \\ m_{21} \end{Bmatrix} = \begin{bmatrix} \dfrac{4EI}{L} & \dfrac{2EI}{L} \\ \dfrac{2EI}{L} & \dfrac{4EI}{L} \end{bmatrix} \begin{Bmatrix} \theta_1 \\ \theta_2 \end{Bmatrix} \tag{3.18}$$

The moment-area theorems can be applied to the element of figure 3.6(c) in a similar manner. In this case the slope change from point 1 to point 2 is zero, while the deflection of point 2 from a tangent at point 1 is equal to y_2. Hence the first moment-area theorem gives

$$0 = \tfrac{1}{2} \cdot m_{12} \cdot \frac{L}{EI} - \tfrac{1}{2} \cdot m_{21} \cdot \frac{L}{EI} \tag{3.19}$$

and the second moment-area theorem gives

$$y_2 = \tfrac{1}{2} \cdot m_{12} \cdot \frac{L}{EI} \cdot \tfrac{2}{3}L - \tfrac{1}{2} \cdot m_{21} \cdot \frac{L}{EI} \cdot \tfrac{1}{3} \cdot L \tag{3.20}$$

From equation (3.19), $m_{12} = m_{21}$ as would be expected from the symmetry of problem. Equation (3.20) thus gives

$$y_2 = \frac{m_{12}L^2}{EI}\left(\tfrac{1}{3} - \tfrac{1}{6}\right)$$

$$= \frac{m_{12}L^2}{6EI} \tag{3.21}$$

The end moments that must be developed to prevent end rotation, while relative end translation occurs, are thus seen to be equal in magnitude to $(6EI/L^2)y_2$. However, they are necessarily of opposite sense to the end moments, defined in figure 3.6(a) as being positive. For this reason the action-displacement relationship for figure 3.6(a) can be written as

$$m_{12}=\frac{4EI}{L}\theta_1+\frac{2EI}{L}\theta_2-\frac{6EI}{L^2}y_2 \tag{3.22}$$

$$m_{21}=\frac{2EI}{L}\theta_1+\frac{4EI}{L}\theta_2-\frac{6EI}{L^2}y_2 \tag{3.23}$$

The end shears are readily introduced into the relationship through the requirements of equilibrium. Taking moments about point 2 gives

$$v_{12}L=m_{12}+m_{21}$$

and for equilibrium in the y direction:

$$v_{21}=-v_{12}$$

The more general relationship applies to the beam element specified with respect to the x–y coordinates as shown in figure 3.7. The displacement y_2 as previously defined in figure 3.6 is now equal to d_2-d_1. Thus

$$\begin{aligned}
m_{12}&=\frac{4EI}{L}\theta_1+\frac{2EI}{L}\theta_2+\frac{6EI}{L^2}d_1-\frac{6EI}{L^2}d_2\\
m_{21}&=\frac{2EI}{L}\theta_1+\frac{4EI}{L}\theta_2+\frac{6EI}{L^2}d_1-\frac{6EI}{L^2}d_2\\
v_{12}&=\frac{6EI}{L^2}\theta_1+\frac{6EI}{L^2}\theta_2+\frac{12EI}{L^3}d_1-\frac{12EI}{L^3}d_2\\
v_{21}&=-\frac{6EI}{L^2}\theta_1-\frac{6EI}{L^2}\theta_2-\frac{12EI}{L^3}d_1+\frac{12EI}{L^3}d_2
\end{aligned} \tag{3.24}$$

The first two of the above set of simultaneous equations are usually referred to as the slope-deflection equations, although they are often written in a form that includes the effect of transverse load on the beam.

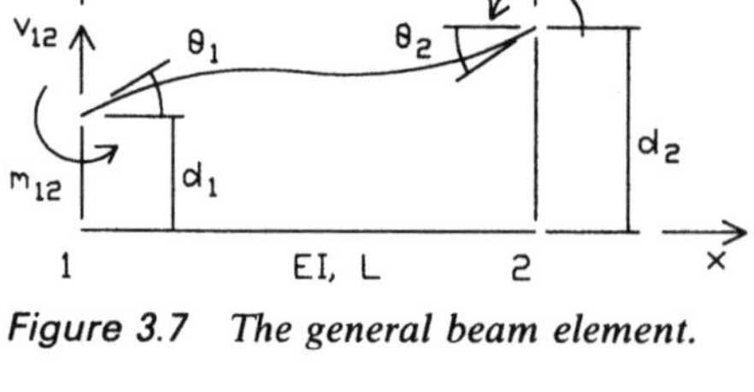

Figure 3.7 The general beam element.

The equations may be written in matrix form as

$$\begin{Bmatrix} v_{12} \\ m_{12} \\ \hline v_{21} \\ m_{21} \end{Bmatrix} = \left[\begin{array}{cc|cc} \frac{12EI}{L^3} & \frac{6EI}{L^2} & -\frac{12EI}{L^3} & \frac{6EI}{L^2} \\ \frac{6EI}{L^2} & \frac{4EI}{L} & -\frac{6EI}{L^2} & \frac{2EI}{L} \\ \hline -\frac{12EI}{L^3} & -\frac{6EI}{L^2} & \frac{12EI}{L^3} & -\frac{6EI}{L^2} \\ \frac{6EI}{L^2} & \frac{2EI}{L} & -\frac{6EI}{L^2} & \frac{4EI}{L} \end{array}\right] \begin{Bmatrix} d_1 \\ \theta_1 \\ \hline d_2 \\ \theta_2 \end{Bmatrix} \tag{3.25}$$

It appears that the matrix expression of equation (3.25) represents four equations with four unknowns. However there are really only three unknowns, namely θ_1, θ_2 and the relative displacement between the beam ends conveniently specified as $(d_2 - d_1)$. It follows then that solutions are only possible when certain restraints are placed on the element, either in its own right or as part of a structure.

The beam element stiffness matrix is given in the action-displacement relationship of equation (3.25) where the actions and displacements are defined by figure 3.7. The matrix is more generally referred to as the continuous beam element stiffness matrix, since it is generally used in that context as will be seen in chapter 4.

3.6 APPLICATION TO SOME SIMPLE BEAM PROBLEMS

As was the case with the moment-area theorems, the slope-deflection equations as presented in equations (3.24) may be applied directly to beam elements to obtain some useful results. However, it is of more interest to consider the equations in the form of the beam element stiffness matrix, using a single element to define a beam and applying the necessary boundary conditions. Figure 3.8 shows a series of such cases where the loads and reactions become equivalent to the element end actions. To satisfy equilibrium there must be a minimum of three independent reaction components and, consistent with the reactions, the displacements there must be zero. It is possible for the boundary displacements to be specified at some non-zero value, as in the case of support settlement, but that will not be considered at this stage.

The effect of specifying certain displacements as zero is to reduce the number of equations to correspond to the number of unknown displacements. Effectively, any coefficients in the matrix related to a zero displacement by matrix multiplication can be ignored. At the same time, any equation which relates the reactions to the displacements is also ignored at this stage.

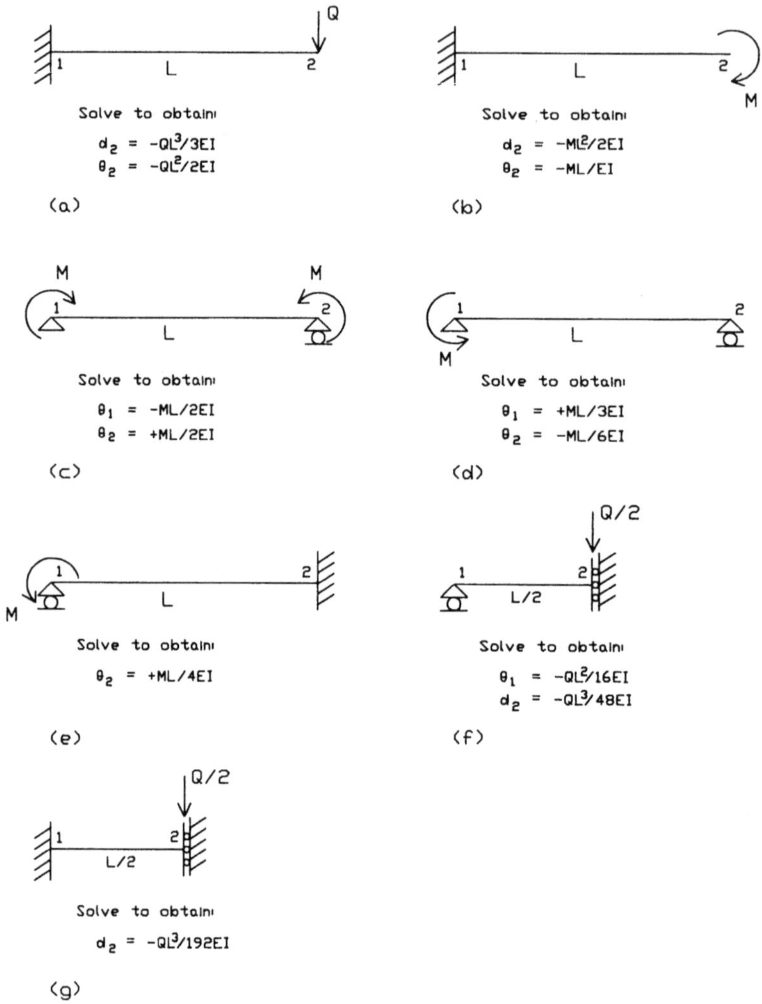

Figure 3.8 Single element beam problems.

The beam problem of figure 3.8(a), where the element is set up as a simple cantilever beam carrying the point load Q at the free end, is solved in example 3.2.

Considering the results of example 3.2, with the displacements of the nodes now known, including the zero values specified as the boundary conditions, the full action-displacement matrix relationship can be used to recover all the actions by matrix multiplication. This operation will give both the applied loads and the reactions and thus allow for a full description of the solution to the beam problems.

Example 3.2: Deflection of a Cantilever Beam (1)

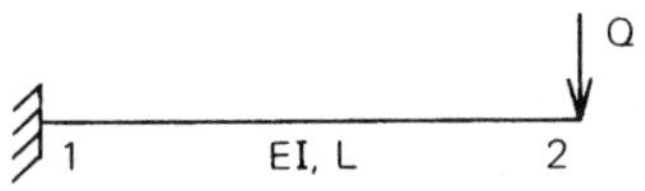

Solve to obtain:

$$d_2 = -\frac{QL^3}{3EI}; \qquad \theta_2 = -\frac{QL^2}{2EI}$$

Given:

$$\begin{Bmatrix} v_{12} \\ m_{12} \\ \hdashline v_{21} \\ m_{21} \end{Bmatrix} = \left[\begin{array}{cc:cc} \frac{12EI}{L^3} & \frac{6EI}{L^2} & -\frac{12EI}{L^3} & \frac{6EI}{L^2} \\ \frac{6EI}{L^2} & \frac{4EI}{L} & -\frac{6EI}{L^2} & \frac{2EI}{L} \\ \hdashline -\frac{12EI}{L^3} & -\frac{6EI}{L^2} & \frac{12EI}{L^3} & -\frac{6EI}{L^2} \\ \frac{6EI}{L^2} & \frac{2EI}{L} & -\frac{6EI}{L^2} & \frac{4EI}{L} \end{array}\right] \begin{Bmatrix} d_1 \\ \theta_1 \\ \hdashline d_2 \\ \theta_2 \end{Bmatrix}$$

Applying the boundary conditions, $d_1 = \theta_1 = 0$, and the known actions at node 2, the remaining equations are given in the relationship

$$\begin{Bmatrix} -Q \\ 0 \end{Bmatrix} = \begin{bmatrix} \frac{12EI}{L^3} & -\frac{6EI}{L^2} \\ -\frac{6EI}{L^2} & \frac{4EI}{L} \end{bmatrix} \begin{Bmatrix} d_2 \\ \theta_2 \end{Bmatrix}$$

Expressed in a non-dimensional form, the matrix equation is

$$QL\begin{Bmatrix} -1 \\ 0 \end{Bmatrix} = \frac{EI}{L}\begin{bmatrix} 12 & -6 \\ -6 & 4 \end{bmatrix}\begin{Bmatrix} d_2/L \\ \theta_2 \end{Bmatrix}$$

and the basis of the solution is seen in

$$B = A \cdot X$$

where

$$B = \begin{Bmatrix} -1 \\ 0 \end{Bmatrix}, \qquad A = \begin{bmatrix} 12 & -6 \\ -6 & 4 \end{bmatrix} \quad \text{and} \quad X = \begin{Bmatrix} X_1 \\ X_2 \end{Bmatrix}$$

with the solution $X_1 = -\frac{1}{3}$ and $X_2 = -\frac{1}{2}$.

Hence

$$\begin{Bmatrix} d_2/L \\ \theta_2 \end{Bmatrix} = QL\frac{L}{EI}\begin{Bmatrix} -\frac{1}{3} \\ -\frac{1}{2} \end{Bmatrix}$$

that is,

$$d_2 = -\frac{QL^3}{3EI} \quad \text{and} \quad \theta_2 = -\frac{QL^2}{2EI}$$

For example 3.2 the appropriate relationship is

$$\begin{Bmatrix} V_1 L \\ M_1 \\ V_2 L \\ M_2 \end{Bmatrix} = \frac{EI}{L} \begin{bmatrix} 12 & 6 & -12 & 6 \\ 6 & 4 & -6 & 2 \\ -12 & -6 & 12 & -6 \\ 6 & 2 & -6 & 4 \end{bmatrix} \begin{Bmatrix} 0 \\ 0 \\ -\dfrac{QL^2}{3EI} \\ -\dfrac{QL^2}{2EI} \end{Bmatrix}$$

Expanding out gives:

$$V_1 L = \frac{EI}{L}\left(\frac{4QL^2}{EI} - \frac{3QL^2}{EI}\right), \qquad \therefore\ V_1 = +Q$$

$$M_1 = \frac{EI}{L}\left(\frac{2QL^2}{EI} - \frac{QL^2}{EI}\right), \qquad \therefore\ M_1 = +QL$$

$$V_2 L = \frac{EI}{L}\left(-\frac{4QL^2}{EI} + \frac{3QL^2}{EI}\right), \qquad \therefore\ V_2 = -Q$$

$$M_2 = \frac{EI}{L}\left(\frac{2QL^2}{EI} - \frac{2QL^2}{EI}\right), \qquad \therefore\ M_2 = 0$$

The results are of course confirmed by the requirements of equilibrium and are obvious since the problem is statically determinate. The procedure is valid, however, irrespective of whether or not the structure is statically determinate. This is because all the requirements of structural behaviour have been met—namely, equilibrium, compatibility, stress-strain laws and boundary conditions.

It is constructive now to consider the above approach in terms of the general stiffness matrix relationship of equation (3.10) of section 3.2, and its subsequent treatment in section 3.3. It may be noted that the load-displacement relationship takes the same form, as indeed it must do, and that the solution is obtained essentially as the solution to $P_F = K_F \cdot d_F$, since d_R is zero.

While many of the results to the problems of figure 3.8 may appear trivial, they represent some important relationships which recur from time to time in the general analysis of beams and frames. It is also useful to consider the results on the beams when d_R is specified not as zero but as some settlement of supports. For the indeterminate beams, such action will result in internal stress even when the beam does not carry any loads. The arrangement of figures 3.8(f) and 3.8(g) are of particular interest, since the boundary conditions, the span and the load have been selected so that the beams model the behaviour of firstly, a simply supported beam under a central load Q, and secondly, a built-in beam under a central load. In further studies in the next chapter, the general question of transverse loads on a beam will be introduced, but at this stage only nodal loads can be considered. This was another reason for the careful selection of figures 3.8(f) and 3.8(g).

3.7 STANDARD SOLUTIONS TO BEAM PROBLEMS

It is now possible to establish some standard solutions for statically indeterminate beams which are, at the same time, kinematically determinate. The standard solutions are of significance in their own right but more especially they are of importance as components in further studies. Although the predominant concern of this chapter is to present the basis of the stiffness method, the results to be presented here are actually based on a flexibility principle rather than a stiffness principle. The flexibility method is considered in detail in chapter 8. Both the moment-area theorems and the results of using the beam element stiffness matrix will be applied in the following problems.

Consider each of the beams of figure 3.9(a) which are restrained against rotation at both ends. The action of the loads is such that end moments

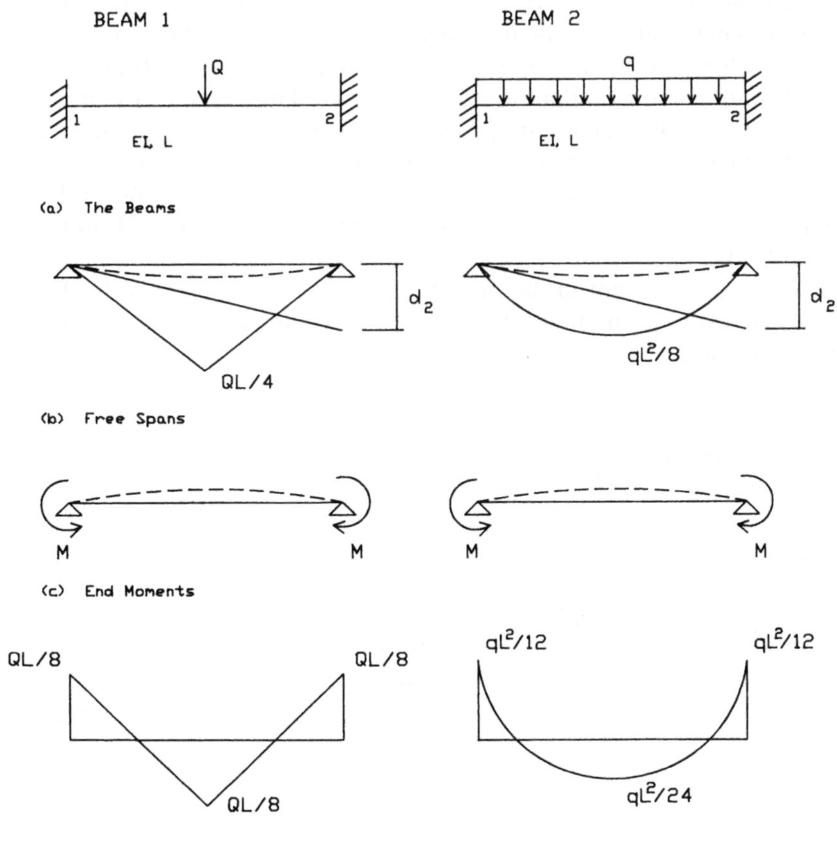

Figure 3.9 Two standard beam problems.

develop at nodes 1 and 2 and effectively prevent the ends from rotating. Suppose each beam were released from the moment restraint and allowed to rotate freely as shown in figure 3.9(b). The beams would now be simply supported beams and the rotations could be calculated using the moment-area theorems. In either case, d_2, being the deflection of 2 from a tangent at 1, is given by the second moment area theorem and the rotation at 1 is given by d_2/L.

For the beam under concentrated load the end rotations are of magnitude $QL^2/16EI$, while for the beam under the uniform load the end rotations are of magnitude $qL^3/24EI$. Now suppose end moments, as shown in figure 3.9(c), were to be applied to the simply supported beams to produce opposite rotations such that the combined effect of the end moments and the applied loads gave the condition of the original beams. The relationship between such moments and the rotation they produce is given as the solution to the beam problem of figure 3.8(c). On this basis the size of the end moments can be calculated taking due regard of sign convention as shown in table 3.1.

For zero net rotation for beam 1, the end moments must be of magnitude $QL/8$; while for beam 2, the end moments required are of magnitude $qL^2/12$. The final bending moment diagram can be obtained in each case by superposition of the bending moment diagrams for the simply supported beam and for the beam under end moments in both cases. The result is shown in figure 3.9(d) and it is also given, along with a range of other standard beam problems, in table B1.3 of the 'Structural Mechanics Students' Handbook' in appendix B.

The remaining beam problems given in table B1.3 of appendix B can be solved in a similar manner, although the algebraic expressions involved in some cases are difficult to handle.

Table 3.1 Beam rotations

	Beam 1		Beam 2	
Action	θ_1	θ_2	θ_1	θ_2
Applied load	$-\dfrac{QL^2}{16EI}$	$\dfrac{QL^2}{16EI}$	$-\dfrac{qL^3}{24EI}$	$\dfrac{qL^3}{24EI}$
Applied moment	$\dfrac{ML}{2EI}$	$-\dfrac{ML}{2EI}$	$\dfrac{ML}{2EI}$	$-\dfrac{ML}{2EI}$

3.8 PROBLEMS FOR SOLUTION

3.1 Analyse the pinjointed plane truss shown in figure P3.1 to find the displacements, element actions and reactions for both load cases. The area of each element is shown in brackets adjacent to the element. $E = 200\ \text{kN/mm}^2$.

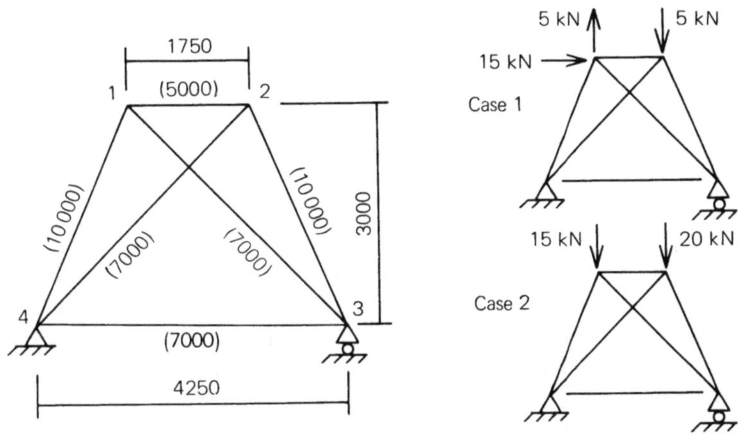

Figure P3.1

3.2 Repeat problem **3.1** for the situation where the roller support at node 3 has been replaced by a pin support, as for node 4, and the element 3–4 has been removed.

3.3 Calculate the forces and displacements in the truss of figure P3.1 due to a vertical settlement of 20 mm at node 3.

3.4 Analyse the truss of figure P2.5(a) of chapter 2, to find the displacements at the nodes, given that the elements 1–2, 2–4, 1–3 and 3–5 have an area of $1800\ \text{mm}^2$; elements 2–3 and 3–4 have an area of $1200\ \text{mm}^2$; and element 4–5 has an area of $2000\ \text{mm}^2$. $E = 200\ \text{kN/mm}^2$.

3.5 Using the beam element stiffness matrix given by equation (3.25), solve each of the beam problems of figure 3.8 to obtain the nodal displacements and actions on the elements.

3.6 Repeat the solutions to the beam problems given in section 3.7 for the case where the right-hand end of the beam is not restrained against rotation.

Chapter 4
The Matrix Stiffness Method—Part 1: Beams and Rectangular Frames

The basic concepts of the matrix stiffness method as presented in chapter 3 can be extended to the analysis of continuous beams and rectangular frames. This group of structures has been selected since it is possible to develop a suitable approach without introducing coordinate transformation. Coordinate transformations are introduced in chapter 6 with what can be described as the general stiffness method. The general stiffness method can be used to analyse all types of skeletal structure and the approach of this chapter will ultimately be seen as a subset of the more general technique.

4.1 THE ANALYSIS OF CONTINUOUS BEAMS

A continuous beam is usually defined as one which continues over more than two supports in such a way that the deflected shape is a continuous curve throughout the beam. From the point of view of modelling a beam, it can be considered as being made up of a series of beam elements connected together at arbitrarily selected nodes. From this definition, even a simply supported beam or a cantilever beam can be considered as a continuous beam if it is considered to be made up of more than one element. This is a definition that will be adopted here, and it will be seen that the same analysis procedure can apply to any beam system, whether or not it is statically determinate. The type of structure covered by this definition is illustrated in figure 4.1.

4.1.1 Forming the Structure Stiffness Matrix

The basic concepts of the stiffness method were introduced in chapter 3, however it is worth while restating some of the points here in introducing the application to continuous beams.

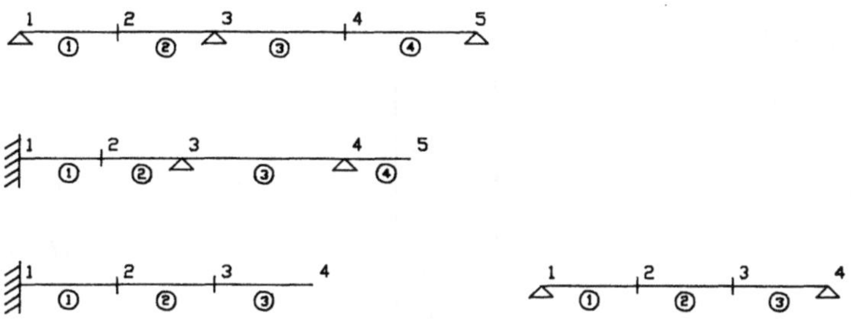

Figure 4.1 *Some continuous beams.*

The basis of the matrix stiffness method is to determine a relationship between the external actions (loads) acting on the structure and the resulting displacements. Since the structure is considered to be an assembly of a number of discrete elements, the external actions and the corresponding displacements are defined at the nodes of the structure so that both form a finite set. It is necessary to introduce a sign convention to define positive loads and displacements with respect to some global axes of reference. This has been done in figure 4.2, where the general terms identifying possible loads and displacements in a continuous beam have been shown.

At this stage, the influence of transverse loads acting between the nodes is not considered. The only loads are thus the nodal loads corresponding to the two degrees of freedom at each node. Boundary conditions will also be treated at a later stage. Both the external loads and the displacements may be written in vector form and they are related by the structure stiffness matrix. Although the structure stiffness matrix is as yet undefined, the general

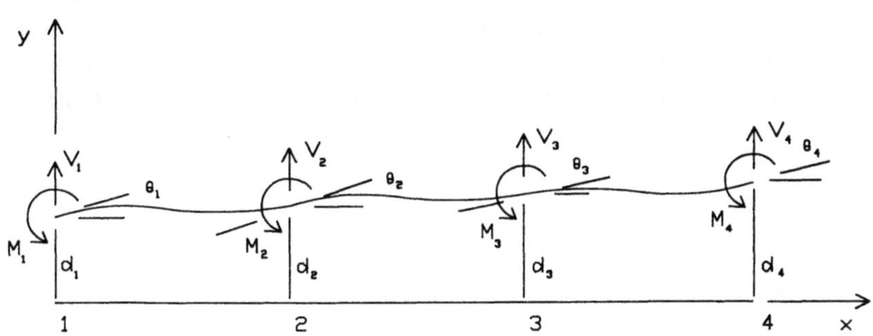

Figure 4.2 *External loads and displacement at nodes.*

form of the relationship is

$$\begin{Bmatrix} V_1 \\ M_1 \\ V_2 \\ M_2 \\ V_3 \\ M_3 \\ V_4 \\ M_4 \end{Bmatrix} = \begin{bmatrix} k_{11} & k_{12} & k_{13} & \cdots & \\ k_{21} & k_{22} & & & \\ k_{31} & & & & \\ k_{41} & & & & \\ \vdots & & & & \\ & & & & \\ & & & & \\ k_{81} & k_{82} & & & k_{88} \end{bmatrix} \begin{Bmatrix} d_1 \\ \theta_1 \\ d_2 \\ \theta_2 \\ d_3 \\ \theta_3 \\ d_4 \\ \theta_4 \end{Bmatrix} \tag{4.1}$$

Equation (4.1) may also be written in the form

$$P = K \cdot d \tag{4.2}$$

where P is a load vector, K is the structure stiffness matrix and d is the displacement vector.

The terms of the load vector can be readily specified and it is the displacements that are sought initially, once the terms of the structure stiffness matrix have been defined. To define these, a knowledge of the behaviour of each beam element and its internal action-displacement relationship is required. The procedure can be compared with physically putting the beam elements together to assemble the continuous structure.

Figure 4.3 shows a free body diagram of node 2 of the beam in figure 4.2, and the details of part of the adjoining elements on either side. For the structure to be in equilibrium, the nodal forces must be in equilibrium, requiring

$$\begin{aligned} V_2 &= v_{21} + v_{23} \\ M_2 &= m_{21} + m_{23} \end{aligned} \tag{4.3}$$

Similar equilibrium equations can be written for all the nodes of the structure, thus linking the external loads to the internal actions. The internal actions of the elements are related to the displacements through the element stiffness matrix as given in equation (3.25), which, in conjunction with the nodal equilibrium equations defines the terms of the structure stiffness

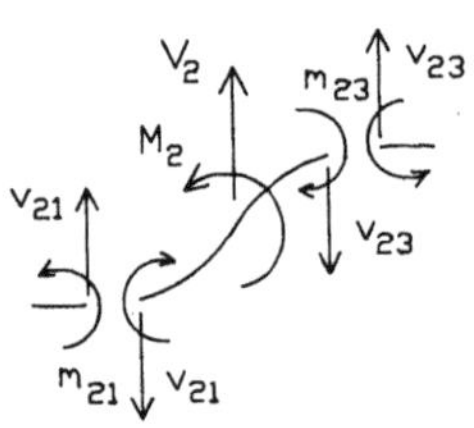

Figure 4.3 Nodal equilibrium at node 2.

matrix, K. This may be demonstrated in algebraic terms as follows; let

$$\begin{Bmatrix} v_{12} \\ m_{12} \\ v_{21} \\ m_{21} \end{Bmatrix} = \begin{bmatrix} a_{11} & a_{12} & a_{13} & a_{14} \\ a_{21} & a_{22} & a_{23} & a_{24} \\ a_{31} & a_{32} & a_{33} & a_{34} \\ a_{41} & a_{42} & a_{43} & a_{44} \end{bmatrix} \begin{Bmatrix} d_1 \\ \theta_1 \\ d_2 \\ \theta_2 \end{Bmatrix} \tag{4.4}$$

and let

$$\begin{Bmatrix} v_{23} \\ m_{23} \\ v_{32} \\ m_{32} \end{Bmatrix} = \begin{bmatrix} b_{11} & b_{12} & b_{13} & b_{14} \\ b_{21} & b_{22} & b_{23} & b_{24} \\ b_{31} & b_{32} & b_{33} & b_{34} \\ b_{41} & b_{42} & b_{43} & b_{44} \end{bmatrix} \begin{Bmatrix} d_2 \\ \theta_2 \\ d_3 \\ \theta_3 \end{Bmatrix} \tag{4.5}$$

where the terms a_{ij} and b_{ij} are known functions of the properties of the elements ① and ② respectively, as previously defined; that is, terms of a beam element stiffness matrix.

From equation (4.4):

$$v_{21} = a_{31}d_1 + a_{32}\theta_1 + a_{33}d_2 + a_{34}\theta_2$$

$$m_{21} = a_{41}d_1 + a_{42}\theta_1 + a_{43}d_2 + a_{44}\theta_2$$

and from equation (4.5)

$$v_{23} = b_{11}d_2 + b_{12}\theta_2 + b_{13}d_3 + b_{14}\theta_3$$

$$m_{23} = b_{21}d_2 + b_{22}\theta_2 + b_{23}d_3 + b_{24}\theta_3$$

Substituting these expressions into equations (4.3) gives

$$V_2 = a_{31}d_1 + a_{32}\theta_1 + (a_{33} + b_{11})d_2 + (a_{34} + b_{12})\theta_2 + b_{13}d_3 + b_{14}\theta_3$$

$$M_2 = a_{41}d_1 - a_{42}\theta_1 + (a_{43} + b_{21})d_2 + (a_{44} + b_{22})\theta_2 + b_{23}d_3 + b_{24}\theta_3$$

Compatibility is expressed through the variables d_2 and θ_2 being common to adjoining spans, so that both equilibrium and compatibility are satisfied at the node. If the operation is carried out for each node, then the complete set of equations defined by equation (4.1) will be developed. However the operation can be seen to be the assembly of the element stiffness matrices according to a strict pattern dictated by the connectivity of the structure. For example, for the beam of figure 4.2, the K matrix is given by

$$\begin{bmatrix} a_{11} & a_{12} & a_{13} & a_{14} & & & & \\ a_{21} & a_{22} & a_{23} & a_{24} & & & & \\ a_{31} & a_{32} & (a_{33}+b_{11}) & (a_{34}+b_{12}) & b_{13} & b_{14} & & \\ a_{41} & a_{42} & (a_{43}+b_{21}) & (a_{44}+b_{22}) & b_{23} & b_{24} & & \\ & & b_{31} & b_{32} & (b_{33}+c_{11}) & (b_{34}+c_{12}) & c_{13} & c_{14} \\ & & b_{41} & b_{42} & (b_{43}+c_{21}) & (b_{44}+c_{22}) & c_{23} & c_{24} \\ & & & & c_{31} & c_{32} & c_{33} & c_{34} \\ & & & & c_{41} & c_{42} & c_{43} & c_{44} \end{bmatrix}$$

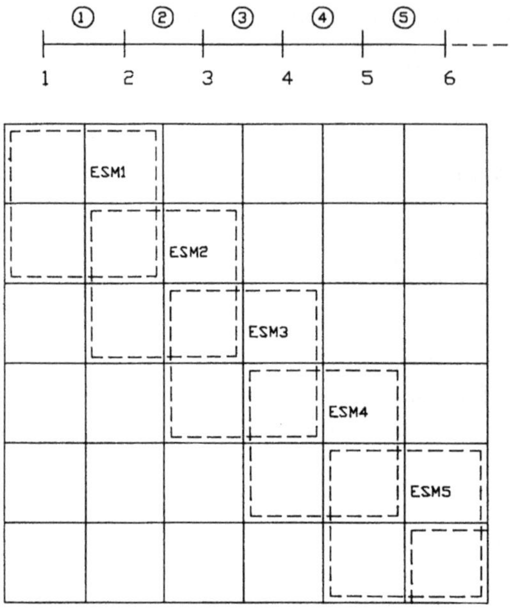

Figure 4.4 *Form of the structure stiffness matrix for continuous beams.*

where the terms c_{ij} are the terms of the element stiffness matrix for the third element. In diagram form, the structure stiffness matrix for a continuous beam of any number of elements is seen to follow the pattern shown in figure 4.4.

4.1.2 Solving for Displacements

The problem is now defined in terms of the relationship expressed by equation (4.2). Once the boundary conditions have been considered, a solution is available in exactly the same way as that presented in section 3.3. For a continuous beam, it is likely that the terms in the load vector representing the reactions will be distributed throughout the load vector, with a corresponding distribution of the restrained displacements in the displacement vector. This means that the equations will not generally be ordered in such a way as to make the partitioning of the matrices immediately obvious. Provided that the restrained displacements are zero, it is a relatively simple matter to reduce the set of equations to the form

$$P_F = K_F \cdot d_F \tag{4.6}$$

This can be done by noting that all the column terms of the stiffness matrix associated, by matrix multiplication, with a zero displacement can be eliminated, along with the row terms which express a relationship between the reactions and the displacements. The remaining terms can be consolidated to give equation (4.6) which can then be solved for d_F. Of course this means that the reactions cannot be so readily recovered although it is still possible.

A more general technique for the consideration of the boundary conditions is presented in chapter 6. As an example of that technique, a continuous beam is analysed to determine its behaviour under support settlement, which is clearly a case of non-zero displacement at a restraint.

4.1.3 Element Actions

With the displacements of the structure known at each of the nodes, including the given displacements at the reactions, it is possible to return to the element action-displacement relationship and recover the internal actions on the element. For each element in turn, the appropriate element stiffness matrix is used in conjunction with the nodal displacements, and the internal actions are found by matrix multiplication. That is, equation (3.25) is applied in a back-substitution role. For ease of reference, equation (3.25) can be written in the form

$$\begin{Bmatrix} f_i \\ \hline f_j \end{Bmatrix} = [ESM] \begin{Bmatrix} d_i \\ \hline d_j \end{Bmatrix} \tag{4.7}$$

where the terms are evident from a comparison with equation (3.25), with the *ESM* being the beam element stiffness matrix.

The procedure is summarised by the analysis flowchart shown in figure 4.5 and illustrated by example 4.1. The same basic data can be used to solve the beam problems of figure 4.6 and it should be noted that it was simply convenient to use beam elements of the same length; this is not a necessary condition.

4.1.4 Consideration of Transverse Loads

In the material presented so far, the external loads have been restricted to moments or forces applied at the nodes of the structure. However a flexural beam element may have transverse loads applied anywhere over the length of the element, including point loads and distributed loads of various types. The question of how such loads are to be considered is resolved through the use of the principle of superposition.

Consider the problem of the cantilever beam modelled as a single beam element. The beam is shown in figure 4.7(a) with a uniformly distributed load applied. If the beam is fully restrained as shown in figure 4.7(b), by applying a restraining clamp and prop to prevent both rotation and translation, the problem is reduced to a standard fixed end beam problem. In this case, the actions on the ends of the element, which are developed by the

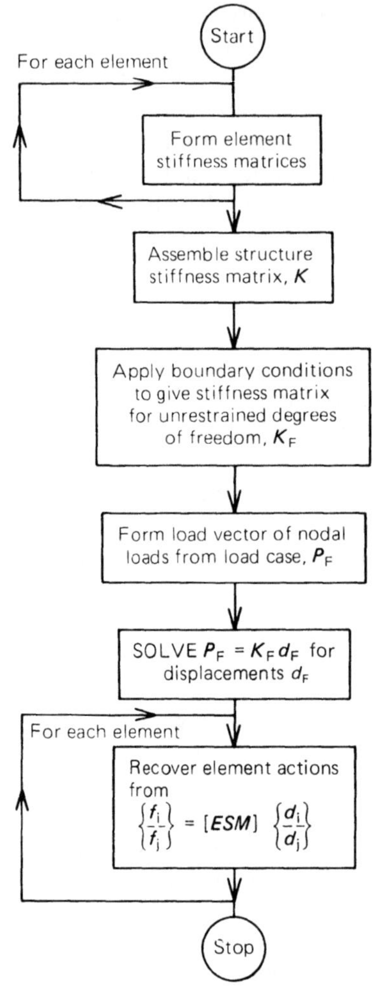

Figure 4.5 *Flowchart for the analysis of continuous beams.*

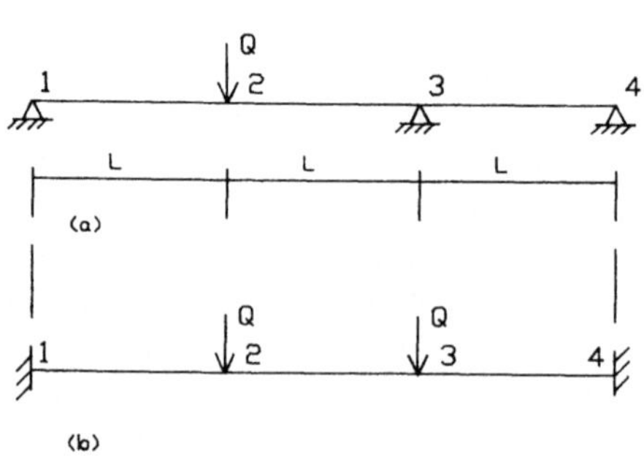

Figure 4.6 *Beam analysis problems.*

Example 4.1: Deflection of a Cantilever Beam (2)

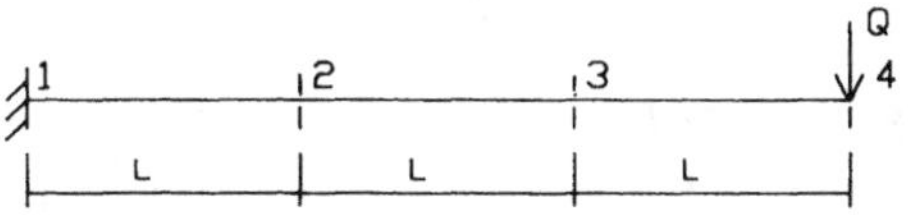

Constant EI

Consider the beam as three uniform elements of length L. Each beam element stiffness matrix is

$$\left[\begin{array}{cc|cc}
\frac{12EI}{L^3} & \frac{6EI}{L^2} & -\frac{12EI}{L^3} & \frac{6EI}{L^2} \\
\frac{6EI}{L^2} & \frac{4EI}{L} & -\frac{6EI}{L^2} & \frac{2EI}{L} \\
\hline
-\frac{12EI}{L^3} & -\frac{6EI}{L^2} & \frac{12EI}{L^3} & -\frac{6EI}{L^2} \\
\frac{6EI}{L^2} & \frac{2EI}{L} & -\frac{6EI}{L^2} & \frac{4EI}{L}
\end{array}\right]$$

The assembled structure stiffness matrix then follows as

$$\begin{bmatrix}
\frac{12EI}{L^3} & \frac{6EI}{L^2} & -\frac{12EI}{L^3} & \frac{6EI}{L^2} & & & & \\
\frac{6EI}{L^2} & \frac{4EI}{L} & -\frac{6EI}{L^2} & \frac{2EI}{L} & & & & \\
-\frac{12EI}{L^3} & -\frac{6EI}{L^2} & \frac{24EI}{L^3} & 0 & -\frac{12EI}{L^3} & \frac{6EI}{L^2} & & \\
\frac{6EI}{L^2} & \frac{2EI}{L} & 0 & \frac{8EI}{L} & -\frac{6EI}{L^2} & \frac{2EI}{L} & & \\
& & -\frac{12EI}{L^3} & -\frac{6EI}{L^2} & \frac{24EI}{L^3} & 0 & -\frac{12EI}{L^3} & \frac{6EI}{L^2} \\
& & \frac{6EI}{L^2} & \frac{2EI}{L} & 0 & \frac{8EI}{L} & -\frac{6EI}{L^2} & \frac{2EI}{L} \\
& & & & -\frac{12EI}{L^3} & -\frac{6EI}{L^2} & \frac{12EI}{L^3} & -\frac{6EI}{L^2} \\
& & & & \frac{6EI}{L^2} & \frac{2EI}{L} & -\frac{6EI}{L^2} & \frac{4EI}{L}
\end{bmatrix}$$

Applying the boundary conditions and forming the load vector gives

$$\begin{Bmatrix} 0 \\ 0 \\ 0 \\ 0 \\ -Q \\ 0 \end{Bmatrix} = \begin{bmatrix} \frac{24EI}{L^3} & 0 & -\frac{12EI}{L^3} & \frac{6EI}{L^2} & 0 & 0 \\ 0 & \frac{8EI}{L} & -\frac{6EI}{L^2} & \frac{2EI}{L} & 0 & 0 \\ -\frac{12EI}{L^3} & -\frac{6EI}{L^2} & \frac{24EI}{L^3} & 0 & -\frac{12EI}{L^3} & \frac{6EI}{L^2} \\ \frac{6EI}{L^2} & \frac{2EI}{L} & 0 & \frac{8EI}{L} & -\frac{6EI}{L^2} & \frac{2EI}{L} \\ 0 & 0 & -\frac{12EI}{L^3} & -\frac{6EI}{L} & \frac{12EI}{L^3} & -\frac{6EI}{L^2} \\ 0 & 0 & \frac{6EI}{L^2} & \frac{2EI}{L} & -\frac{6EI}{L^2} & \frac{4EI}{L} \end{bmatrix} \begin{Bmatrix} d_2 \\ \theta_2 \\ d_3 \\ \theta_3 \\ d_4 \\ \theta_4 \end{Bmatrix}$$

which can be written as:

$$QL \begin{Bmatrix} 0 \\ 0 \\ 0 \\ 0 \\ -1 \\ 0 \end{Bmatrix} = \frac{EI}{L} \begin{bmatrix} 24 & 0 & -12 & 6 & 0 & 0 \\ 0 & 8 & -6 & 2 & 0 & 0 \\ -12 & -6 & 24 & 0 & -12 & 6 \\ 6 & 2 & 0 & 8 & -6 & 2 \\ 0 & 0 & -12 & -6 & 12 & -6 \\ 0 & 0 & 6 & 2 & -6 & 4 \end{bmatrix} \begin{Bmatrix} d_2/L \\ \theta_2 \\ d_3/L \\ \theta_3 \\ d_4/L \\ \theta_4 \end{Bmatrix} \quad \text{(a)}$$

Solving equation (a) with a suitable solution routine such as that available in MATOP gives

$$\begin{Bmatrix} d_2/L \\ \theta_2 \\ d_3/L \\ \theta_3 \\ d_4/L \\ \theta_4 \end{Bmatrix} = \frac{QL^2}{EI} \begin{Bmatrix} -1.333 \\ -2.5 \\ -4.666 \\ -4.0 \\ -9.0 \\ -4.5 \end{Bmatrix}$$

so that

$$d_2 = -1.333\frac{QL^3}{EI}; \qquad \theta_2 = -2.5\frac{QL^2}{EI}$$

$$d_3 = -4.666\frac{QL^3}{EI}; \qquad \theta_3 = -4.0\frac{QL^2}{EI}$$

$$d_4 = -9.0\frac{QL^3}{EI}; \qquad \theta_4 = -4.5\frac{QL^2}{EI}$$

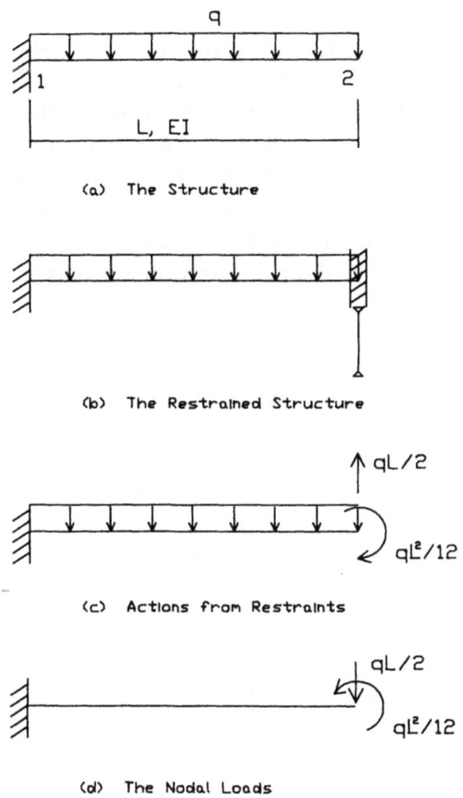

Figure 4.7 Transverse loads on a cantilever beam.

imposed restraints, can be calculated to be as shown in figure 4.7(c). If the reverse actions were to be applied to the otherwise unloaded beam as shown in figure 4.7(d), and the solution to this problem were combined with the solution of the problem specified by figure 4.7(c), then the solution to the given beam problem of figure 4.7(a) would be obtained.

The solution to the beam of figure 4.7(b) can be described as a fixed end action solution, while the solution to the beam of figure 4.7(d) can be described as the analysis of a beam under nodal loads only. The latter solution is therefore available by matrix analysis. For this reason the matrix analysis of structures is often described as a two-part solution problem. The fixed end action solution can be described as a Part 1 solution, to be followed by the nodal load solution as Part 2. The final solution is of course a combination of both parts. It may be noted that there are no nodal displacements in a Part 1 solution so that the nodal displacements given in Part 2 are the final nodal displacements. The flowchart of figure 4.5 does not reflect this requirement since its emphasis is on the matrix analysis from the nodal loads. The preliminary step of considering the fixed end action solution in order to establish the nodal loads, and the final step of combining that solution with the nodal load solution, must be added. The same procedure can be applied when a series of elements are used. In this case the nodal loads are the sum of the reversed end actions which are applied to the ends of the elements connected to the node, as a result of the fixed end actions. This is illustrated in figure 4.8 using numerical data and that beam is fully analysed as example 4.2.

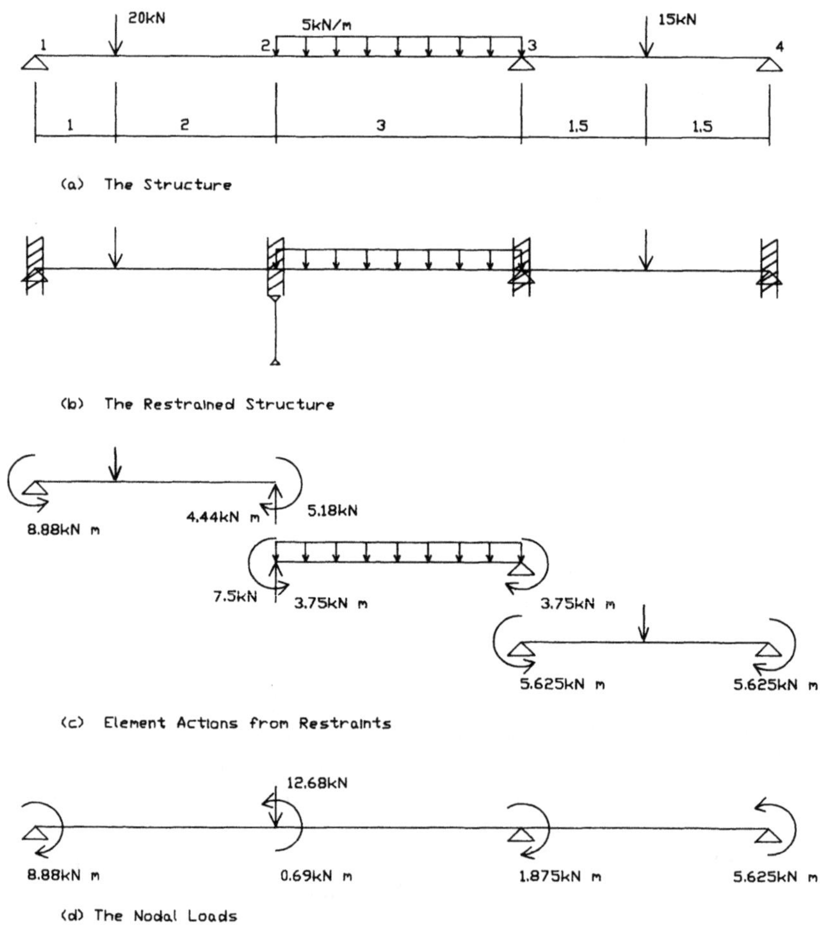

Figure 4.8 *Transverse loads on a typical beam.*

Example 4.2: Analysis of a Continuous Beam

Given data:

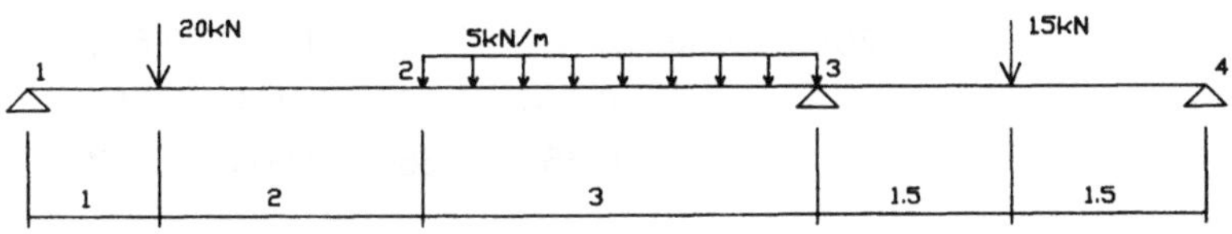

$E = 200\ \text{kN/mm}^2$

$I = 10 \times 10^6\ \text{mm}^4$

Nodes and Elements:

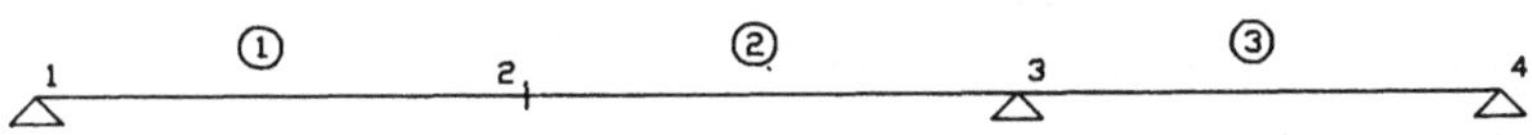

Procedure: Form beam element stiffness matrices, assemble to give structure stiffness matrix. Apply boundary conditions and solve to obtain displacements. Recover element actions.

Element Stiffness Matrices (Units: kN and m):

All elements have:

$$ESM = EI \begin{bmatrix} 0.444 & 0.666 & -0.444 & 0.666 \\ 0.666 & 1.333 & -0.666 & 0.666 \\ -0.444 & -0.666 & 0.444 & -0.666 \\ 0.666 & 0.666 & -0.666 & 1.333 \end{bmatrix}$$

Structure Stiffness Matrix:

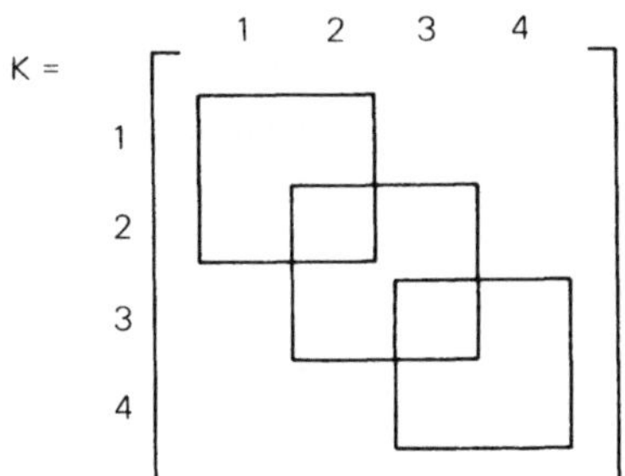

$$\therefore \quad K = EI \begin{bmatrix} 0.444 & 0.666 & -0.444 & 0.666 & & & & \\ 0.666 & 1.333 & -0.666 & 0.666 & & & & \\ -0.444 & -0.666 & 0.888 & 0 & -0.444 & 0.666 & & \\ 0.666 & 0.666 & 0 & 2.666 & -0.666 & 0.666 & & \\ & & -0.444 & -0.666 & 0.888 & 0 & -0.444 & 0.666 \\ & & 0.666 & 0.666 & 0 & 2.666 & -0.666 & 0.666 \\ & & & & -0.444 & -0.666 & 0.444 & -0.666 \\ & & & & 0.666 & 0.666 & -0.666 & 1.333 \end{bmatrix}$$

Applying the boundary conditions gives $P_F = K_F \cdot d_F$, as

$$\begin{Bmatrix} M_1 \\ V_2 \\ M_2 \\ M_3 \\ M_4 \end{Bmatrix} = EI \begin{bmatrix} 1.333 & -0.666 & 0.666 & 0 & 0 \\ -0.666 & 0.888 & 0 & 0.666 & 0 \\ 0.666 & 0 & 2.666 & 0.666 & 0 \\ 0 & 0.666 & 0.666 & 2.666 & 0.666 \\ 0 & 0 & 0 & 0.666 & 1.333 \end{bmatrix} \begin{Bmatrix} \theta_1 \\ d_2 \\ \theta_2 \\ \theta_3 \\ \theta_4 \end{Bmatrix}$$

Load Vector: From Part 1 Solution Fixed End Actions

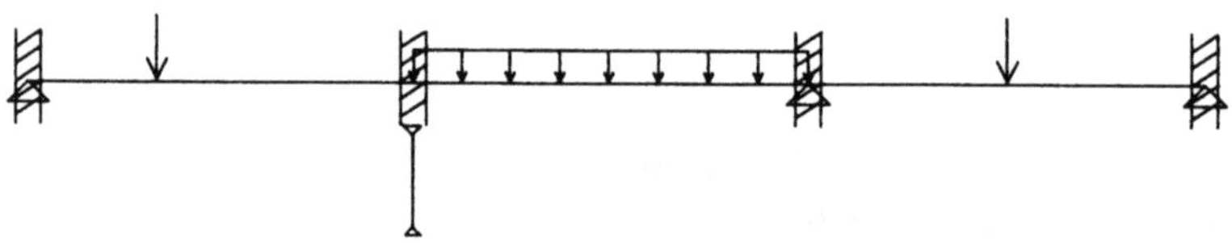

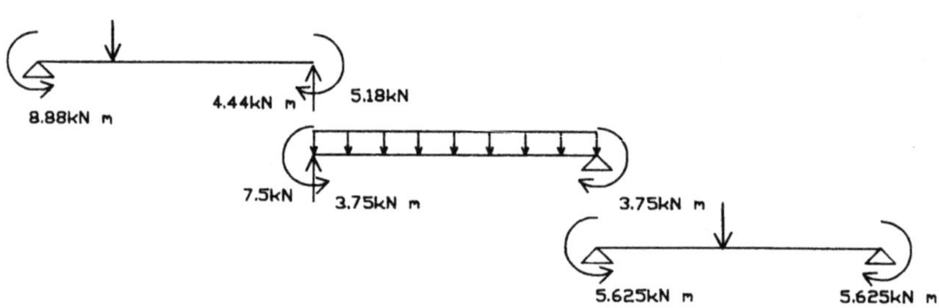

Hence nodal loads are

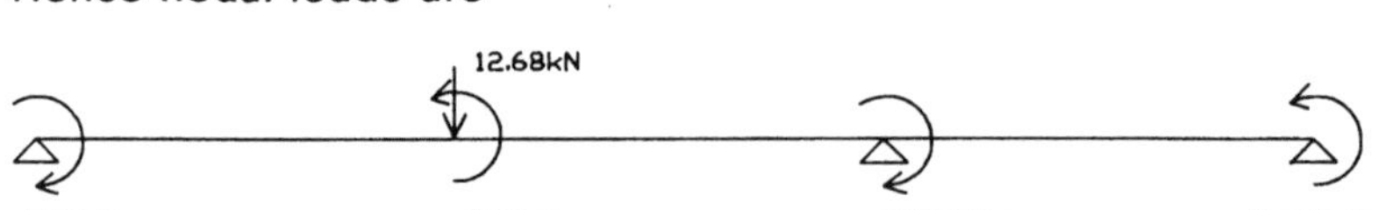

Solution: Using MATOP, the following results were obtained with the displacements being interpreted as shown in the table.

```
LOAD.K.5.5
SCALE.K.2000
LOAD.P.5.1
SOLVE.K.P
PRINT.P  DISPLACEMENTS
-.162437E-01
-.227981E-01
0.302937E-02
0.464374E-02
-.212494E-03
SCALE.P.2000
PRINT.P  SCALED DISPLACEMENTS
-.324875E+02
-.455962E+02
0.605874E+01
0.928748E+01
-.424988E+00
End of File
```

Node	d (mm)	θ (radian)
1	0	−0.0162
2	−22.8	0.0031
3	0	0.0046
4	0	−0.0002

Element Actions: Using equation (4.7), the element actions can be recovered noting in particular the additional zero terms in the displacement vector. The results from matrix multiplication are shown in the following table.

	Element		
	1–2	2–3	3–4
v_{ij} (kN)	2.65	−10.03	5.91
m_{ij} (kN m)	−8.88	−16.13	12.09
v_{ji} (kN)	−2.65	10.03	−5.91
m_{ji} (kN m)	16.82	−13.97	5.62

Final Actions

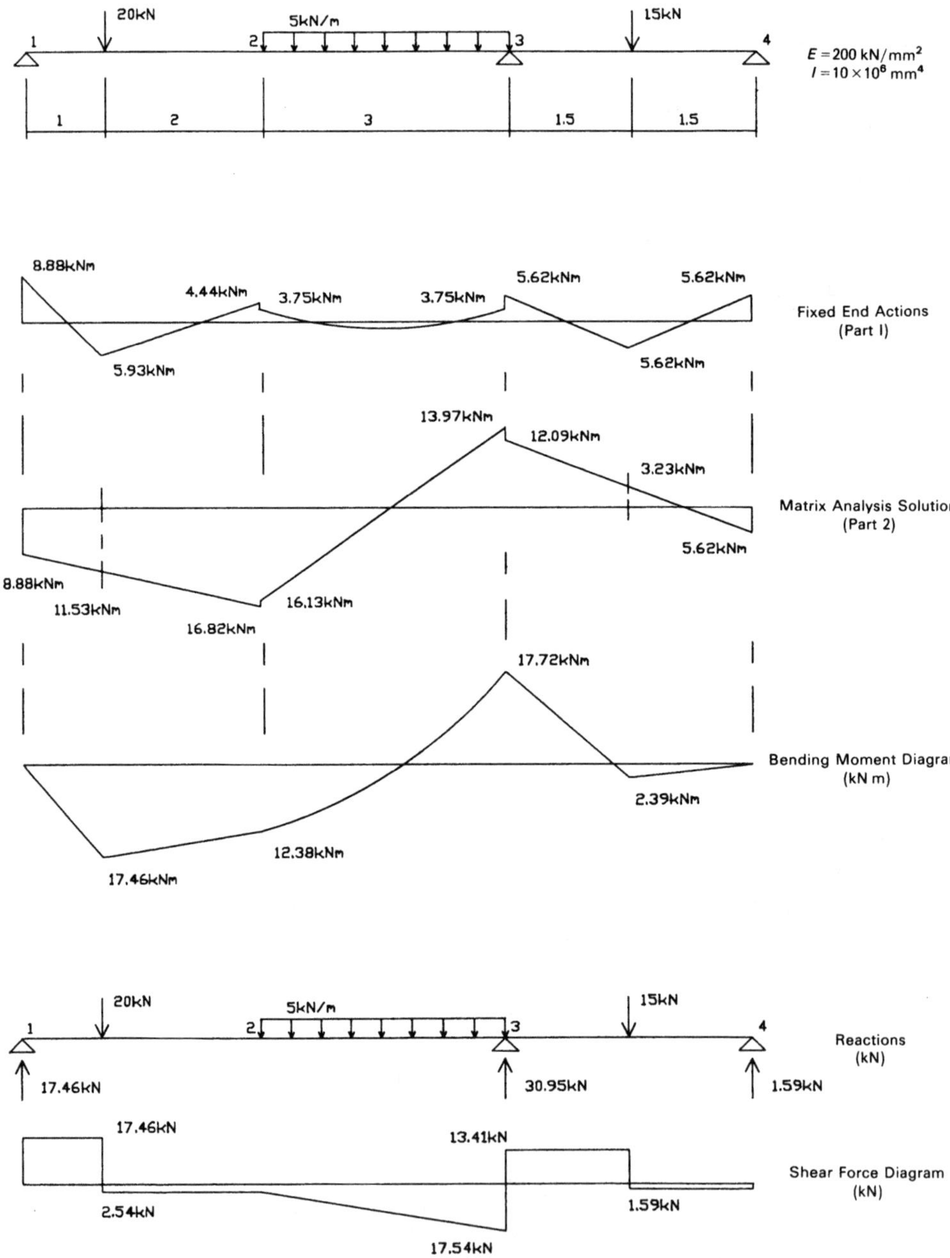

4.2 THE ANALYSIS OF RECTANGULAR FRAMES

The matrix stiffness method as applied to continuous beams can be extended to the analysis of rigid jointed plane frames of a rectangular nature. The principal action of such frames is the flexural bending of the beams and columns, and horizontal forces are resisted through this action. The rigid joint is an essential part of this action since this is how the moments are transmitted through the frame. Under horizontal or lateral forces, a frame may deflect horizontally at the level of the beams and this is known as lateral sway, or simply a sway deflection. A typical set of frames to which this method of analysis applies is shown in figure 4.9. It may be noted that some of the frames are actually restrained against lateral sway by the supports, but in general this must be considered.

A fundamental difference between the continuous beam analysis and the frame analysis lies in the fact that the elements do not have the same orientation with respect to each other. For this reason it is necessary to introduce some limitations on the nature of the elements. The limitations can be overcome in a more general analysis procedure that will be presented in chapter 6.

The limitations are:

(a) The column is considered to be the fundamental element having the actions and displacements as indicated in figure 4.10, where each action and displacement is shown in the positive sense.

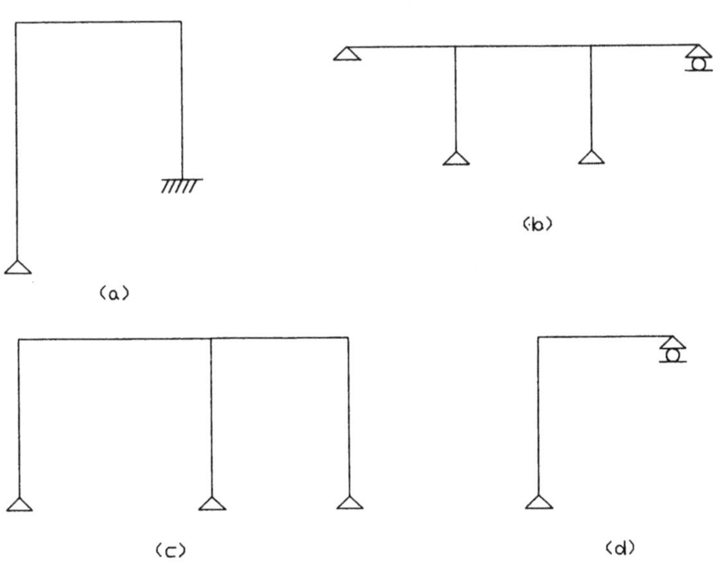

Figure 4.9 Some rectangular rigid-jointed frames.

(b) Each beam is considered to be a single element subjected to end moments and corresponding end rotations only.

(c) Although the elements of the structure are subjected to axial load, axial deformation of the elements is ignored. This means that the only translation considered is the lateral sway of the frame at the level of the beams. The general degrees of freedom of the frame are thus reduced by this assumption. The assumption is reasonably valid since the axial stiffness of most frames is significantly greater than the flexural stiffness which dominates the sway characteristics.

One of the consequences of these limitations is that the degrees of freedom of the frame are reduced to the rotation at each node and the sway translations at each beam level, plus any translations that may be admitted at the boundary nodes. Of course the nodal loads are correspondingly reduced and particular attention should be paid to the fact that only a single horizontal force, or sway force, is applied at a given beam level. This follows from the fact that the beams (and columns) are regarded as axially rigid so that all horizontal forces evaluated at a given beam level may be lumped together.

4.2.1 The Column Element Stiffness Matrix

The slope–deflection equations expressed in equation (3.24) may be applied directly to the column element of figure 4.10 to give

$$m_{12}=\frac{4EI}{L}\theta_1+\frac{2EI}{L}\theta_2-\frac{6EI}{L^2}d_1+\frac{6EI}{L^2}d_2$$

$$m_{21}=\frac{2EI}{L}\theta_1+\frac{4EI}{L}\theta_2-\frac{6EI}{L^2}d_1+\frac{6EI}{L^2}d_2$$

where it should be noted that the moments induced by translation are now positive moments.

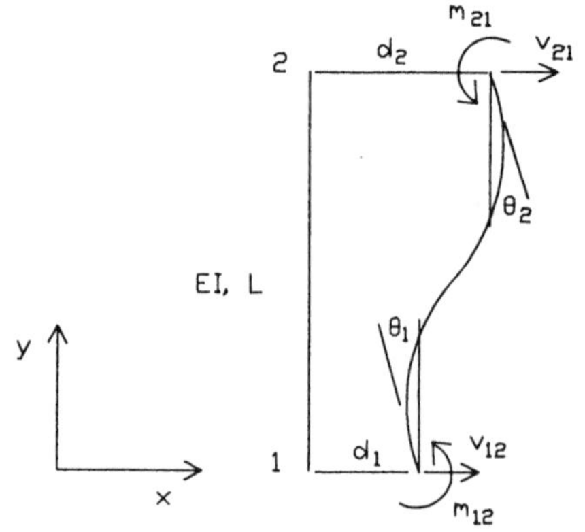

Figure 4.10 The column element.

Taking moments about node 2 results in

$$v_{12} = -\frac{(m_{12}+m_{21})}{L}$$

$$\therefore \quad v_{12} = -\frac{6EI}{L^2}\theta_1 - \frac{6EI}{L^2}\theta_2 + \frac{12EI}{L^3}d_1 - \frac{12EI}{L^3}d_2$$

and since

$v_{21} = -v_{12}$, then

$$v_{21} = \frac{6EI}{L^2}\theta_1 + \frac{6EI}{L^2}\theta_2 - \frac{12EI}{L^3}d_1 + \frac{12EI}{L^3}d_2$$

The above equations may be written in matrix form to define the column element stiffness matrix in the relationship:

$$\begin{Bmatrix} v_{12} \\ m_{12} \\ \hdashline v_{21} \\ m_{21} \end{Bmatrix} = \left[\begin{array}{cc:cc} \frac{12EI}{L^3} & \frac{-6EI}{L^2} & -\frac{12EI}{L^3} & -\frac{6EI}{L^2} \\ -\frac{6EI}{L^2} & \frac{4EI}{L} & \frac{6EI}{L^2} & \frac{2EI}{L} \\ \hdashline -\frac{12EI}{L^3} & \frac{6EI}{L^2} & \frac{12EI}{L^3} & \frac{6EI}{L^2} \\ -\frac{6EI}{L^2} & \frac{2EI}{L} & \frac{6EI}{L^2} & \frac{4EI}{L} \end{array}\right] \begin{Bmatrix} d_1 \\ \theta_1 \\ \hdashline d_2 \\ \theta_2 \end{Bmatrix} \tag{4.8}$$

4.2.2 Assembly of the Structure Stiffness Matrix

The structure stiffness matrix is assembled according to the same principles as those used in the analysis of continuous beams. However, in this case parallel column elements are linked by beam elements and are not directly connected. The consequences of this are that while the nodal equilibrium equations, with respect to moments, have a precise parallel with those used in the continuous beam analysis, the equivalent expression for the transverse loads is different. In this case the transverse loads are acting horizontally and the equilibrium equations arise from a consideration of horizontal shear taken right through the frame. The procedure is best understood with the help of an example.

Consider the portal frame shown in figure 4.11(a). The node numbering of the frame is significant, since it is important that the columns are numbered such that each column can be addressed as low node number, high node number, consistent with the column element of figure 4.10. The alternative to this is to introduce coordinate transformation as will be seen in chapter 6. The possible external loads and displacements are those defined in figure 4.11(b). As has been previously mentioned, axial deformation is

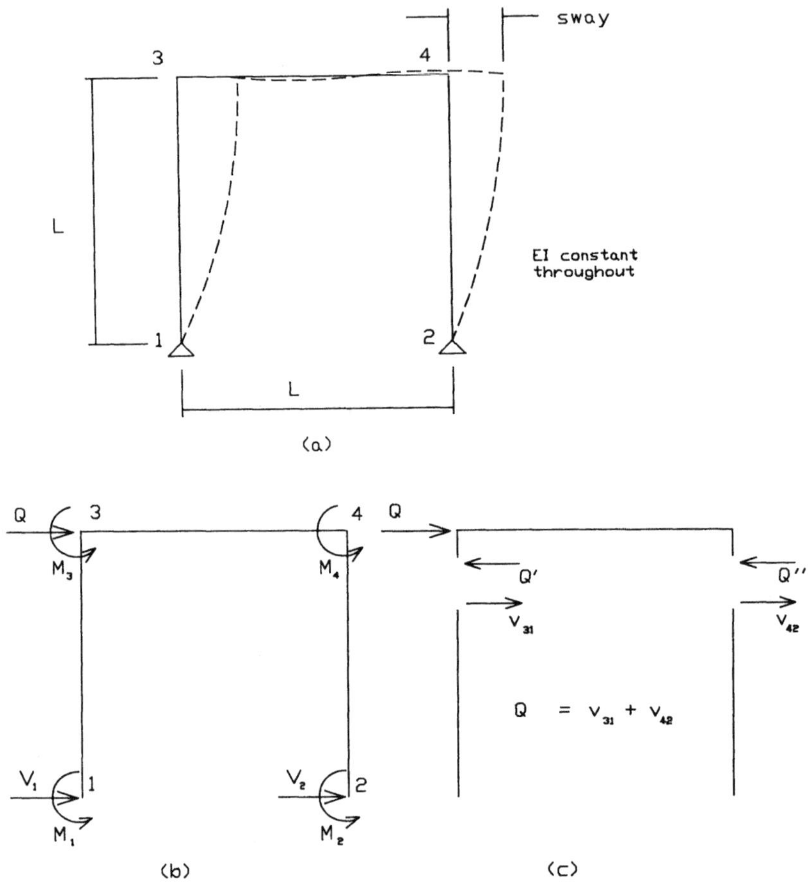

Figure 4.11 *Details of a swaying frame.*

ignored, so that there are no translations in the y direction. Further, any loads applied directly in the line of the columns will be transmitted directly to the reactions at the column bases. To avoid subscripts in the interests of clarity, each element has the same length and flexural rigidity but this is not a necessary condition.

The sway load, Q, is distributed into the columns according to their capacity to resist sway. This is illustrated in figure 4.11(c), which also demonstrates the horizontal shear equilibrium requirement. In general, at any beam level the horizontal load is distributed into the column lines and is taken up by the shear in the column elements at that node. It is convenient to introduce the notation Q', and Q'' to describe the distribution so that

$$Q = Q' + Q''$$

This allows the assembly of the structure stiffness matrix to proceed on similar lines to that used in continuous beam analysis. However the resulting matrix must then be compacted by combining the equations relating to Q' and Q'' as will be seen.

The element stiffness matrix for the column element 1-3 is thus defined in the relationship

$$\begin{Bmatrix} v_{13} \\ m_{13} \\ 0 \\ 0 \\ v_{31} \\ m_{31} \end{Bmatrix} = \begin{bmatrix} \frac{12EI}{L^3} & -\frac{6EI}{L^2} & 0 & 0 & -\frac{12EI}{L^3} & -\frac{6EI}{L^2} \\ -\frac{6EI}{L^2} & \frac{4EI}{L} & 0 & 0 & \frac{6EI}{L^2} & \frac{2EI}{L} \\ 0 & 0 & 0 & 0 & 0 & 0 \\ 0 & 0 & 0 & 0 & 0 & 0 \\ -\frac{12EI}{L^3} & \frac{6EI}{L^2} & 0 & 0 & \frac{12EI}{L^3} & \frac{6EI}{L^2} \\ -\frac{6EI}{L^2} & \frac{2EI}{L} & 0 & 0 & \frac{6EI}{L^2} & \frac{4EI}{L} \end{bmatrix} \begin{Bmatrix} d_1 \\ \theta_1 \\ 0 \\ 0 \\ d_3 \\ \theta_3 \end{Bmatrix}$$

while that for column 2-4 is given by

$$\begin{Bmatrix} v_{24} \\ m_{24} \\ 0 \\ 0 \\ v_{42} \\ m_{42} \end{Bmatrix} = \begin{bmatrix} \frac{12EI}{L^3} & -\frac{6EI}{L^2} & 0 & 0 & -\frac{12EI}{L^3} & -\frac{6EI}{L^2} \\ -\frac{6EI}{L^2} & \frac{4EI}{L} & 0 & 0 & \frac{6EI}{L^2} & \frac{2EI}{L} \\ 0 & 0 & 0 & 0 & 0 & 0 \\ 0 & 0 & 0 & 0 & 0 & 0 \\ -\frac{12EI}{L^3} & \frac{6EI}{L^2} & 0 & 0 & \frac{12EI}{L^3} & \frac{6EI}{L^2} \\ -\frac{6EI}{L^2} & \frac{2EI}{L} & 0 & 0 & \frac{6EI}{L^2} & \frac{4EI}{L} \end{bmatrix} \begin{Bmatrix} d_2 \\ \theta_2 \\ 0 \\ 0 \\ d_4 \\ \theta_4 \end{Bmatrix}$$

In both cases the element stiffness matrices have been expanded to accommodate the non-sequential node numbering so that the assembly procedure is more obvious.

The element stiffness matrix for the beam element 3-4 is given in the relationship

$$\begin{Bmatrix} m_{34} \\ m_{43} \end{Bmatrix} = \begin{bmatrix} \frac{4EI}{L} & \frac{2EI}{L} \\ \frac{2EI}{L} & \frac{4EI}{L} \end{bmatrix} \begin{Bmatrix} \theta_3 \\ \theta_4 \end{Bmatrix}$$

which is a subset of the continuous beam element stiffness matrix given by equation (3.25). The structure stiffness matrix, which can now be assembled from the element stiffness matrices as defined, has the form indicated by the block diagram of figure 4.12.

The structure stiffness matrix, which satisfies nodal equilibrium and compatibility, is then initially given by

$$
\begin{Bmatrix} V_1 \\ M_1 \\ V_2 \\ M_2 \\ Q' \\ M_3 \\ Q'' \\ M_4 \end{Bmatrix} =
\begin{bmatrix}
\frac{12EI}{L^3} & -\frac{6EI}{L^2} & & & -\frac{12EI}{L^3} & -\frac{6EI}{L^2} & & \\
-\frac{6EI}{L^2} & \frac{4EI}{L} & & & \frac{6EI}{L^2} & \frac{2EI}{L} & & \\
 & & \frac{12EI}{L^3} & -\frac{6EI}{L^2} & & & -\frac{12EI}{L^3} & -\frac{6EI}{L^2} \\
 & & -\frac{6EI}{L^2} & \frac{4EI}{L} & & & \frac{6EI}{L^2} & \frac{2EI}{L} \\
-\frac{12EI}{L^3} & \frac{6EI}{L^2} & & & \frac{12EI}{L^3} & \frac{6EI}{L^2} & & \\
-\frac{6EI}{L^2} & \frac{2EI}{L} & & & \frac{6EI}{L^2} & \left(\frac{4EI}{L}+\frac{4EI}{L}\right) & & \frac{2EI}{L} \\
 & & -\frac{12EI}{L^3} & \frac{6EI}{L^2} & & & \frac{12EI}{L^3} & \frac{6EI}{L^2} \\
 & & -\frac{6EI}{L^2} & \frac{2EI}{L} & & \frac{2EI}{L} & \frac{6EI}{L^2} & \left(\frac{4EI}{L}+\frac{4EI}{L}\right)
\end{bmatrix}
\begin{Bmatrix} d_1 \\ \theta_1 \\ d_2 \\ \theta_2 \\ \Delta \\ \theta_3 \\ \Delta \\ \theta_4 \end{Bmatrix}
$$

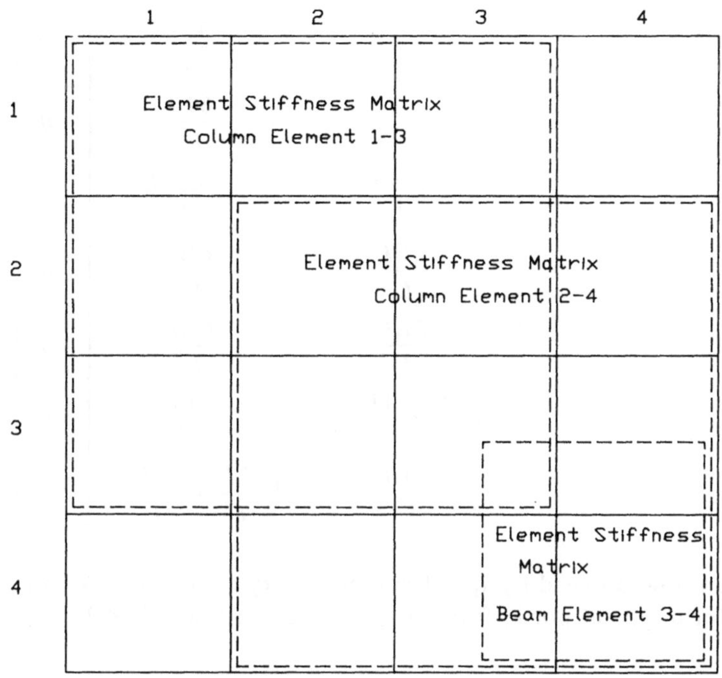

Figure 4.12 *Form of the structure stiffness matrix for the frame of figure 4.11.*

Combining the fifth and seventh equations and coefficients of Δ, which is a repeated displacement term, results in the following relationship which describes the structure stiffness,

$$\left\{\begin{matrix} V_1 \\ M_1 \\ V_2 \\ M_2 \\ M_3 \\ M_4 \\ Q \end{matrix}\right\} = \left[\begin{matrix} \frac{12EI}{L^3} & -\frac{6EI}{L^2} & & & -\frac{6EI}{L^2} \\ -\frac{6EI}{L^2} & \frac{4EI}{L} & & & \frac{2EI}{L} \\ & & \frac{12EI}{L^3} & -\frac{6EI}{L^2} & \\ & & -\frac{6EI}{L^2} & \frac{4EI}{L} & \\ -\frac{6EI}{L^2} & \frac{2EI}{L} & & & \left(\frac{4EI}{L}+\frac{4EI}{L}\right) \\ & & -\frac{6EI}{L^2} & \frac{2EI}{L} & \frac{2EI}{L} \\ -\frac{12EI}{L^3} & \frac{6EI}{L^2} & -\frac{12EI}{L^3} & \frac{6EI}{L^2} & \frac{6EI}{L^2} \end{matrix}\right.$$

(*continued*)

$$
\begin{bmatrix}
 & -\dfrac{12EI}{L^3} \\[2ex]
 & \dfrac{6EI}{L^2} \\[2ex]
-\dfrac{6EI}{L^2} & -\dfrac{12EI}{L^3} \\[2ex]
\dfrac{2EI}{L} & \dfrac{6EI}{L^2} \\[2ex]
\dfrac{2EI}{L} & \dfrac{6EI}{L^2} \\[2ex]
\left(\dfrac{4EI}{L}+\dfrac{4EI}{L}\right) & \dfrac{6EI}{L^2} \\[2ex]
\dfrac{6EI}{L^2} & \left(\dfrac{12EI}{L^3}+\dfrac{12EI}{L^3}\right)
\end{bmatrix}
\begin{Bmatrix} d_1 \\ \theta_1 \\ d_2 \\ \theta_2 \\ \theta_3 \\ \theta_4 \\ \Delta \end{Bmatrix}
$$

The solution can now proceed by specifying boundary conditions. A complete solution for a typical portal frame is presented as example 4.3.

Example 4.3: Analysis of a Rectangular Plane Frame

Given data:

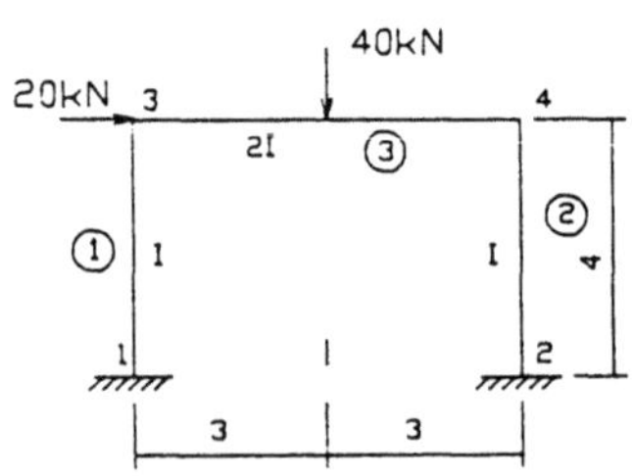

$E = 200\ \text{kN/mm}^2$

$I = 200 \times 10^6\ \text{mm}^4$

Case (a)—Fixed Bases

Procedure: Form column stiffness matrices and beam element stiffness matrix. Assemble to give structure stiffness matrix. Apply boundary conditions; form load vector from Fixed End Action solution; solve for displacements; recover element actions.

Degrees of Freedom:

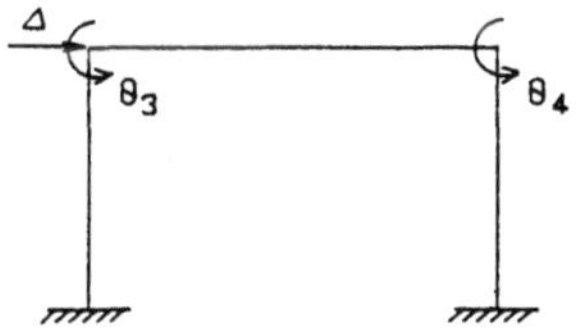

Ignore axial deformation

DOF = 3

Element Stiffness Matrices: (Units: kN and m)

Column Elements ① and ②

$$EI\begin{bmatrix} 0.1875 & -0.375 & -0.1875 & -0.375 \\ -0.375 & 1 & 0.375 & 0.5 \\ -0.1875 & 0.375 & 0.1875 & 0.375 \\ -0.375 & 0.5 & 0.375 & 1 \end{bmatrix}$$

Beam Element ③

$$EI\begin{bmatrix} 1.33\dot{3} & 0.66\dot{6} \\ 0.66\dot{6} & 1.33\dot{3} \end{bmatrix}$$

Structure Stiffness Matrix: From Figure 4.12, initially with Q' and Q''

$$\begin{Bmatrix} V_1 \\ M_1 \\ V_2 \\ M_2 \\ Q' \\ M_3 \\ Q'' \\ M_4 \end{Bmatrix} = EI \begin{bmatrix} 0.1875 & -0.375 & & & -0.1875 & -0.375 & & \\ -0.375 & 1 & & & 0.375 & 0.5 & & \\ & & 0.1875 & -0.375 & & & -0.1875 & -0.375 \\ & & -0.375 & 1 & & & 0.375 & 0.5 \\ -0.1875 & 0.375 & & & 0.1875 & 0.375 & & \\ -0.375 & 0.5 & & & 0.375 & 2.333 & & 0.666 \\ & & -0.1875 & 0.375 & & & 0.1875 & 0.375 \\ & & -0.375 & 0.5 & & 0.66\dot{6} & 0.375 & 2.333 \end{bmatrix} \begin{Bmatrix} d_1 \\ \theta_1 \\ d_2 \\ \theta_2 \\ \Delta \\ \theta_3 \\ \Delta \\ \theta_4 \end{Bmatrix}$$

Combining Q' and Q'' to give Q and combining coefficients of Δ gives

$$\begin{Bmatrix} V_1 \\ M_1 \\ V_2 \\ M_2 \\ M_3 \\ M_4 \\ Q \end{Bmatrix} = EI \begin{bmatrix} 0.1875 & -0.375 & & & -0.375 & & -0.1875 \\ -0.375 & 1 & & & 0.5 & & 0.375 \\ & & 0.1875 & -0.375 & & -0.375 & -0.1875 \\ & & -0.375 & 1 & & 0.5 & 0.375 \\ -0.375 & 0.5 & & & 2.333 & 0.666 & 0.375 \\ & & -0.375 & 0.5 & 0.666 & 2.333 & 0.375 \\ -0.1875 & 0.375 & -0.1875 & 0.375 & 0.375 & 0.375 & 0.375 \end{bmatrix} \begin{Bmatrix} d_1 \\ \theta_1 \\ d_2 \\ \theta_2 \\ \theta_3 \\ \theta_4 \\ \Delta \end{Bmatrix}$$

Applying the boundary conditions to give $P_F = K_F \cdot d_F$ gives

$$\begin{Bmatrix} M_3 \\ M_4 \\ Q \end{Bmatrix} = EI \begin{bmatrix} 2.33\dot{3} & 0.66\dot{6} & 0.375 \\ 0.66\dot{6} & 2.33\dot{3} & 0.375 \\ 0.375 & 0.375 & 0.375 \end{bmatrix} \begin{Bmatrix} \theta_3 \\ \theta_4 \\ \Delta \end{Bmatrix}$$

Load Vector: Fixed end actions

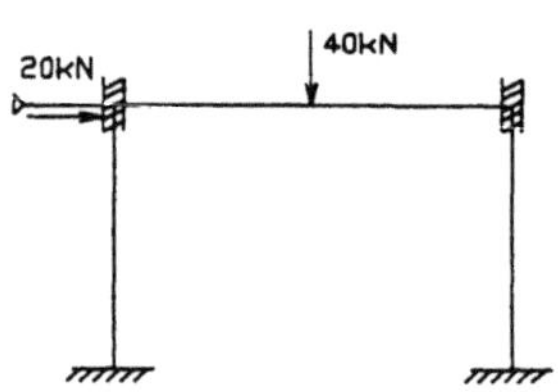

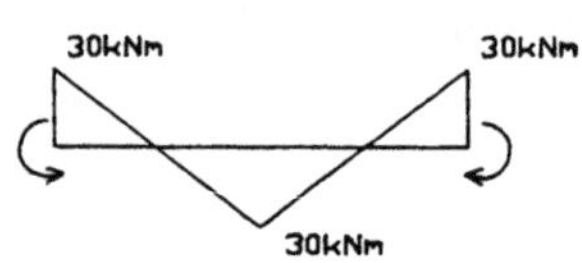

Hence

$$\begin{Bmatrix} M_3 \\ M_4 \\ Q \end{Bmatrix} = \begin{Bmatrix} -30 \\ 30 \\ 20 \end{Bmatrix}$$

Solution: The following results were obtained from MATOP

```
LOAD.K.3.3
SCALE.K.40000
LOAD.P.3.1
SOLVE.K.P
PRINT.P  DISPLACEMENTS
-.672222E-03
0.227778E-03
0.177778E-02
SCALE.P.40000
PRINT.P  SCALED DISPLACEMENTS
-.268889E+02
0.911110E+01
0.711111E+02
End of File
```

Node	x (mm)	y (mm)	rot. (radian)
1	—	—	—
2	—	—	—
3	1.78	—	−0.0007
4	1.78	—	0.0002

Element Actions: Using the known displacements and the element stiffness matrices in the relationship of the form of equation (4.7), the internal actions can be found. For example

$$\begin{Bmatrix} v_{13} \\ m_{13} \\ v_{31} \\ m_{31} \end{Bmatrix} = EI \begin{bmatrix} 0.1875 & -0.375 & -0.1875 & -0.375 \\ -0.375 & 1 & 0.375 & 0.5 \\ -0.1875 & 0.375 & 0.1875 & 0.375 \\ -0.375 & 0.5 & 0.375 & 1 \end{bmatrix} \frac{1}{EI} \begin{Bmatrix} 0 \\ 0 \\ 71.11 \\ -26.88 \end{Bmatrix}$$

$$= \begin{Bmatrix} -3.25 \\ 13.22 \\ 3.25 \\ -0.22 \end{Bmatrix}$$

Similarly

$$\begin{Bmatrix} v_{24} \\ m_{24} \\ v_{42} \\ m_{42} \end{Bmatrix} = \begin{Bmatrix} -16.75 \\ 31.22 \\ 16.75 \\ 35.77 \end{Bmatrix} \quad \text{and} \quad \begin{Bmatrix} m_{34} \\ m_{43} \end{Bmatrix} = \begin{Bmatrix} -29.77 \\ -5.77 \end{Bmatrix}$$

Final Actions:

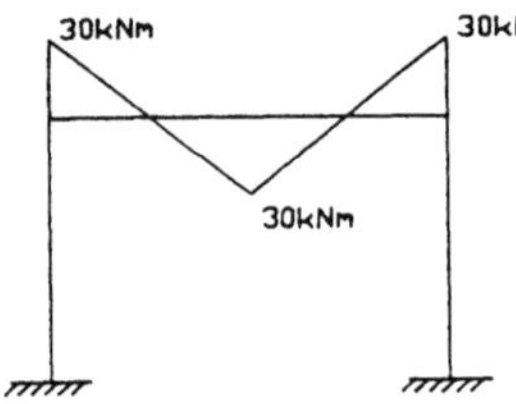

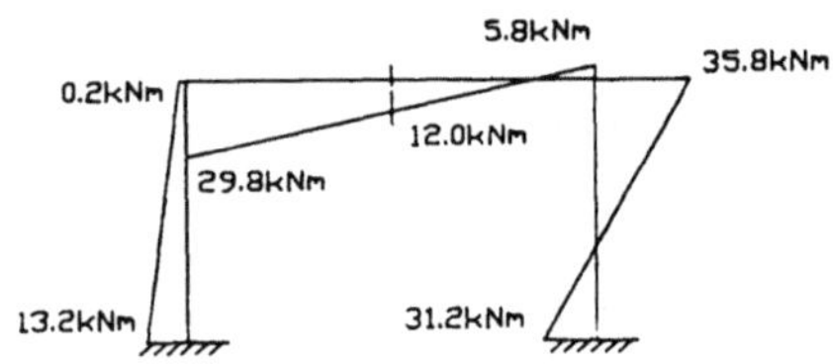

Fixed End Actions (Part 1)

Matrix Analysis Solution (Part 2)

Frame Bending Moment Diagram

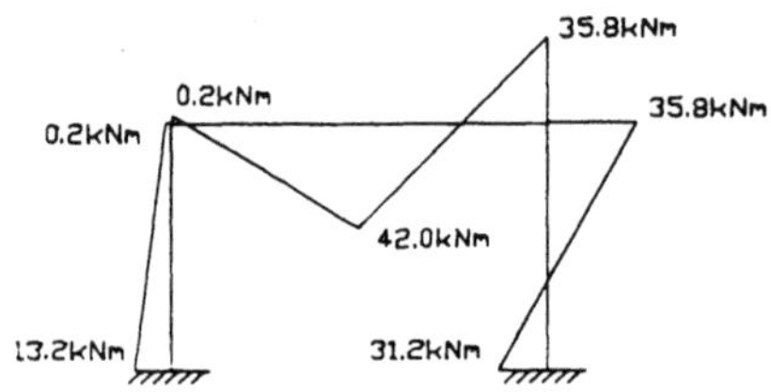

Reactions

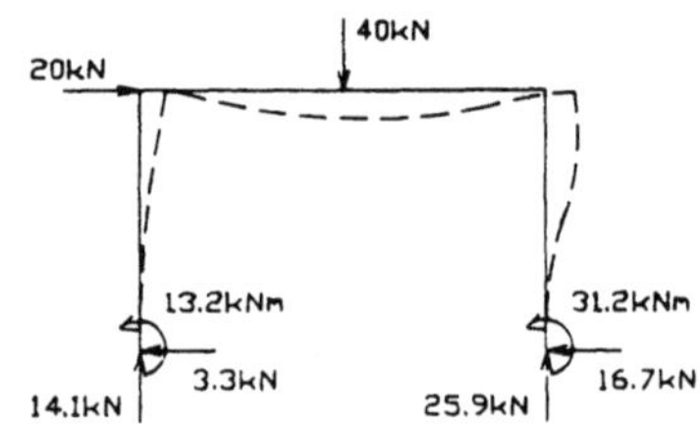

The same frame can now be analysed for the case when the column bases are pinned.

Case (b)—Pinned Bases

Other data as for Case (a)

Degrees of Freedom:

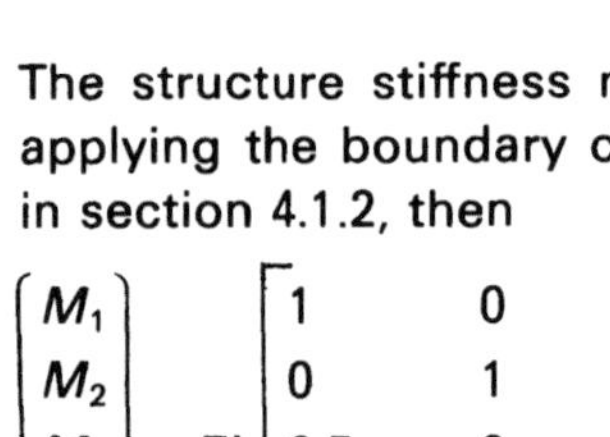

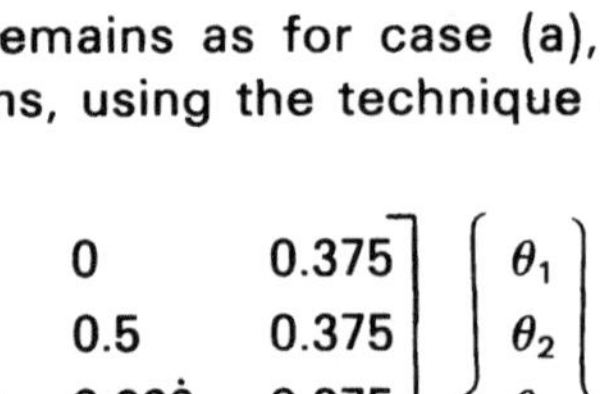

DOF = 5

The structure stiffness matrix remains as for case (a), but after applying the boundary conditions, using the technique described in section 4.1.2, then

$$\begin{Bmatrix} M_1 \\ M_2 \\ M_3 \\ M_4 \\ Q \end{Bmatrix} = EI \begin{bmatrix} 1 & 0 & 0.5 & 0 & 0.375 \\ 0 & 1 & 0 & 0.5 & 0.375 \\ 0.5 & 0 & 2.33\dot{3} & 0.66\dot{6} & 0.375 \\ 0 & 0.5 & 0.66\dot{6} & 2.333 & 0.375 \\ 0.375 & 0.375 & 0.375 & 0.375 & 0.375 \end{bmatrix} \begin{Bmatrix} \theta_1 \\ \theta_2 \\ \theta_3 \\ \theta_4 \\ \Delta \end{Bmatrix}$$

Load Vector: Fixed End Actions

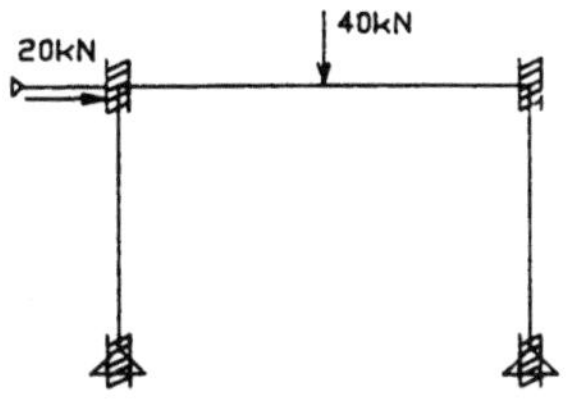

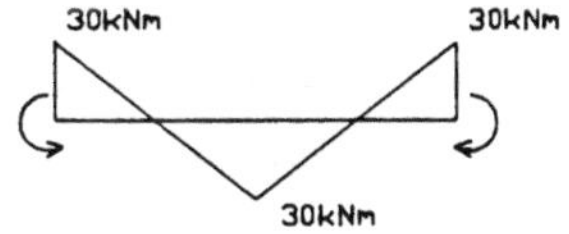

Hence

$$\begin{Bmatrix} M_1 \\ M_2 \\ M_3 \\ M_4 \\ Q \end{Bmatrix} = \begin{Bmatrix} 0 \\ 0 \\ -30 \\ 30 \\ 20 \end{Bmatrix}$$

Solution: The following results were obtained from MATOP

```
LOAD.K.5.5
SCALE.K.40000
LOAD.P.5.1
SOLVE.K.P
PRINT.P  DISPLACEMENTS
-.223529E-02
-.276471E-02
-.102941E-02
0.294114E-04
0.733333E-02
SCALE.P.40000
PRINT.P  SCALED DISPLACEMENTS
-.894118E+02
-.110588E+03
-.411765E+02
0.117645E+01
0.293333E+03
End of File
```

Node	x (mm)	y (mm)	rot. (radian)
1	—	—	−0.0022
2	—	—	−0.0028
3	7.33	—	−0.0010
4	7.33	—	0.0000

Element Actions: Using the now known displacements, the following internal actions were found using the element stiffness matrices.

	Element		
Action	1–3	2–4	3–4
v_{ij} (kN)	−6.03	−13.97	
m_{ij} (kN m)	0	0	−54.10
v_{ji} (kN)	+6.03	13.97	
m_{ji} (kN m)	24.13	55.88	−25.88

Final Actions:

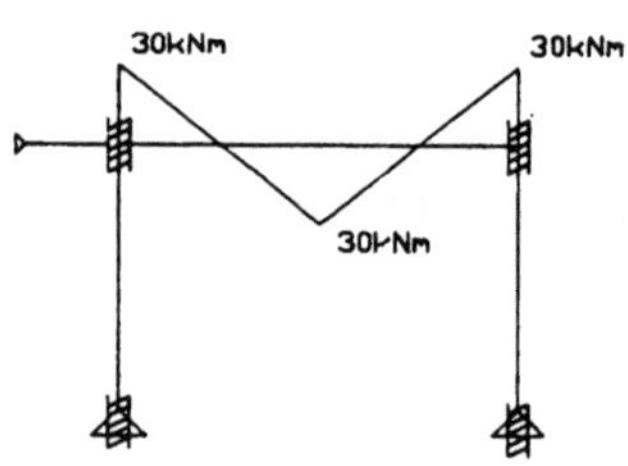

Fixed End Actions
(Part 1)

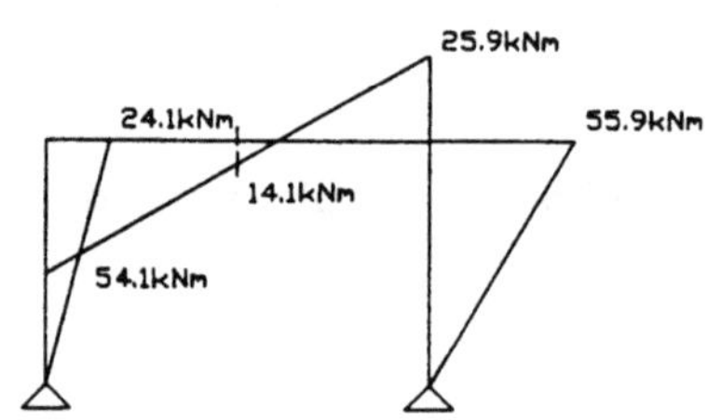

Matrix Analysis Solution
(Part 2)

Frame Bending Moment Diagram

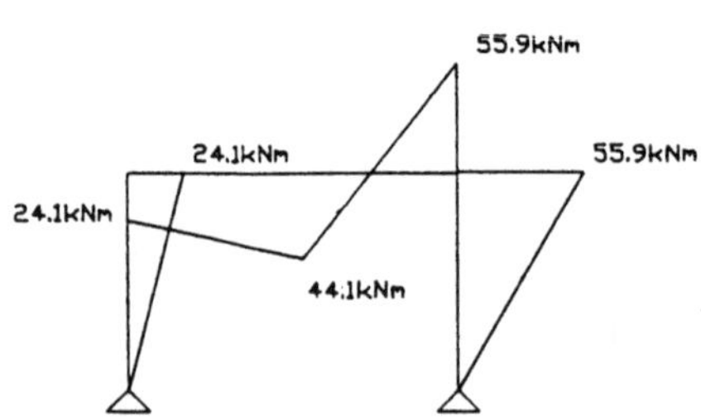

Reactions:

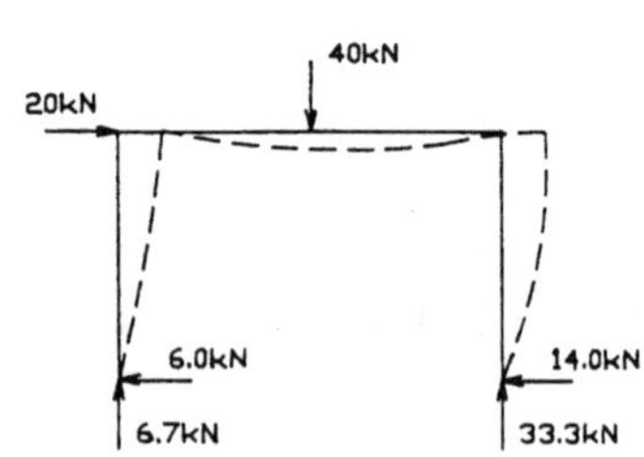

4.3 THE DIRECT STIFFNESS METHOD

The routine of forming the structure stiffness matrix by assembling a series of element stiffness matrices can be eliminated in a procedure which may be referred to as the direct stiffness method. The process may be applied to any matrix equation describing an action-displacement relationship.

The general form of such an equation may be written as

$$\begin{Bmatrix} P_1 \\ P_2 \\ P_3 \\ \vdots \\ P_n \end{Bmatrix} = \begin{bmatrix} k_{11} & k_{12} & k_{13} & \cdots & \\ k_{21} & k_{22} & & & \\ k_{31} & & & & \\ \vdots & & & & \\ k_{n1} & \cdots & & & k_{nn} \end{bmatrix} \begin{Bmatrix} d_1 \\ d_2 \\ d_3 \\ \vdots \\ d_n \end{Bmatrix} \tag{4.9}$$

If a nominated term d_j in the displacement vector were given a unit value, while all the remaining terms were specified as zero, then the expansion of the equation (4.9) would give

$$\begin{Bmatrix} P_1 \\ P_2 \\ P_3 \\ \vdots \\ P_n \end{Bmatrix} = \begin{Bmatrix} k_{1j} \\ k_{2j} \\ k_{3j} \\ \vdots \\ k_{nj} \end{Bmatrix} \tag{4.10}$$

It may be seen from equation (4.10) that the load or action vector which imposes the set of displacements to be such that d_j equals unity while all other displacements are zero, is equal to the jth column of the coefficients of the matrix linking the actions and the displacements.

The technique may be effectively used to check the continuous beam element stiffness matrix previously defined. The series of specific actions to create the displacement patterns required is shown in figure 4.13. It may be noted that in each case the required actions (which may be determined from the slope-deflection equations and equilibrium considerations) are the coefficients taken from the appropriate column of the continuous beam element stiffness matrix. In applying the technique to the formation of a structure stiffness matrix, the actions are the nodal actions, or loads, that would have to be applied to cause the unique set of displacements required. The magnitude of nodal loads can be readily determined once the internal element actions are identified from the imposed displacements. This is illustrated in the series of diagrams shown in figure 4.14.

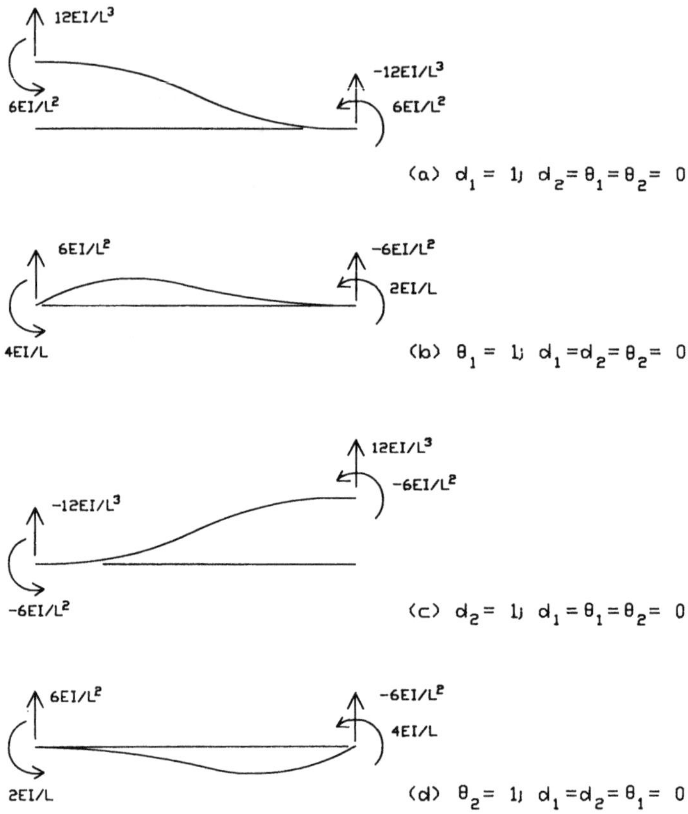

Figure 4.13 *Beam actions under specified displacements.*

For a given structure, the appropriate number of degrees of freedom should be nominated, excluding the restrained degrees of freedom at boundary nodes unless the total structure stiffness matrix is required, and the action-displacement relationship written out in a general form similar to that given in equation (4.9). Each column of the coefficients of the stiffness matrix can then be determined in turn, by sketching the appropriate displacement pattern and noting the necessary nodal loads that would have to be applied to give the required displacements. With the structure stiffness matrix formed, the analysis can proceed on the same basis as the earlier study. In particular, once the matrix equation has been solved to determine the displacements, the element actions can be calculated using the element stiffness matrices in the usual way. Example 4.3 (Case (a)—Fixed Bases) can now be repeated as example 4.4 using the direct stiffness method.

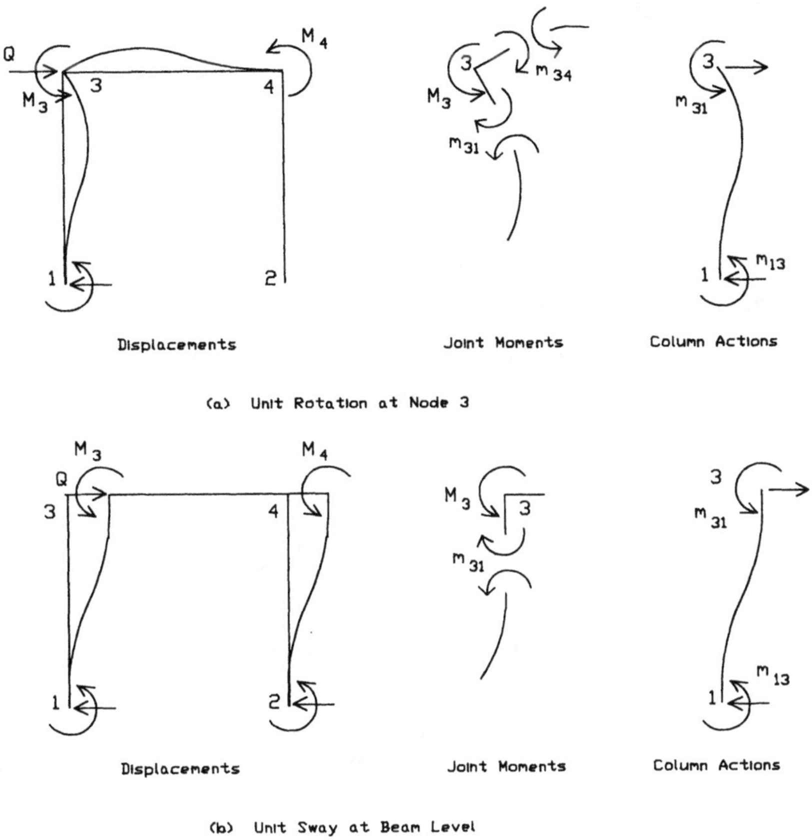

Figure 4.14 Rectangular frame displacements.

Example 4.4: Frame Analysis using the Direct Stiffness Method

---

Example 4.3 repeated using the direct stiffness method of forming the structure stiffness matrix.

Case (a)—Fixed Bases

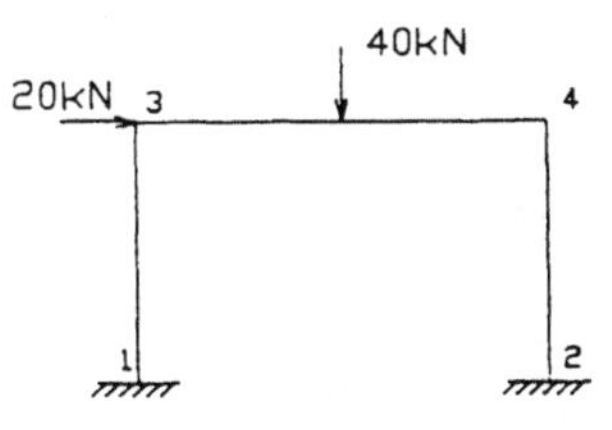

Given data:

$E = 200\ \text{kN/mm}^2$

$I = 200 \times 10^6\ \text{mm}^4$

Degrees of Freedom:

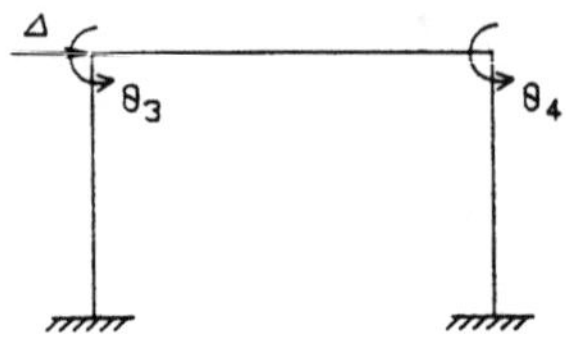

Structure Stiffness Matrix:

Require K_F of the form:

$$\begin{Bmatrix} M_3 \\ M_4 \\ Q \end{Bmatrix} = \begin{bmatrix} k_{11} & k_{12} & k_{13} \\ k_{21} & k_{22} & k_{23} \\ k_{31} & k_{32} & k_{33} \end{bmatrix} \begin{Bmatrix} \theta_3 \\ \theta_4 \\ \Delta \end{Bmatrix}$$

$\theta_3 = 1; \qquad \theta_4 = \Delta = 0$

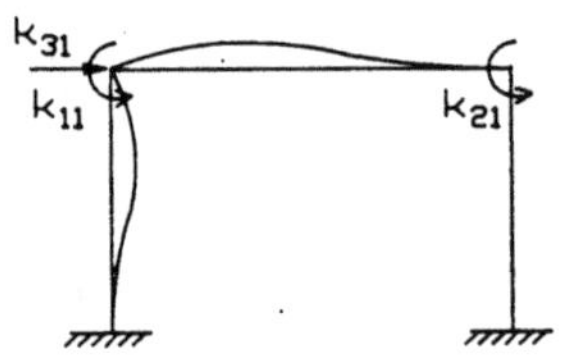

$$k_{11} = \frac{4EI}{4} + \frac{4E(2I)}{6} = 2.33\dot{3}EI$$

$$k_{21} = \frac{2E(2I)}{6} = 0.66\dot{6}EI$$

$$k_{31} = \frac{6EI}{(4)^2} = 0.375EI$$

$\theta_4 = 1; \qquad \theta_3 = \Delta = 0$

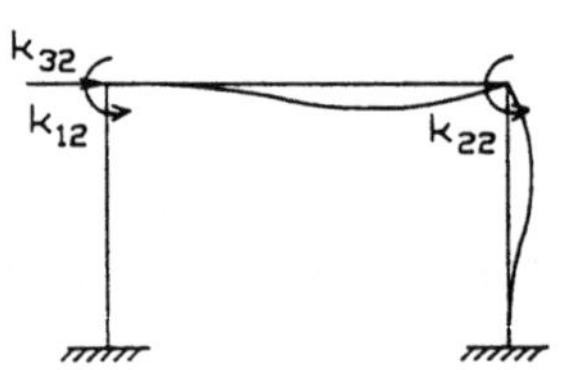

$$k_{12} = \frac{2E(2I)}{6} = 0.66\dot{6}EI$$

$$k_{22} = \frac{4E(2I)}{6} + \frac{4EI}{4} = 2.33\dot{3}EI$$

$$k_{32} = \frac{6EI}{(4)^2} = 0.375EI$$

$\Delta = 1; \qquad \theta_3 = \theta_4 = 0$

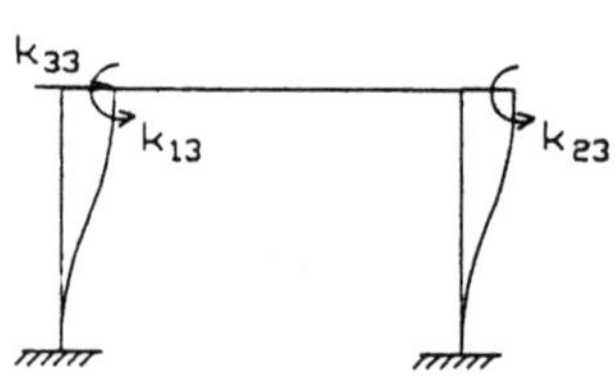

$$k_{13} = \frac{6EI}{(4)^2} = 0.375EI$$

$$k_{23} = \frac{6EI}{(4)^2} = 0.375EI$$

$$k_{33} = \frac{12EI}{(4)^3} + \frac{12EI}{(4)^3} = 0.375EI$$

Hence K_F has the form of example 4.3 and the analysis proceeds from there as before.

Case (b)—Pinned Bases

Degrees of Freedom:

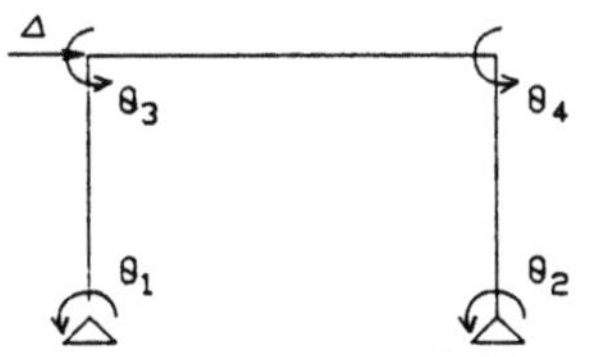

Structure Stiffness Matrix:

$$\begin{Bmatrix} M_1 \\ M_2 \\ M_3 \\ M_4 \\ Q \end{Bmatrix} = \begin{bmatrix} k_{11} & k_{12} & k_{13} & k_{14} & k_{15} \\ k_{21} & k_{22} & k_{23} & k_{24} & k_{25} \\ k_{31} & k_{32} & k_{33} & k_{34} & k_{35} \\ k_{41} & k_{42} & k_{43} & k_{44} & k_{45} \\ k_{51} & k_{52} & k_{53} & k_{54} & k_{55} \end{bmatrix} \begin{Bmatrix} \theta_1 \\ \theta_2 \\ \theta_3 \\ \theta_4 \\ \Delta \end{Bmatrix}$$

Unit Displacement Diagrams:

$\theta_1 = 1; \qquad \theta_2 = \theta_3 = \theta_4 = \Delta = 0$

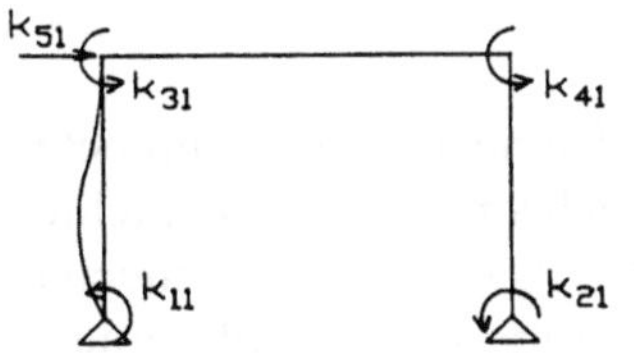

$$k_{11} = \frac{4EI}{4} = 1\,EI$$

$$k_{21} = 0 = 0$$

$$k_{31} = \frac{2EI}{4} = 0.5EI$$

$$k_{41} = 0 = 0$$

$$k_{51} = \frac{6EI}{4^2} = 0.375EI$$

(Similarly for the second column)

$\theta_3 = 1; \qquad \theta_1 = \theta_2 = \theta_4 = \Delta = 0$

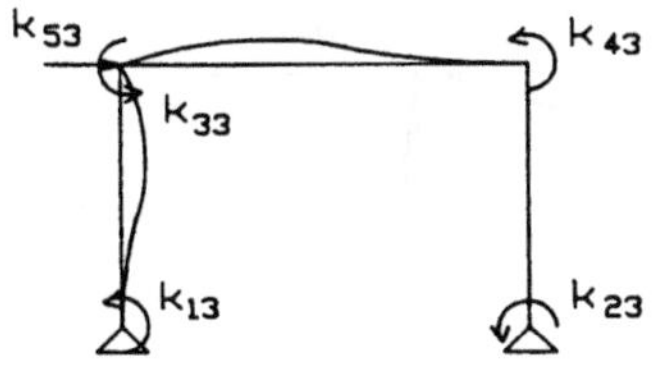

$$k_{13} = \frac{2EI}{4} = 0.5EI$$

$$k_{23} = 0 = 0$$

$$k_{33} = \frac{4EI}{4} + \frac{4E(2I)}{6} = 2.333EI$$

$$k_{43} = \frac{2E(2I)}{6} = 0.666EI$$

$$k_{53} = \frac{6EI}{4^2} = 0.375EI$$

(Similarly for the fourth column)

$\Delta=1;\qquad \theta_1=\theta_2=\theta_3=\theta_4=0\qquad k_{15}=\frac{6EI}{4^2}=0.375EI$

$$k_{25}=\frac{6EI}{4^2}=0.375EI$$

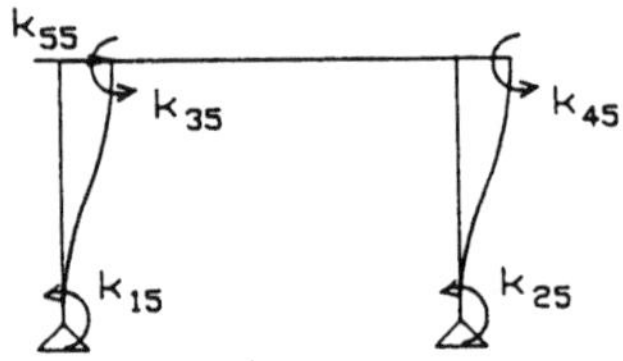

$$k_{35}=\frac{6EI}{4^2}=0.375EI$$

$$k_{45}=\frac{6EI}{4^2}=0.375EI$$

$$k_{55}=\frac{12EI}{4^3}+\frac{12EI}{4^3}=0.375EI$$

which leads to the form of K_F in example 4.3, Case (b).

4.4 MODIFICATION TO ELEMENT STIFFNESS MATRICES FOR END MOMENT RELEASE

While the elements of a continuous beam or plane frame are often connected in such a way as to provide for moment transfer, such structures may include connections where the elements should be regarded as being pin-connected. In the general case of three or more elements framing into a node, one of which is pin-connected, there may still be moments at the node and a nodal rotation, although the local moment at the pin-ended element must be zero and its end rotation is independent. Such a condition is shown in the frame of figure 4.15(a), along with a detail of a possible rotation response of node 5 in figure 4.15(b). Clearly the pin-connected element has a different stiffness and it is necessary to define appropriate element stiffness matrices for these circumstances.

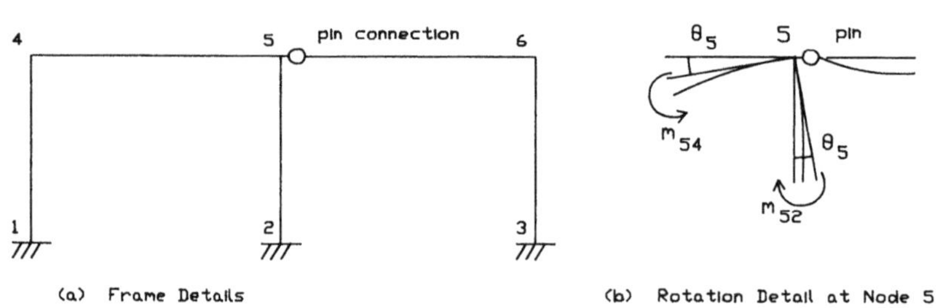

Figure 4.15 Element end moment release.

Case (a): Pinned Left-hand End of the Continuous Beam Element

Consider again the general beam element as shown in figure 3.8 of chapter 3. From the element stiffness matrix, the general equations describing the element behaviour can be written as

$$v_{12}=\frac{12EI}{L^3}d_1+\frac{6EI}{L^2}\theta_1-\frac{12EI}{L^3}d_2+\frac{6EI}{L^2}\theta_2 \qquad \text{(i)}$$

$$m_{12}=\frac{6EI}{L^2}d_1+\frac{4EI}{L}\theta_1-\frac{6EI}{L^2}d_2+\frac{2EI}{L}\theta_2 \qquad \text{(ii)}$$

$$v_{21}=-\frac{12EI}{L^3}d_1-\frac{6EI}{L^2}\theta_1+\frac{12EI}{L^3}d_2-\frac{6EI}{L^2}\theta_2 \qquad \text{(iii)}$$

$$m_{21}=\frac{6EI}{L^2}d_1+\frac{2EI}{L}\theta_1-\frac{6EI}{L^2}d_2+\frac{4EI}{L}\theta_2 \qquad \text{(iv)}$$

However in this case, m_{12} is zero, so that equation (ii) may be used to express θ_1 in terms of the remaining displacements. Equation (ii) is then

$$0=\frac{6EI}{L^2}d_1+\frac{4EI}{L}\theta_1-\frac{6EI}{L^2}d_2+\frac{2EI}{L}\theta_2$$

so that

$$\frac{4EI}{L}\theta_1=-\frac{6EI}{L^2}d_1+\frac{6EI}{L^2}d_2-\frac{2EI}{L}\theta_2$$

and

$$\theta_1=-\frac{3}{2L}d_1+\frac{3}{2L}d_2-\frac{1}{2}\theta_2$$

This expression can be substituted back into the remaining equations to eliminate θ_1, to give

$$v_{12}=\frac{3EI}{L^3}d_1+0-\frac{3EI}{L^3}d_2+\frac{3EI}{L^2}\theta_2$$

$$m_{12}=0$$

$$v_{21}=-\frac{3EI}{L^3}d_1-0+\frac{3EI}{L^3}d_2+\frac{3EI}{L^2}\theta_2$$

$$m_{21}=\frac{3EI}{L^2}d_1+0-\frac{3EI}{L^2}d_2+\frac{3EI}{L}\theta_2$$

Writing the coefficients in matrix form gives

$$\left[\begin{array}{cc|cc} \frac{3EI}{L^3} & 0 & -\frac{3EI}{L^3} & \frac{3EI}{L^2} \\ 0 & 0 & 0 & 0 \\ \hline -\frac{3EI}{L^3} & 0 & \frac{3EI}{L^3} & -\frac{3EI}{L^2} \\ \frac{3EI}{L^2} & 0 & -\frac{3EI}{L^2} & \frac{3EI}{L} \end{array}\right]$$

which is the required element stiffness matrix.

Case (b): Pinned Right-hand End of the Continuous Beam Element

In the same way, considering the moment m_{21} as zero, the continuous beam element stiffness matrix when the right-hand end of the element is pinned is found to be

$$\left[\begin{array}{cc|cc} \frac{3EI}{L^3} & \frac{3EI}{L^2} & -\frac{3EI}{L^3} & 0 \\ \frac{3EI}{L^2} & \frac{3EI}{L} & -\frac{3EI}{L^3} & 0 \\ \hline -\frac{3EI}{L^3} & -\frac{3EI}{L^2} & \frac{3EI}{L^3} & 0 \\ 0 & 0 & 0 & 0 \end{array}\right]$$

Case (c): Pinned Base of the Column Element Stiffness Matrix

Again, following considerations similar to those outlined for Case (a), the appropriate element stiffness matrix is found to be

$$\left[\begin{array}{cc|cc} \frac{3EI}{L^3} & 0 & -\frac{3EI}{L^3} & -\frac{3EI}{L^2} \\ 0 & 0 & 0 & 0 \\ \hline -\frac{3EI}{L^3} & 0 & \frac{3EI}{L^3} & \frac{3EI}{L^2} \\ -\frac{3EI}{L^2} & 0 & \frac{3EI}{L^2} & \frac{3EI}{L} \end{array}\right]$$

When an element is pin-connected at a boundary node, the analyst has to decide which way this is considered. The first approach is to ignore the specific boundary conditions initially and to form the structure stiffness matrix in a general way. The boundary conditions can then be applied to accommodate the pin connection. This approach has been used in all of the examples presented to this stage. However, since the specific behaviour of a pin-ended element is known, the rotational degree of freedom at the boundary node need not be admitted in defining the behaviour of the structure. This leads to an overall simplification of the problem as is explained in more detail in section 4.5 and illustrated in example 4.5.

Example 4.5: Analysis of a Continuous Beam

Given data:

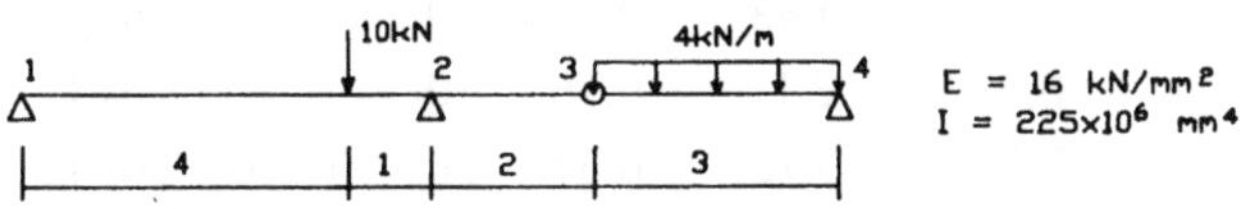

Nodes and Elements:

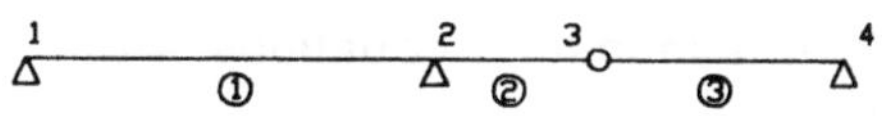

Procedure: Form beam element stiffness matrices and assemble to give structure stiffness matrix. Recognize left-hand end pinned for element ①; right-hand end pinned element ②; both ends pinned element ③, therefore zero stiffness. Apply boundary conditions and solve to obtain displacements. Recover element actions.

Element Stiffness Matrices (Units: kN and m):

$$ESM1 = EI\begin{bmatrix} 0.024 & 0 & -0.024 & 0.12 \\ 0 & 0 & 0 & 0 \\ -0.024 & 0 & 0.024 & -0.12 \\ 0.12 & 0 & -0.12 & 0.6 \end{bmatrix}$$

$$ESM2 = EI\begin{bmatrix} 0.375 & 0.75 & -0.375 & 0 \\ 0.75 & 1.5 & -0.75 & 0 \\ -0.375 & -0.75 & 0.375 & 0 \\ 0 & 0 & 0 & 0 \end{bmatrix}$$

Structure Stiffness Matrix

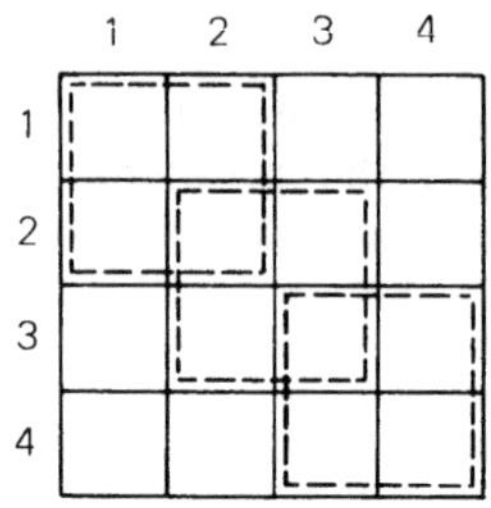

$$\begin{Bmatrix} V_1 \\ M_1 \\ V_2 \\ M_2 \\ V_3 \\ M_3 \\ V_4 \\ M_4 \end{Bmatrix} = EI \begin{bmatrix} 0.024 & 0 & -0.024 & 0.12 & & & & \\ 0 & 0 & 0 & 0 & & & & \\ -0.024 & 0 & 0.399 & 0.63 & -0.375 & 0 & & \\ & & 0.63 & 2.1 & -0.75 & 0 & & \\ & & -0.375 & -0.75 & 0.375 & 0 & 0 & 0 \\ & & 0 & 0 & 0 & 0 & 0 & 0 \\ & & & & 0 & 0 & 0 & 0 \\ & & & & 0 & 0 & 0 & 0 \end{bmatrix} \begin{Bmatrix} d_1 \\ \theta_1 \\ d_2 \\ \theta_2 \\ d_3 \\ \theta_3 \\ d_4 \\ \theta_4 \end{Bmatrix}$$

Applying the boundary conditions and eliminating equations with zero coefficients gives $P_F = K_F \cdot d_F$ as

$$\begin{Bmatrix} M_2 \\ V_3 \end{Bmatrix} = EI \begin{bmatrix} 2.1 & -0.75 \\ -0.75 & 0.375 \end{bmatrix} \begin{Bmatrix} \theta_2 \\ d_3 \end{Bmatrix}$$

The same result can be obtained by the direct stiffness method.

Consider:

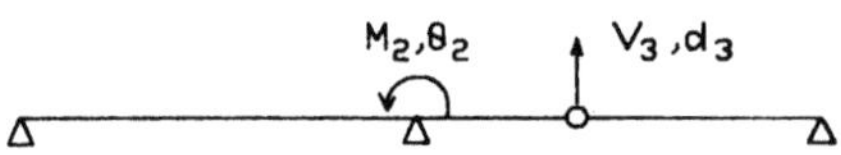

Degrees of Freedom = 2

$$\begin{Bmatrix} M_2 \\ V_3 \end{Bmatrix} = \begin{bmatrix} k_{11} & k_{12} \\ k_{21} & k_{22} \end{bmatrix} \begin{Bmatrix} \theta_2 \\ d_3 \end{Bmatrix}$$

$\theta_2 = 1;\ d_3 = 0$

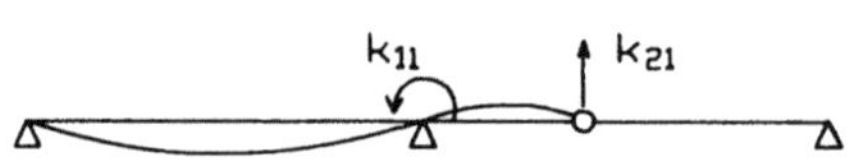

$$k_{11} = \frac{3EI}{5} + \frac{3EI}{2} = 2.1\,EI$$

$$k_{21} = -\frac{3EI}{2^2} = -0.75EI$$

$\theta_2 = 0;\ d_3 = 1$

$$k_{12} = -\frac{3EI}{2^2} = -0.75EI$$

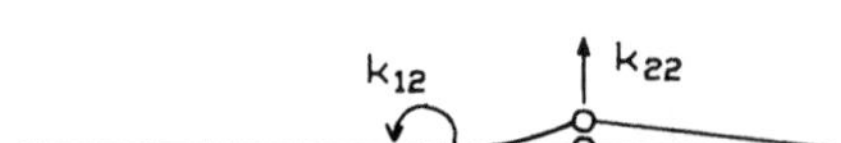

$$k_{22} = \frac{3EI}{2^3} = 0.375EI$$

which gives the same result, as required.

Load Vector: From Part 1 Solution—Fixed end actions, noting pins

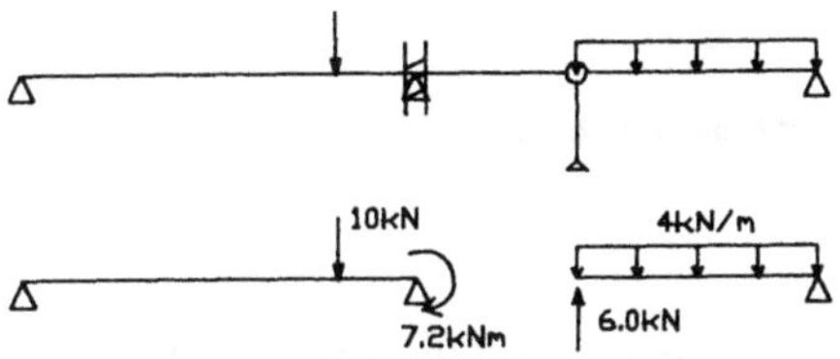

$M_2 = 7.2$

$V_3 = -6$

Solution: Solve

$$\begin{Bmatrix} 7.2 \\ -6 \end{Bmatrix} = EI \begin{bmatrix} 2.1 & -0.75 \\ -0.75 & 0.375 \end{bmatrix} \begin{Bmatrix} \theta_2 \\ d_3 \end{Bmatrix}$$

$$\begin{Bmatrix} \theta_2 \\ d_3 \end{Bmatrix} = \frac{1}{EI} \cdot \frac{1}{0.225} \begin{bmatrix} 0.375 & 0.75 \\ 0.75 & 2.1 \end{bmatrix} \begin{Bmatrix} 7.2 \\ -6 \end{Bmatrix} = \frac{1}{EI} \begin{Bmatrix} -8.0 \\ -32.0 \end{Bmatrix}$$

$\therefore \quad \theta_2 = -0.0022$ radian and $d_3 = -8.9$ mm

Element Actions: Using equation (4.7), then

	Element		
	1–2	2–3	3–4
v_{ij} (kN)	−0.96	6.0	0
m_{ij} (kN m)	0	12.0	0
v_{ji} (kN)	0.96	−6.0	0
m_{ji} (kN m)	−4.80	0	0

Final Actions:

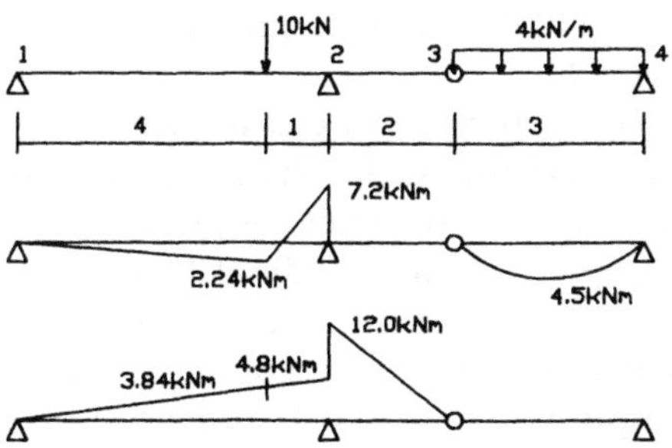

Fixed End Actions
(Part 1)

Matrix Analysis Solution
(Part 2)

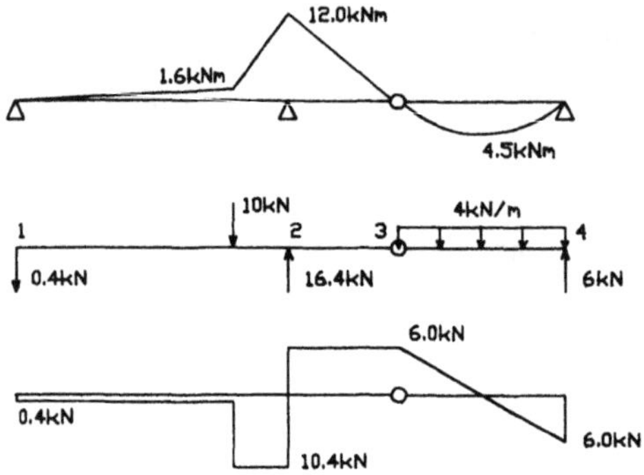

4.5 APPLICATION OF THE STIFFNESS METHOD TO BEAMS AND RECTANGULAR FRAMES

The matrix stiffness method, as applied to beams and rectangular frames, may be consolidated through a consideration of the application of the method to the structures shown in figure 4.16. It should now be apparent that there are several distinct approaches available in the analysis of such structures using the stiffness method. The variations can be clarified through a general discussion of how the analysis may proceed. For convenience in this discussion, the node and element numbering has already been selected for the structures shown in figure 4.16, although this would clearly be the starting point of any analysis.

In figure 4.16(b), a node point has been nominated at the mid-point of the second span. This ensures that the deflection and other data at that point are obtained as part of the immediate results of the analysis, although such nodes are optional. The other node points on the continuous beams are essential since the boundary conditions are expressed through them. For the frames of figure 4.16, there is little flexibility in the selection of nodes or in the node numbering. Because of the restrictions placed on the method presented in section 4.2, nodes must be confined to the supports and the beam-column intersections. These restrictions do not apply to the more general approach offered in chapter 6.

A feature of the method is that axial deformation of the elements is ignored. This is of no consequence to the continuous beams since they are considered to be under transverse loads only, and the effect of pin supports is exactly the same as that of roller supports. If longitudinal loads exist they can always be considered in a separate analysis, and the distribution of longitudinal forces will then depend on the precise nature of the supports. For the rectangular frames, the assumption that axial deformations are negligible introduces an approximation which is overcome in the more

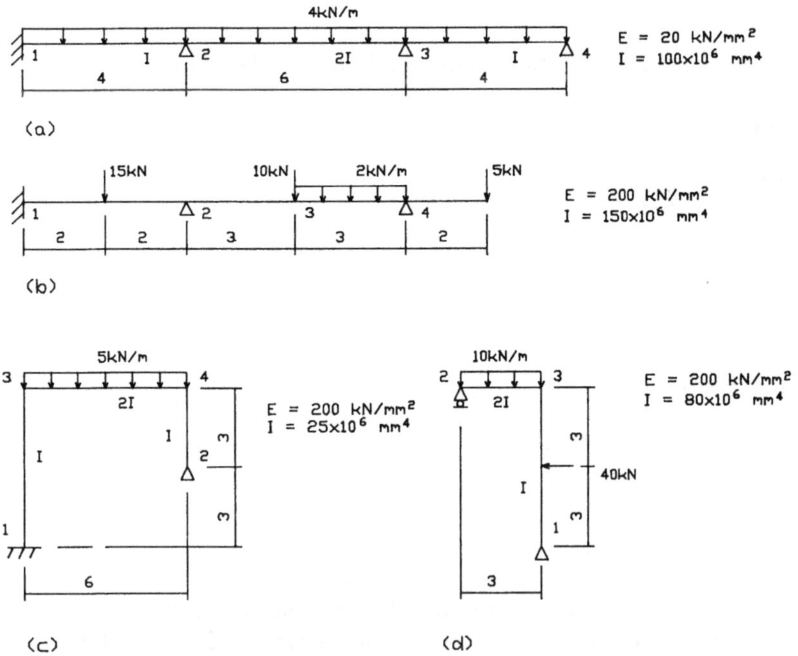

Figure 4.16 *Typical beam and frame problems.*

general technique of chapter 6. The omission of axial deformation for the frames leads to a reduction of the number of degrees of freedom so that it is only necessary to introduce one sway term at each level. Although it has not been particularly discussed, the method can of course be readily extended to multi-storey and multi-bay rectangular frames.

Either the direct stiffness method or the more general method of assembly from element stiffness matrices can be used for forming a structure stiffness matrix. Further, as was pointed out in section 4.4, decisions on handling boundary conditions must be taken. All of this means that the analyst must be specific about the details of the method of solution. While much of this is irrelevant to the use of fully developed structural analysis computer programs, it is highly relevant to an understanding of what is going on. In any case, one objective of this study is to show that quite complex problems can be reduced to some routine calculations quite within the scope of very modest computational aids, as is illustrated with example 4.6.

In a final comment on the application of the matrix stiffness method of this chapter, it is worth noting that the same approach applies irrespective of the determinacy of the structure. If the structure happens to be statically determinate, then the method can still be applied to advantage since relevant deflection data can be obtained. Further, the same basic data may be used

Example 4.6: Analysis of a Non-swaying Rectangular Frame

Given data:

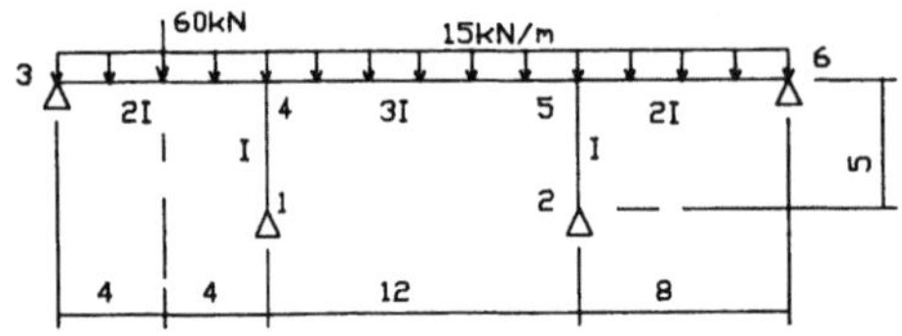

E = 20 kN/mm²
I = 1000×10⁶ mm⁴

Nodes and Elements:

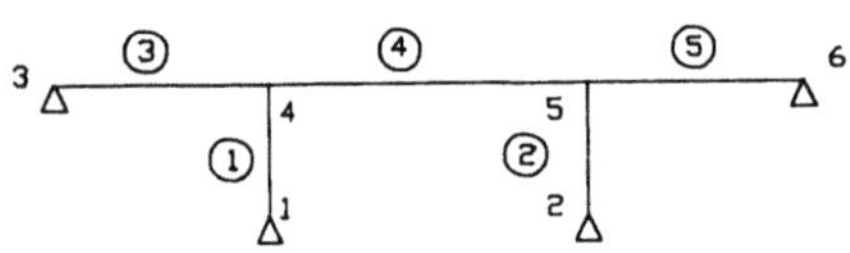

Procedure: Form column and beam element stiffness matrices, recognising the pin-ended conditions of elements ①, ②, ③ and ⑤ immediately. Use direct stiffness method to form structure stiffness matrix, K_F. Solve for displacements and recover element actions.

Element Stiffness Matrices (Units: kN and m):

Column Elements ① and ②

$$EI\begin{bmatrix} 0.024 & 0 & -0.024 & -0.12 \\ 0 & 0 & 0 & 0 \\ -0.024 & 0 & 0.024 & 0.12 \\ -0.12 & 0 & 0.12 & 0.60 \end{bmatrix}$$

Beam Element ③

$$EI\begin{bmatrix} 0.01172 & 0 & -0.01172 & 0.09375 \\ 0 & 0 & 0 & 0 \\ -0.01172 & 0 & 0.01172 & -0.09375 \\ 0.09375 & 0 & -0.09375 & 0.75 \end{bmatrix}$$

Beam Element ④

$$EI\begin{bmatrix} 0.02083 & 0.125 & -0.02083 & 0.125 \\ 0.125 & 1.0 & -0.125 & 0.5 \\ -0.02083 & -0.125 & 0.02083 & -0.125 \\ 0.125 & 0.5 & -0.125 & 1.0 \end{bmatrix}$$

Beam Element ⑤

$$EI\begin{bmatrix} 0.01172 & 0.09375 & -0.01172 & 0 \\ 0.09375 & 0.75 & -0.09375 & 0 \\ -0.01172 & -0.09375 & 0.01172 & 0 \\ 0 & 0 & 0 & 0 \end{bmatrix}$$

Structure Stiffness Matrix K_F based on:

Degrees of Freedom:

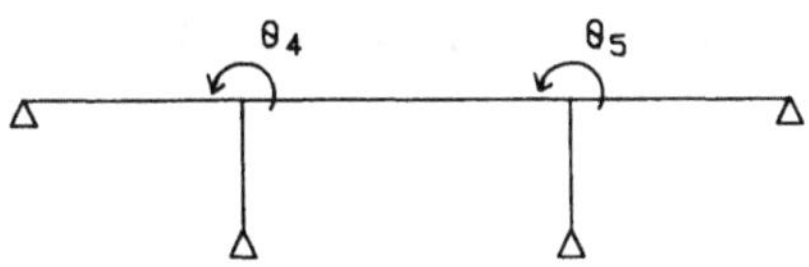

$$\begin{Bmatrix} M_4 \\ M_5 \end{Bmatrix} = \begin{bmatrix} k_{11} & k_{12} \\ k_{21} & k_{22} \end{bmatrix} \begin{Bmatrix} \theta_4 \\ \theta_5 \end{Bmatrix}$$

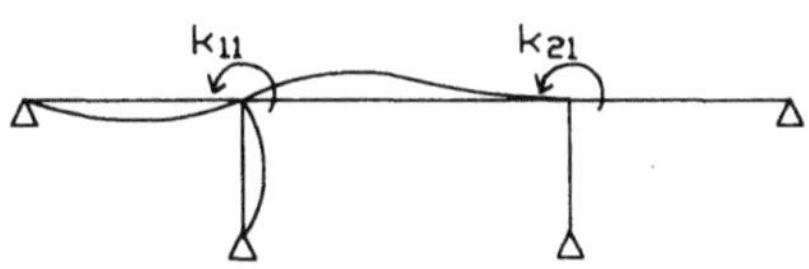

$$k_{11} = \frac{3EI}{5} + \frac{3E(2I)}{8} + \frac{4E(3I)}{12}$$

$$= 2.35EI$$

$$k_{21} = \frac{2E(3I)}{12} = 0.5EI$$

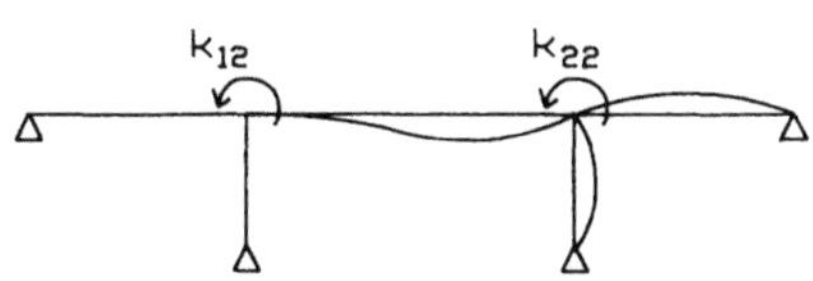

Similarly:

$$k_{12} = 0.5EI$$

$$k_{22} = 2.35EI$$

Load Vector: From Part 1 Solution—Fixed End Actions, noting pins

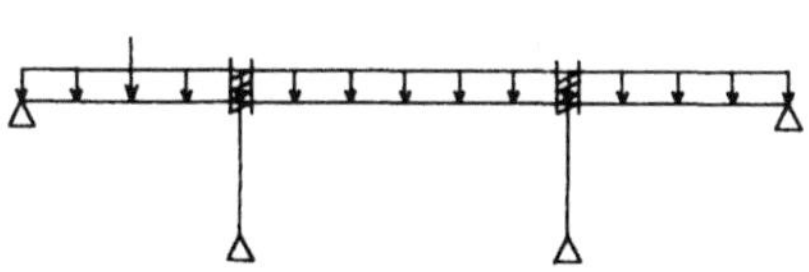

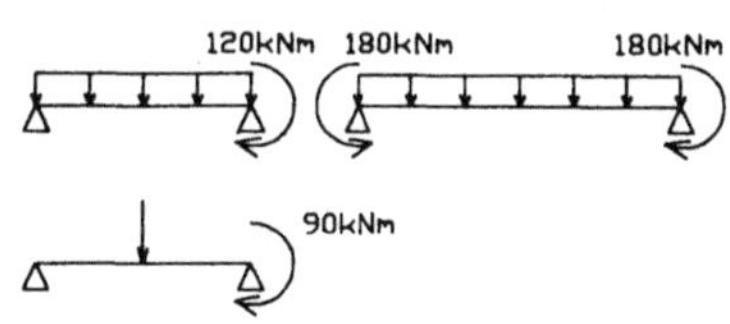

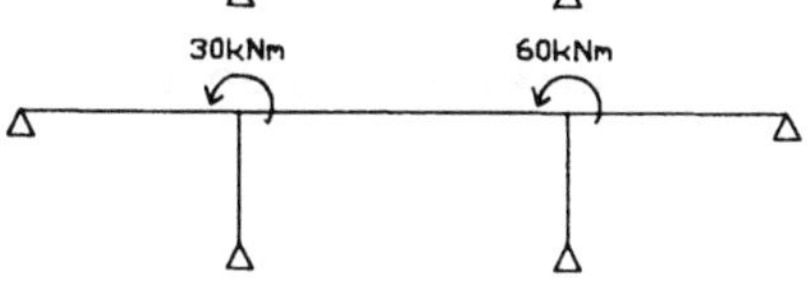

Solution: Solve

$$\begin{Bmatrix} 30 \\ 60 \end{Bmatrix} = EI \begin{bmatrix} 2.35 & 0.5 \\ 0.5 & 2.35 \end{bmatrix} \begin{Bmatrix} \theta_4 \\ \theta_5 \end{Bmatrix}$$

$$\begin{Bmatrix} \theta_4 \\ \theta_5 \end{Bmatrix} = \frac{1}{EI} \cdot \frac{1}{5.2725} \begin{bmatrix} 2.35 & -0.5 \\ -0.5 & 2.35 \end{bmatrix} \begin{Bmatrix} 30 \\ 60 \end{Bmatrix} = \frac{1}{EI} \begin{Bmatrix} 7.6814 \\ 23.8976 \end{Bmatrix}$$

$\therefore \quad \theta_4 = 0.004$ radian and $\theta_5 = 0.0012$ radian

Element Actions: Using equation (4.7), then

	Element				
	1–4	2–5	3–4	4–5	5–6
v_{ij} (kN)	−0.92	−2.87	0.72	3.95	2.24
m_{ij} (kN m)	0	0	0	19.63	17.92
v_{ji} (kN)	+0.92	2.87	−0.72	−3.95	−2.24
m_{ji} (kN m)	4.61	14.34	5.76	27.74	0

Final Actions:

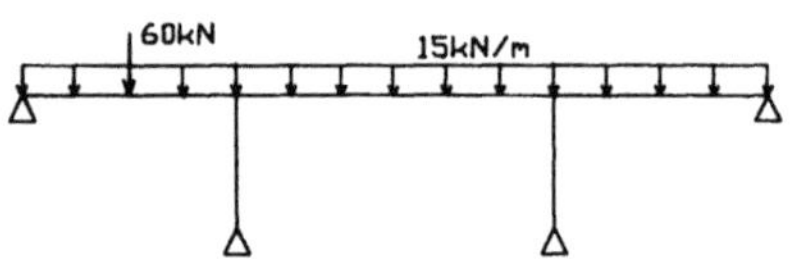

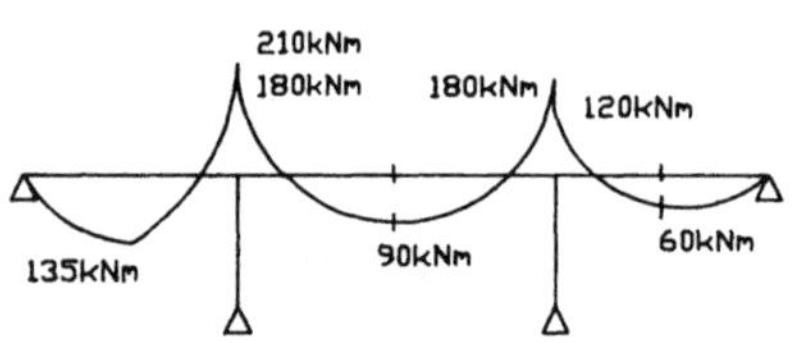

Fixed End Actions
(Part 1)

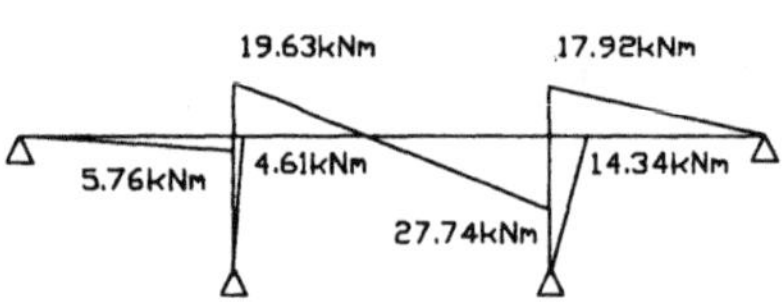

Matrix Analysis Solution
(Part 2)

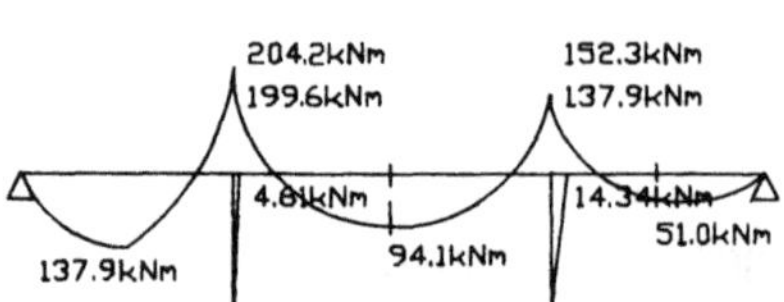

Bending Moment Diagram
(kN m)

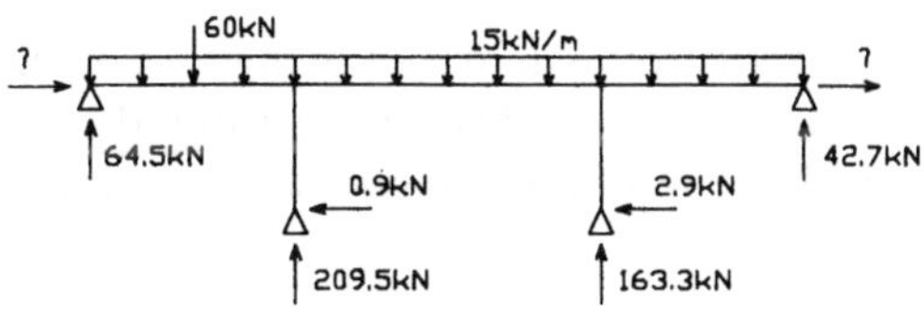

Reactions (kN)

Check Vertical Equilibrium:

$60+28(15)=480$

$64.5+209.5+163.3+42.7=480$

---

in variations on the structure, though different boundary conditions and multiple load cases can be readily considered. In examples 4.5 and 4.6, there is considerable choice available in respect of the modification to the degrees of freedom based on the pin-connected boundary nodes. The degrees of freedom have in fact been selected to minimise the size of the stiffness matrix K, and it would be useful to consider the problems again on the basis of additional degrees of freedom in order to understand the differences. It can be seen that in the final results for the reactions of the frame of example 4.6, the distribution of the horizontal reaction at the beam level is unknown. This is because axial deformation is ignored and there is no data on the longitudinal stiffness of the beam elements. Of course the force is a small one and it could be proportionally distributed between the supports with little error. If the support at node 6 had been a roller support, then the analysis as presented would not be any different (again because of the assumption of zero axial deformation), and the nett horizontal force required for equilibrium would be at node 3.

4.6 PROBLEMS FOR SOLUTION

4.1 Considering the cantilever beam of figure P4.1 as two beam elements, calculate the deflection and rotation at nodes 2 and 3. Complete the analysis by matrix methods to confirm the bending moment diagram that is given by equilibrium alone.

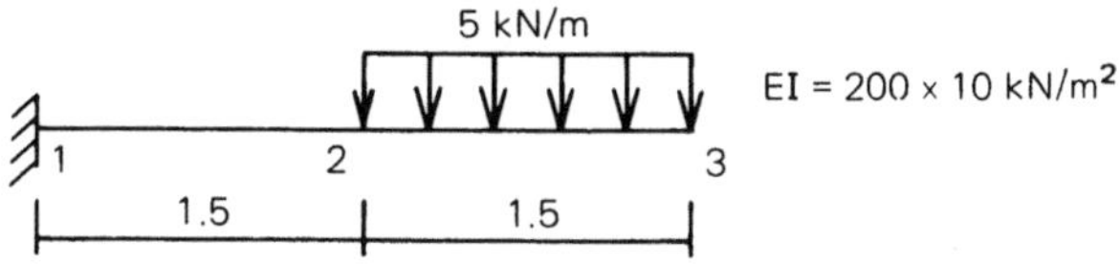

Figure P4.1.

4.2 The built-in end beam of figure P4.2 is to be analysed using two beam elements to find the end moments due to a uniformly distributed load acting on half the beam. Show that the displacements at node 2 are

$$d_2 = -\frac{qL^4}{768EI} \quad \text{and} \quad \theta_2 = -\frac{qL^3}{768EI}$$

and that

$$M_{AB} = \frac{5}{192} qL^2 \quad \text{and} \quad M_{BA} = -\frac{11}{192} qL^2$$

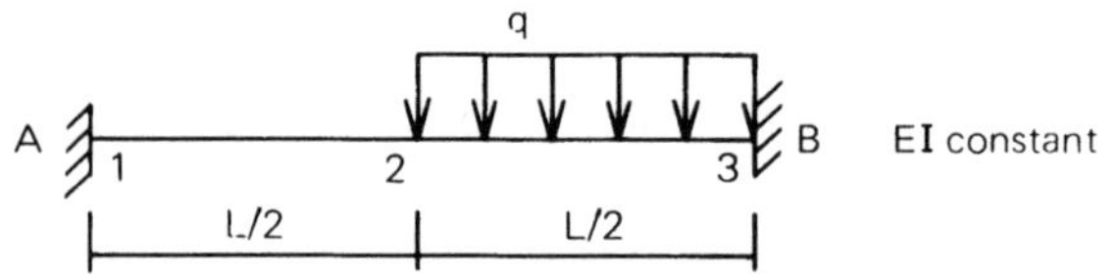

Figure P4.2.

4.3 The three element cantilever beam of figure P4.3 is an approximation to a tapered beam. Calculate the deflections of the beam for both the load shown and when the beam is under a uniformly distributed load of 5 kN/m.

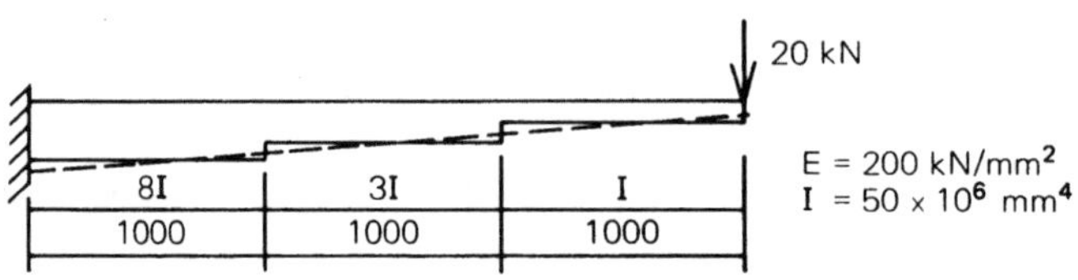

Figure P4.3.

4.4 Analyse the structures shown in figure 4.16 of chapter 4. Sketch the deflected shapes and draw the bending moment diagrams showing all principal ordinates.

Chapter 5
The Moment Distribution Method

It is a characteristic of the structural analysis problem that most formulations lead to the need to solve a set of simultaneous equations. Without using a computer or programmable calculator, solving four or more equations is a tedious, if not daunting, task that most people would prefer to avoid. The general basis of the matrix stiffness method as presented in earlier chapters was well understood prior to the development of the digital computer, so that analysts were left with having to find the solution to a large number of simultaneous equations by hand calculation, or to avoid them.

Attention was focused on the latter option and, in the 1930s, Professor H. Cross published a method of moment distribution (Cross, 1936) which circumvented the actual solution to the set of equations. The essential elements of that method are presented in this chapter, and it is considered important since there will always be a need for hand calculations in the design office. A few simple hand calculations, even using gross assumptions, are invaluable either in a checking situation or in the preliminary design phase. This point is expanded in chapter 9 in a study of approximate methods.

The moment distribution method is based on the philosophy of the stiffness method of structural analysis. While the technique can be used on more extensive structures, this presentation will be restricted to a study of beams and frames similar to the applications of chapter 4.

5.1 AN ITERATIVE SOLUTION TO A SET OF SIMULTANEOUS EQUATIONS

Before proceeding to a study of the moment distribution method, it is appropriate to consider an iterative method for the solution to a system of linear equations. The method of moment distribution is itself an iterative method based on the following technique, although the actual solution for the primary unknowns is usually by-passed. Any iteration technique relies on an iteration formula or recursive equation, in which the current value of the $(n+1)$th variable is expressed in terms of the previously determined variables from 1 to n.

Consider the general set of n simultaneous equations:

$$\begin{aligned} a_{11}x_1+a_{12}x_2+\cdots a_{1n}x_n &= b_1 \\ a_{21}x_1+a_{22}x_2+\cdots a_{2n}x_n &= b_2 \\ &\vdots \\ a_{n1}x_1+a_{n2}x_2+\cdots a_{nn}x_n &= b_n \end{aligned} \tag{5.1}$$

Equations (5.1) can be rewritten in the form:

$$\begin{aligned} x_1 &= \frac{1}{a_{11}}(b_1-a_{12}x_2-a_{13}x_3\cdots-a_{1n}x_n) \\ x_2 &= \frac{1}{a_{22}}(b_2-a_{21}x_1-a_{23}x_3\cdots-a_{2n}x_n) \\ &\vdots \\ x_n &= \frac{1}{a_{nn}}(b_n-a_{n1}x_1-a_{n2}x_2\cdots-a_{nn-1}x_{n-1}) \end{aligned} \tag{5.2}$$

A first approximation to the solution is given by assuming that all the x_i values in the right-hand side of equation (5.2) are zero. In practical terms, the coefficients along the leading diagonal of the matrix A, representing the coefficients a_{ij} of equation (5.1), often dominate the other coefficients so that the first approximation may in fact be close to the solution.

Equation (5.2) is a set of recursive equations so that substitution of the first approximation into equation (5.2) gives the second approximation. The process can be repeated up to the kth iteration, where the results are not significantly altered by further iterations. The method is generally known as Gauss–Seidel Iteration and it is important to stress that it is not an approximate method, since any desired accuracy can be achieved by increasing the number of iterations.

The convergence of the procedure is usually improved by modifying the technique to the extent that in any one iteration, to find a new value, next-best estimates of the lower variables are used. That is, as opposed to using the values of the ith iteration in a substitution into the recursive equations for the $(i+1)$th iteration, the currently improved values are used. For example, the set of equations used in the solution to example 4.3 of chapter 4 can be written as

$$\begin{aligned} 2.333x_1+0.666x_2+0.375x_3 &= -30 \\ 0.666x_1+2.333x_2+0.375x_3 &= 30 \\ 0.375x_1+0.375x_2+0.375x_3 &= 20 \end{aligned} \tag{5.3}$$

ignoring the scalar effect of the flexural rigidity *EI*. Applying equations (5.2) to equations (5.3) gives

$$
\begin{aligned}
x_1 &= \frac{1}{2.333}(-30-0.666x_2-0.375x_3) \\
x_2 &= \frac{1}{2.333}(30-0.666x_1-0.375x_3) \\
x_3 &= \frac{1}{0.375}(20-0.375x_1-0.375x_2)
\end{aligned}
\tag{5.4}
$$

The first approximation is then

$$x_1 = -12.86; \qquad x_2 = 16.53; \quad \text{and} \quad x_3 = 49.66$$

In the second iteration, x_1 is given by the first of the recursive equations as −25.56, and it is this value which is used with the first approximation to x_3 to give the second approximation to x_2, found to be 12.17. Now both second approximation values of x_1 and x_2 are used to find the second approximation of x_3. The results from three iterations are recorded in table 5.1, along with the exact solution given by Gaussian Elimination from the program MATOP.

The iterative method is simply a numerical analysis technique and it does little to relieve the computational effort associated with large numbers of equations. However the process can be interpreted in a physical sense in association with the analysis of a continuous beam. In a three-span continuous beam taking the length of each element as equal to the span, the structure has four degrees of freedom expressed by the rotations θ_1 to θ_4 over the supports. The rotations are the unknown variables in the resulting equations that express the moment-rotation relationship for the structure through the structure stiffness matrix. An iterative solution would start with all the rotations held at zero. The next step would allow the first node to rotate to give a first approximation to θ_1 while all the other rotations remained at zero. For the first approximation to θ_2, θ_1 would be held at its current value while θ_2 was released with θ_3 and θ_4 both remaining at zero. In one iteration all of the nodes would be progressively released as the process moved from one node to the next. Subsequent iterations would

Table 5.1 Iterative solution to equations

k	x_1	x_2	x_3
1	−12.86	16.53	49.66
2	−25.56	12.17	66.72
3	−27.06	9.86	70.53
Exact	−26.89	9.11	71.11

start again at the first node and move across the structure. The actions of first setting the rotation to zero, and subsequently allowing a rotation and holding it, can be further interpreted physically if the nodes are assumed to have imaginary clamps applied. The iteration procedure can then be seen as one of progressively relaxing the restrained form of the structure to the free deflected shape.

Element rotations are related to the end moments acting, and the moment distribution method exploits this by considering the effects of the rotation rather than the actual rotation. By carrying out the iterative operations in a systematic and tabular fashion, much of the computational effort in the solution of simultaneous equations is reduced.

5.2 THE ELEMENTS OF THE MOMENT DISTRIBUTION METHOD

The method will be presented initially for structures where the degrees of freedom are restricted to nodal rotation only, such as continuous beams and non-swaying frames with nodes selected at the joints. In the matrix stiffness method, the problem is generalised to one of nodal loads of applied moments producing a set of nodal rotations as the displacement vector. The applied moments generally derive from the effects of transverse loads on the beams and columns, although that is of little consequence at this stage.

Figure 5.1 shows three examples of structures with nodal rotation only—in fact, restricted to one rotational degree of freedom so that the load-deflection relationship is given by

$$M_2 = K \cdot \theta_2 \tag{5.5}$$

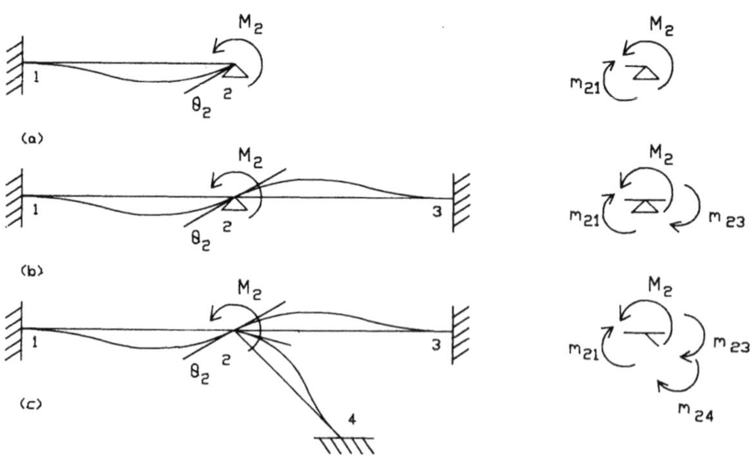

Figure 5.1 Structural systems with one rotational DOF.

where equation (5.5) is actually a scalar relationship with the matrix K reduced to one term.

The element stiffness matrix for prismatic beam elements subjected to end rotations only is a subset of the continuous beam element stiffness matrix and takes the form

$$\begin{Bmatrix} m_{ij} \\ m_{ji} \end{Bmatrix} = \begin{bmatrix} \dfrac{4EI}{L} & \dfrac{2EI}{L} \\ \dfrac{2EI}{L} & \dfrac{4EI}{L} \end{bmatrix} \begin{Bmatrix} \theta_i \\ \theta_j \end{Bmatrix} \tag{5.6}$$

Taking case (b) as a specific example from figure 5.1, the moment–rotation relationship can be found as follows.

From joint moment equilibrium:

$$M_2 = m_{21} + m_{23} \tag{5.7}$$

However values of m_{21} and m_{23} can be found using equation (5.6) applied to each element in turn, noting particularly that the far end of each element is fixed, so that

$$m_{21} = \frac{4EI_1}{L_1} \cdot \theta_2 \quad \text{and} \quad m_{23} = \frac{4EI_2}{L_2} \cdot \theta_2$$

with the subscript on the second moment of area, I, and the length, L, used to refer to the specific element.

Substituting for m_{21} and m_{23} into equation (5.7) and noting the compatibility requirement that θ_2 is common to both beam ends, then

$$M_2 = \left(\frac{4EI_1}{L_1} + \frac{4EI_2}{L_2}\right) \cdot \theta_2$$

hence

$$\theta_2 = \frac{1}{\left(\dfrac{4EI_1}{L_1} + \dfrac{4EI_2}{L_2}\right)} \cdot M_2$$

The internal beam end moments m_{21} and m_{23} can now be recovered by back-substitution into equation (5.7) and they are found to be

$$\begin{aligned} m_{21} &= \frac{\dfrac{4EI_1}{L_1}}{\left(\dfrac{4EI_1}{L_1} + \dfrac{4EI_2}{L_2}\right)} \cdot M_2 \\ m_{23} &= \frac{\dfrac{4EI_2}{L_2}}{\left(\dfrac{4EI_1}{L_1} + \dfrac{4EI_2}{L_2}\right)} \cdot M_2 \end{aligned} \tag{5.8}$$

The results given by equation (5.8) can be interpreted in a general way to describe the influence of a moment applied to a node. The coefficient $4EI_k/L_k$ used in the moment–rotation relationship can be defined as a stiffness factor k_{2j}, in this case appropriate to a prismatic element subjected to end rotation with the far end fixed. Other stiffness factors can be evaluated for non-prismatic elements and for other forms of displacement, although they will not be considered here.

With this definition of element stiffness in mind, it can be seen that the applied moment, M_2, has been distributed into the beam elements in proportion to the relative stiffness of each beam, and a simple distribution factor, r, can be defined.

Equation (5.8) may be written as

$$m_{21} = r_{21} \cdot M_2$$
$$m_{23} = r_{23} \cdot M_2 \tag{5.9}$$

where

$$r_{21} = \frac{k_{21}}{(k_{21} + k_{23})}$$

and

$$r_{23} = \frac{k_{23}}{(k_{21} + k_{23})}$$

In general, the distribution factor applicable to the beam end ji is

$$r_{ji} = \frac{k_{ji}}{\sum_{m=1}^{n} k_{jm}} \tag{5.10}$$

where n is equal to the number of beams framing into the node j.

At the same time as the beam end moments m_{21} and m_{23} develop due to the rotation θ_2, end moments m_{12} and m_{32} are induced at the far ends of the beam which are held against rotation. These far end or carry-over moments can be found from equation (5.6) which gives

$$m_{j2} = \frac{2EI_k}{L_k} \cdot \theta_2 \tag{5.11}$$

where $j = 1, 3$ and $k = 1, 2$ for case (b) of figure 5.1. This shows that for a prismatic beam, the carry-over moments are one-half of the distributed moments and of the same sign.

The preceding analysis has followed the guidelines of the direct stiffness method as developed in chapter 4 and it has introduced the concepts behind the moment distribution method. It remains for these concepts to be introduced in a systematic procedure for the analysis of a complete structure.

5.3 APPLICATION OF THE MOMENT DISTRIBUTION METHOD

The three-span continuous beam of figure 5.2(a) will be analysed both by the matrix stiffness method and by moment distribution, in order to emphasise the relationship between the two techniques. Ignoring for the present the modification that can be made for the pin-ended condition of the exterior spans, the matrix stiffness method proceeds by restraining all the nodal rotations to develop the fixed end moments shown in figure 5.2(c). This leads to the set of applied moments of figure 5.2(d), representing the load vector for the matrix analysis.

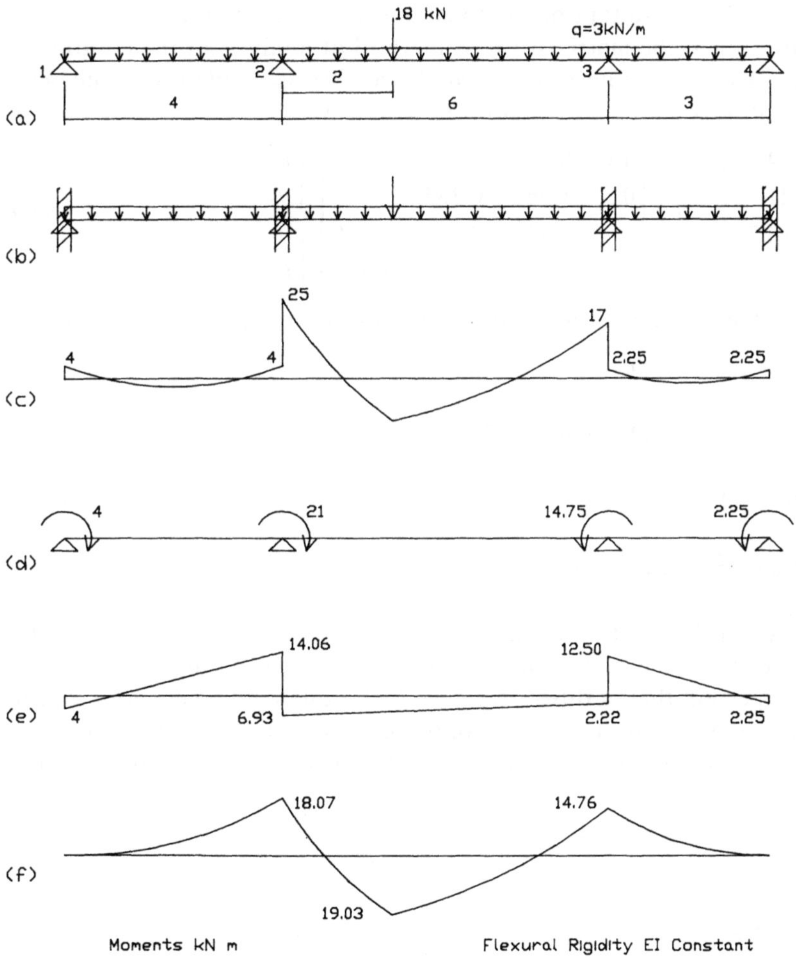

Figure 5.2 *Analysis of a continuous beam.*

Clarification may be needed for the sign of the moments acting on the node. A consistent sign convention has been used throughout this text and the convention itself embodies nodal moment equilibrium. It may be seen that anticlockwise internal moments acting on an element are positive along with the conjugate clockwise internal moment acting on the node. The external moment is necessarily of opposite sense to the resultant internal moments in order to satisfy the equilibrium of moments at the node. The positive external moment has therefore been taken in an anticlockwise sense. The sign convention is illustrated in figure 5.3 where all the moments shown are positive. Restraining moments represent external moments maintaining moment equilibrium and, since the nodal loads must negate the effects of the restraints, the sign of the applied moments of figure 5.2(d) is opposite to the sign of the resultant of the fixed end moments acting at the node.

Returning to the matrix stiffness method of analysis of the continuous beam, the structure stiffness matrix can be readily assembled from the beam element stiffness matrix of equation (5.6) to give

$$\begin{Bmatrix} -4 \\ -21 \\ 14.75 \\ 2.25 \end{Bmatrix} = EI \begin{bmatrix} 1.0 & 0.5 & & \\ 0.5 & 1.666 & 0.333 & \\ & 0.333 & 2.0 & 0.666 \\ & & 0.666 & 1.333 \end{bmatrix} \begin{Bmatrix} \theta_1 \\ \theta_2 \\ \theta_3 \\ \theta_4 \end{Bmatrix} \tag{5.12}$$

The recursive equations for an iterative solution to equation (5.12) are

$$\begin{aligned} EI\theta_1 &= \tfrac{1}{1}(-4-0.5\theta_2) \\ EI\theta_2 &= \frac{1}{1.666}(-21-0.5\theta_1-0.333\theta_3) \\ EI\theta_3 &= \tfrac{1}{2}(14.75-0.333\theta_2-0.666\theta_4) \\ EI\theta_4 &= \frac{1}{1.333}(2.25-0.666\theta_3) \end{aligned} \tag{5.13}$$

The results for the solution using both strict Gauss–Seidel Iteration and the modified technique outlined in section 5.1 are given in table 5.2.

As expected, the modified iteration technique converges more rapidly and while the iterations could continue, sufficient accuracy has been achieved with the cycles shown. The beam end moments due to the nodal

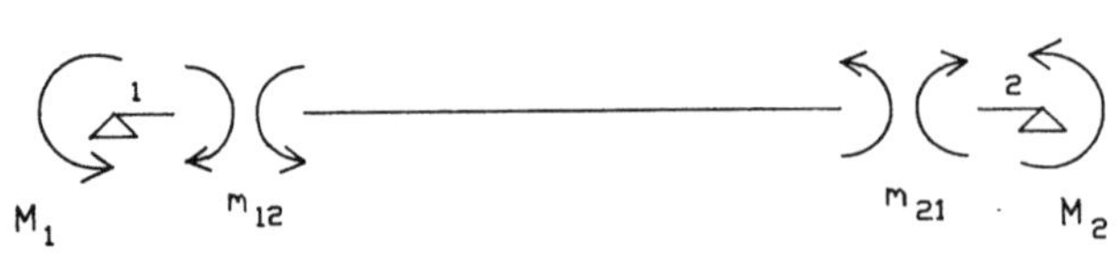

Figure 5.3 *Positive moments.*

Table 5.2 Iterative solutions for beam rotations

k	$EI\theta_1$	$EI\theta_2$	$EI\theta_3$	$EI\theta_4$
Gauss-Seidel Iteration				
1	−4.00	−12.60	7.38	1.69
2	2.30	−12.88	8.91	−2.00
3	2.44	−15.07	10.91	−2.77
4	3.54	−15.37	10.81	−3.41
5	3.69	−15.82	11.07	−3.72
6	3.91	−15.82	11.25	−3.85
7	3.96	−16.02	11.31	−3.94
8	4.01	−16.05	11.36	−3.97
Modified Gauss-Seidel Iteration				
1	−4.00	−11.40	9.27	−2.95
2	1.70	−14.97	10.85	−3.74
3	3.49	−15.82	11.26	−3.94
4	3.91	−16.03	11.36	−3.99

loads are recovered by back-substitution and these are shown in figure 5.2(e). Typically, the results for the interior span are given as

$$\begin{Bmatrix} m_{23} \\ m_{32} \end{Bmatrix} = EI \begin{bmatrix} 0.666 & 0.333 \\ 0.333 & 0.666 \end{bmatrix} \begin{Bmatrix} \dfrac{-16.03}{EI} \\ \dfrac{11.36}{EI} \end{Bmatrix}$$

$$= \begin{Bmatrix} -6.91 \\ 2.22 \end{Bmatrix}$$

The final moments in the beam are given by the combination of the moment diagrams of figures 5.2(c) and 5.2(e), as shown in figure 5.2(f).

The moment distribution method uses a similar iteration technique based on the distribution of moments into the beam ends and the influence of the subsequent carry-over moments. A physical interpretation of the operations is shown in figure 5.4, starting with all the nodes clamped against rotation. Each node is then released in turn, with all the remaining nodes still restrained, resulting in the series of diagrams of figure 5.4. Applied to the continuous beam of figure 5.2(a), the nodal moments of figure 5.2(d) are distributed into the beam ends according to the relative stiffness of the elements. A distribution should be seen as a release followed by a replacement of the restraining clamp. Initially, the release relaxes the node so that the end moments are balanced, consistent with the applied moment. The results at this stage represent the first approximation of an iterative solution. This operation is shown as cycle 1 in table 5.3 where each column is headed by the subscripts appropriate to the element end moment. The distribution of the nodal moments is according to the distribution factors calculated

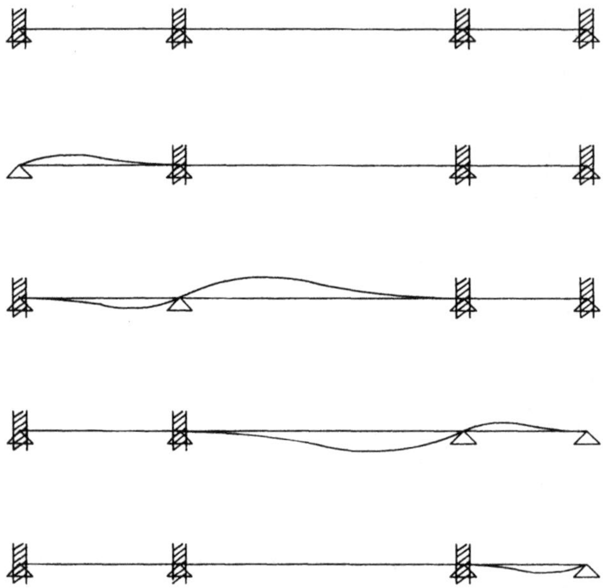

Figure 5.4 Progressive nodal release.

from equation (5.12) and shown immediately underneath the subscripts in the table. For example, at node 2 the applied moment of −21.00 kN m results in internal moments m_{21} of 0.6 (−21.00) [that is, −12.60] and m_{23} of 0.4 (−21.00) [that is, −8.40] at the node.

As was pointed out previously, the distribution of moments induces carry-over moments at the far end of a restrained element, causing a balancing moment to develop in the restraint. The carry-over moments are shown immediately after the distribution cycle and are derived as one-half of the opposite end distributed moment. Again using node 2 as an example, the carry-over moments are to m_{12} equal to −6.30 kN m and to m_{32} equal to −4.20 kN m, with the arrows in table 5.3 indicating the initial carry over for the first cycle only. In the second cycle of operation each node must again be relaxed to distribute the unbalanced moment produced by the carry-over moments and given as the sum of the carry-over moments at the node. Since the balancing moment in the restraint is always of opposite sense to the resultant internal moments acting on the node, the distributed moments are of opposite sign to the unbalanced moment. This cycle also produces carry-over moments and the process must be repeated until the carry-over moments are small enough to be neglected, in which case the node is fully relaxed from any rotational restraint. The fact that a balance is achieved after each distribution is indicated by the bars drawn in table 5.3

Table 5.3 Moment distribution table for beam of figure 5.2

Subscript	12	21	23	32	34	43
Dist. F	1	0.6	0.4	0.33	0.67	1
Cycle Number						
1	−4.00	−12.60	−8.40	4.92	9.83	2.25
	−6.30	−2.0	2.46	−4.20	1.13	4.92
2	6.30	−0.28	−0.18	1.02	2.05	−4.92
	−0.14	3.15	0.51	−0.09	−2.46	1.03
3	0.14	−2.20	−1.46	0.85	1.70	−1.03
	−1.10	0.07	0.43	−0.73	−0.52	0.85
4	1.10	−0.30	−0.20	0.42	0.83	−0.85
	−0.15	0.55	0.21	−0.10	−0.43	0.42
5	0.15	−0.45	−0.30	0.18	0.35	−0.42
	−0.23	0.08	0.09	−0.15	−0.21	0.18
6	0.23	−0.10	−0.07	0.12	0.24	−0.18
	−0.05	0.11	0.06	−0.03	−0.09	0.12
7	0.05	−0.10	−0.07	0.04	0.08	−0.12
	−0.05	0.02	0.02	−0.03	−0.06	0.04
8	0.05	−0.02	−0.02	0.03	0.06	−0.04
	−4.00	−14.07	−6.93	2.24	12.52	2.25

and it should be recalled that in the first cycle, the moments are balanced with the applied moment. For the given beam, eight cycles are considered to be sufficient and the element end moments are then given as the sum of the end moment entries in each column. The results agree substantially with those obtained by Gauss–Seidel Iteration.

Since the modified Gauss–Seidel Iteration produced more rapid convergence, it is pertinent to seek a parallel procedure in the moment distribution method. This can be done by considering the carry-over moment as soon as a distribution has been performed. This is equivalent to taking the next-best estimate of the rotations instead of using all the values of rotation from the previous cycle. The process is shown in table 5.4, in which it should be noted that the value of the fixed end moments is included at the head of each column. These fixed end moments are responsible for the initial applied nodal moments which are of opposite sign to the unbalanced moment, as previously explained.

At node 1, the initial unbalanced moment is 4.00 kN m and the distribution factor is 1, so that a moment of −4.00 kN m is distributed into the end '12' of the first span. This immediately induces a carry-over moment of −2.00 kN m at the end '21' of the span, and this is considered in determining the unbalanced moment of 19.00 kN m now acting at node 2. Moment distribution is now applied at node 2 to give the values of −11.40 and −7.60

Table 5.4 Moment distribution—immediate carry-over

Subscript	12		21	23		32	34		43
Dist. F	1		0.6	0.4		0.33	0.67		1
Cycle Number	4.00		−4.00	25.00		−17.00	2.25		−2.25
1	−4.00	→	−2.00						
	−5.70	←	−11.40	−7.60	→	−3.80			
2	5.70	→	2.85	3.09	←	6.18	12.37	→	6.18
	−1.78	←	−3.57	−2.38	→	−1.19	−1.97	←	−3.93
3	1.78	→	0.89	0.53	←	1.05	2.10	→	1.05
	−0.43	←	−0.85	−0.57	→	−0.28	−0.53	←	−1.05
4	0.43	→	0.21	0.14	←	0.27	0.54	→	0.27
			−0.21	−0.14	→	−0.07	−0.14	←	−0.27
						0.07	0.14	→	0.07
									−0.07
Σm	0.0		−18.07	18.07		−14.77	14.77		0.0
$\Sigma d \cdot m$	3.91		−16.02	−10.68		7.57	15.15		−5.32
$L/4\Sigma d \cdot m$	3.91		−16.02	−16.02		11.36	11.36		−3.99

to balance that node, and the operations continue throughout the structure. Four iterations are shown in table 5.4 but since the carry-over operation is considered at the same time as the distribution of moments, the iterations are staggered throughout the table. To give emphasis to this and to indicate that the node is currently balanced, a bar is drawn below the moment entries for each distribution. Further, the source of the carry-over moments, which impinge on the previously achieved balance, is shown by the arrows in the table. After four iterations, the carry-over moments are considered small enough to be neglected and the end moments, including the fixed end moments, can be summed to give the end moments corresponding to those of figure 5.2(f).

The relationship between the modified Gauss–Seidel Iteration and the moment distribution method as applied in table 5.4 can be demonstrated more clearly when the rotations of the nodes are identified from table 5.4. Each node relaxes or rotates under a distributed moment and, from equation (5.7), the amount of rotation is $L/4EI$ times the distributed moment. The sum of the distributed moments of each column is shown in table 5.4 along with these values multiplied by $L/4$, which can be compared with the last line of entries in table 5.2. Intermediate values of rotation can also be calculated at the end of any iteration and these values will be found to be consistent with the corresponding terms in table 5.2. A similar comparison can be made between table 5.3 and the strict Gauss–Seidel Iterations of table 5.2. It is clear that the approach to moment distribution taken in table 5.4 is a preferable one and no further consideration will be given to the procedure used in table 5.3.

5.3.1 Modification for Pin-ended Elements

At this stage the computational effort of moment distribution is still quite considerable but a simple modification is available for handling pin-ended elements. This modification, which is in line with that used in the matrix stiffness method, again reduces the computation.

The end moments at nodes 1 and 4 of the continuous beam of figure 5.2(a) are known to be zero because of the nature of the pin supports. A beam has a reduced rotational stiffness, with respect to a moment applied at one end, when the far end is free to rotate. Under these conditions the element stiffness matrix for a prismatic beam is given in the relationship

$$\left\{\begin{array}{c} m_{ij} \\ \hline m_{jL} \end{array}\right\} = \left[\begin{array}{c|c} 0 & 0 \\ \hline 0 & \dfrac{3EI}{L} \end{array}\right] \left\{\begin{array}{c} \theta_i \\ \hline \theta_j \end{array}\right\} \tag{5.14}$$

so that the appropriate stiffness factor is now $3EI/L$, and this may be used to calculate the distribution factors. There is of course no carry-over moment

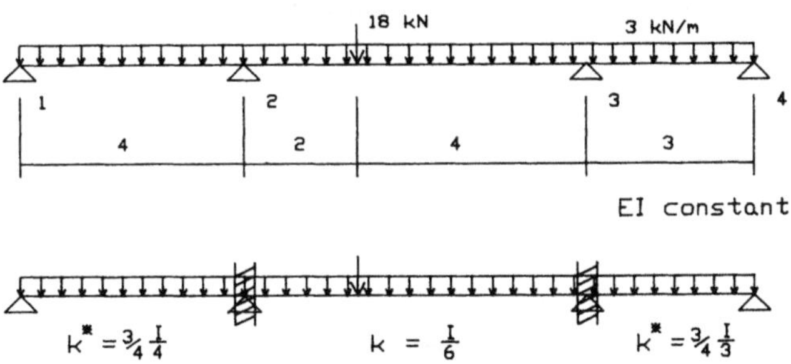

21	23	32	34
0.53	0.47	0.40	0.60
-6.00	25.00	-17.00	3.38
-10.07	-8.93	-4.47	
	3.61	7.23	10.86
-1.91	-1.70	-0.85	
	0.17	0.34	0.51
-0.09	-0.08	-0.04	
		0.02	0.02
-18.07	18.07	-14.77	14.77

Figure 5.5 Moment distribution using modified stiffness.

to the pinned end. The pinned ends of the elements are not considered to be clamped at any stage, and this must be noted when calculating the fixed end moments. The continuous beam of figure 5.2(a) can now be analysed again, this time using the modified stiffness for the exterior spans.

The results of the analysis and the moment distribution table are shown in figure 5.5, together with the beam and its initial restrained form. Figure 5.5 introduces a practical notion of the modified stiffness factor that is convenient to use with prismatic elements. With the standard stiffness factor expressed as $4EI/L$, and assuming all the elements have the modulus of elasticity, E, the term $4E$ can be cancelled from the expression of the distribution factor r_{ji} given by equation (5.10). The standard stiffness factor might well then be defined simply as I/L. The modified stiffness factor, denoted as k^*, must now be defined as $\frac{3}{4}$ of the standard stiffness factor to preserve the cancellation of the term $4E$ in calculating the distribution factors. This approach is demonstrated in the solution shown in figure 5.5.

A further study of the use of the moment distribution method is given in example 5.1, where a rigid jointed frame superstructure of a bridge system is analysed.

Example 5.1: Moment Distribution of a Non-swaying Frame

Given data:

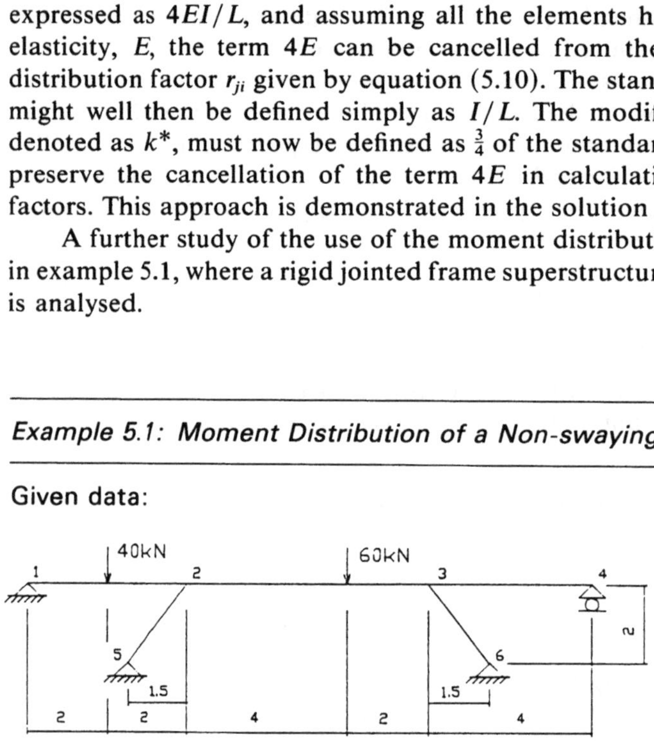

E constant throughout. Relative *I* values: beams 4*I*; columns *I*.

Calculate distribution factors and fixed end moments from

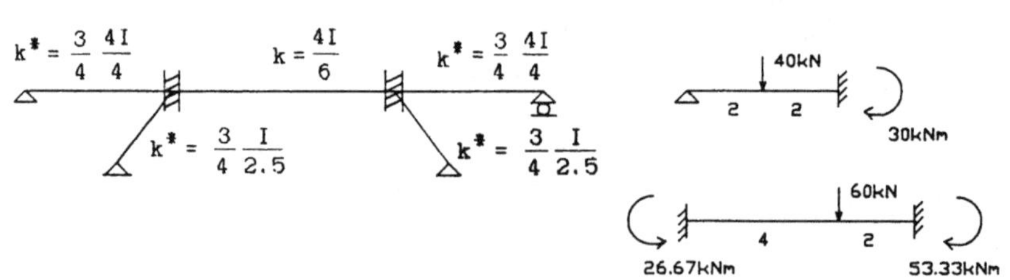

Moment Distribution Table:

21	25	23	32	36	34	
0.44	0.17	0.39	0.39	0.17	0.44	Distribution Factors
−30		26.67	−53.33			Fixed End Moments
1.47	0.56	1.30	0.65			
		10.28	20.55	8.96	23.18	
−4.52	−1.75	−4.01	−2.00			
		0.39	0.78	0.34	0.88	
−0.17	−0.07	−0.15	−0.08			
		. . .	+0.03	0.01	0.04	
−33.20	−1.26	34.48	−33.40	9.30	24.10	Final End Moments (kN m)

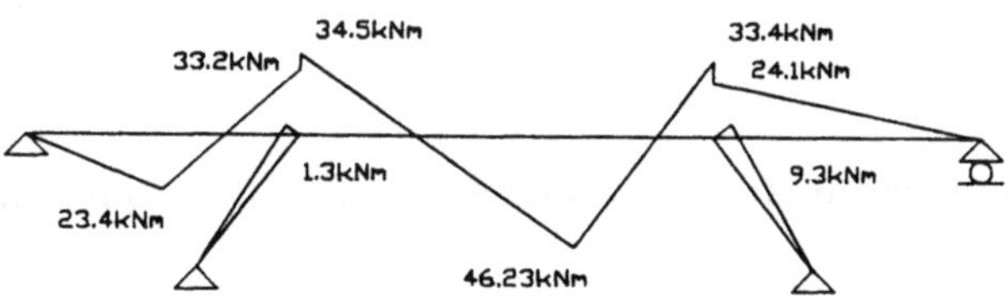

Frame Bending Moment Diagram

5.4 MOMENT DISTRIBUTION APPLIED TO SWAYING RECTANGULAR FRAMES

In extending the application of the moment distribution method to swaying frames, the more general principles of the stiffness method of analysis are introduced. Although the technique can be used for non-rectangular frames, the additional computational effort erodes the advantages of simple hand calculations, and the general stiffness method of matrix analysis or the flexibility method of analysis is probably to be preferred.

5.4.1 Beam Element Behaviour under Transverse Displacement

Before proceeding to the analysis, some general observations about the displacement of a beam element need to be made. The general behaviour of a beam element was presented in chapter 3 and summarised by the beam element stiffness matrix given in equation (3.25). Two particular cases are of interest here and they are shown in figure 5.6.

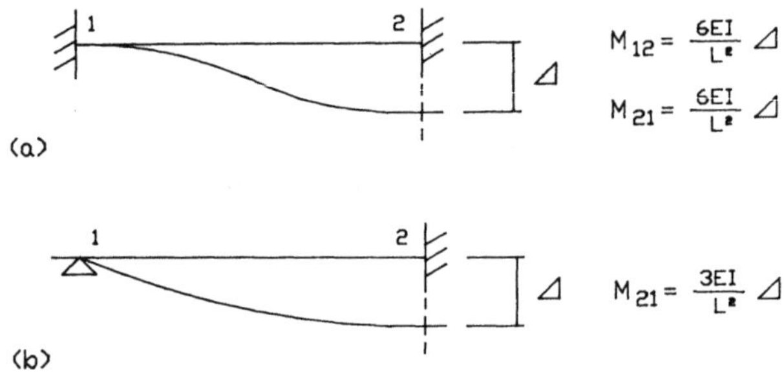

Figure 5.6 *Fixed end moments under transverse displacements.*

For case (a) of figure 5.6, d_1, θ_1 and θ_2 are zero, while $d_2 = -\Delta$, so that equation (3.25) gives

$$m_{12} = \frac{6EI}{L^2}\Delta$$

$$m_{21} = \frac{6EI}{L^2}\Delta \qquad (5.15)$$

For case (b) of figure 5.6, d_1 and θ_2 are zero while $d_2 = -\Delta$, so that equation (3.25) gives

$$m_{12} = \frac{4EI}{L}\theta_1 + \frac{6EI}{L^2}\Delta \qquad \text{(i)}$$

$$m_{21} = \frac{2EI}{L}\theta_1 + \frac{6EI}{L^2}\Delta \qquad \text{(ii)}$$

Since m_{12} is zero, equation (i) allows θ_1 to be expressed in terms of Δ as

$$\theta_1 = -\frac{6}{4L}\Delta$$

Substituting for θ_1 into equation (ii) results in

$$m_{21} = \frac{3EI}{L^2}\Delta \qquad (5.16)$$

Equations (5.15) and (5.16) represent values of fixed end moments due to translation of one end of a beam element, and this information is relevant to the analysis of swaying frames.

5.4.2 Frames with One Sway Degree of Freedom

The behaviour of rectangular frames with sway displacement was introduced in chapter 4. It may be recalled that with the nodes nominated at the joints, and ignoring the effects of axial deformation, the degrees of freedom of a single-storey frame can be described by the rotations at the nodes and the sway of the frame at the level of the beam. Using the principle of superposition, the behaviour of such a frame can be considered in two parts: firstly with the sway prevented by a restraining prop and secondly with the effects of sway taken into account. This is illustrated in the diagrams of figure 5.7 where the frame sways by an amount Δ_1. In figure 5.7(b), the frame is restrained against sway by the prop which, because of the loads, develops a reactive force that can be designated as F_{10}. In figure 5.7(c) a sway force F_1 is assumed to be applied to the structure to produce the sway Δ_1. Since the behaviour of the frame is given by the superposition of these two systems, equilibrium requires F_{10} plus F_1 to be zero. This provides an equilibrium equation which is the key to the solution.

The restraining force F_{10} can be readily calculated since it involves the analysis of a non-swaying frame. In terms of the moment distribution method, each node can be clamped as required and the fixed end moments can be calculated. The nodes can then be progressively released to give the end moments from which the bending moment diagram can be drawn and the reactions, including the propping force, can be found.

Sway moment distribution can be applied to determine a relationship between the sway deflection and the sway force. Rather than calculate the sway Δ_1 from the known value of F_1 of figure 5.7(c), it is more convenient to introduce an arbitrary sway, Δ', and determine the force necessary to produce it. For sway moment distribution, the frame is again assumed to be clamped at the nodes as necessary to prevent rotation and the arbitrary sway is applied. This induces sway fixed end moments in accordance with equations (5.15) and (5.16) as appropriate, and the moment distribution process can be applied again to give the bending moment diagrams and reactions. The arbitrary sway force, say F_{11}, is then found from equilibrium. Since the sway stiffness of the frame is expressed by F_{11}/Δ', the equilibrium

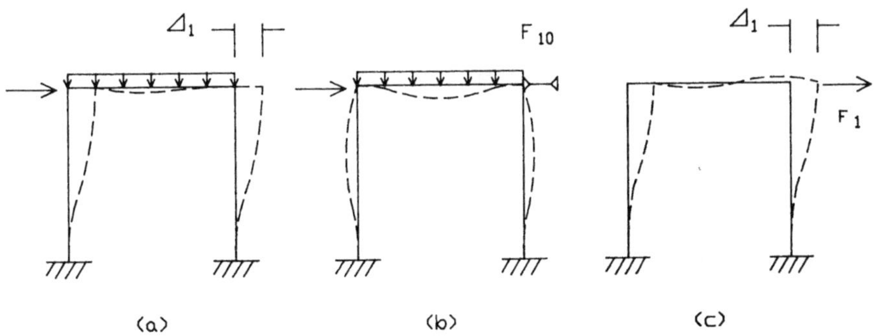

Figure 5.7 Behaviour of a swaying frame.

equation satisfying the superposition requirements of the two part study is then,

$$F_{10}+\frac{\Delta_1}{\Delta'}F_{11}=0 \tag{5.17}$$

Equation (5.17) may be used to calculate the sway Δ_1 as a ratio of the assumed sway. The same factor would have to be applied to the assumed sway moments, before they were combined with the no-sway moment results to give the bending moments in the frame. The method is demonstrated in example 5.2 for a typical single-bay frame. The different column heights and the base conditions have been deliberately chosen to illustrate the nature of the sway fixed end moments in relationship to the assumed sway. The final bending moment diagram and reactions of the frame are not shown but these can be deduced from the final end moments given by combination in the sway moment distribution table of example 5.2.

The technique can be immediately applied to multi-bay rectangular frames with one sway degree of freedom, and it is also appropriate for the analysis of continuous beams and frames with support movements. While most of the routine structural analysis assumes unyielding supports, it is possible to estimate foundation settlement due to the reactive loads. The effect of settlement or other support movements on statically indeterminate structures can be readily analysed by the moment distribution method. If the structure is initially displaced, without nodal rotation or other translation, to accommodate the support movements, then this action will set up transverse fixed end moments again in accordance with equations (5.15) and (5.16). The nodal restraints can then be relaxed through moment distribution to give the final condition of the structure.

Example 5.2: Moment Distribution of a Swaying Frame

Given data:

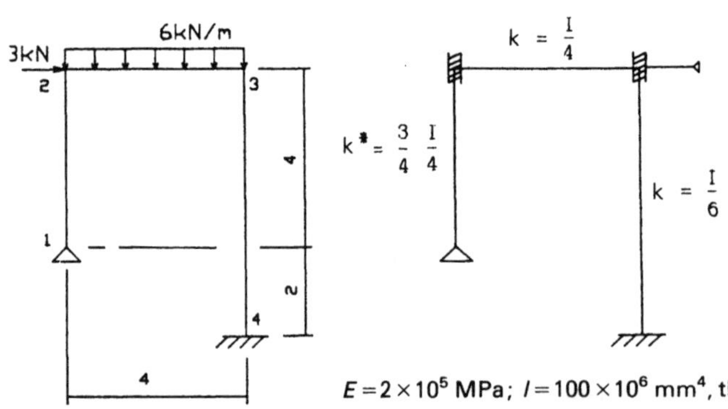

$E=2\times10^5$ MPa; $I=100\times10^6$ mm^4, throughout

Part A—No-sway Moment Distribution

21	23	32	34	43
0.43	0.57	0.60	0.40	—
	+8.00	−8.00		
−3.44	−4.56	−2.28		
	3.09	6.17	4.11	2.06
−1.33	−1.76	−0.88		
	0.27	0.53	0.35	0.18
−0.11	−0.15	−0.08		
	. . .	0.05	0.03	0.01
−4.88	4.89	−4.49	4.49	2.25

Moments (kN m)

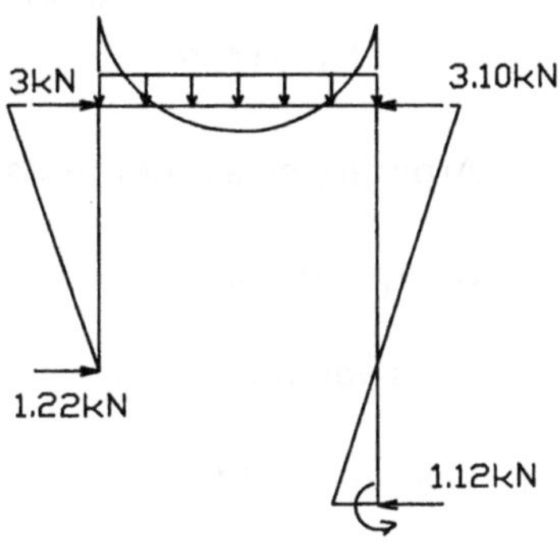

$F_{10} = -3.10$ kN

Part B—Sway Moment Distribution

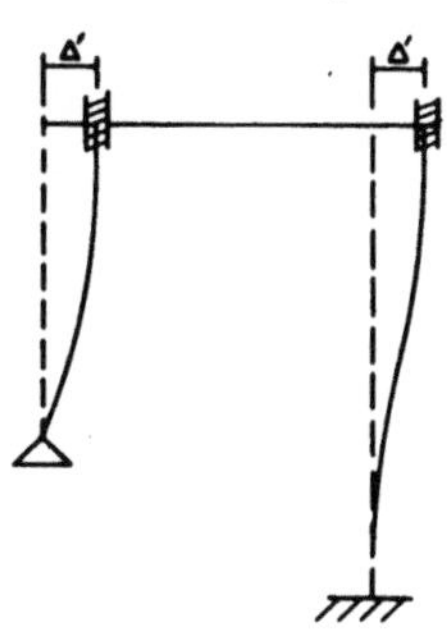

Sway Fixed End Moments:

$$m_{21} = \frac{3EI}{4^2}\Delta'$$

$$m_{34} = m_{43} = \frac{6EI}{6^2}\Delta'$$

Select Δ' such that

$$\frac{3EI}{16}\Delta' = 10 \qquad \therefore \quad \Delta' = \frac{160}{3EI}$$

then $m_{21} = 10$; $m_{34} = m_{43} = 8.89$

21	23	32	34	43	
0.43	0.57	0.60	0.40	—	
10.00			8.89	8.89	
−4.30	−5.70	−2.85			
	−1.81	−3.62	−2.42	−1.21	
0.78	1.03	0.52			
	−0.16	−0.31	−0.21	−0.10	
0.07	0.09	0.05			
	. . .	−0.03	−0.02	. . .	
6.55	−6.55	−6.24	6.24	7.58	Arbitrary sway moments
5.17	−5.17	−4.93	4.93	5.99	Sway moments
−4.88	4.88	−4.49	4.49	2.25	No-sway moments
0.29	−0.29	−9.42	9.42	8.24	Final moments

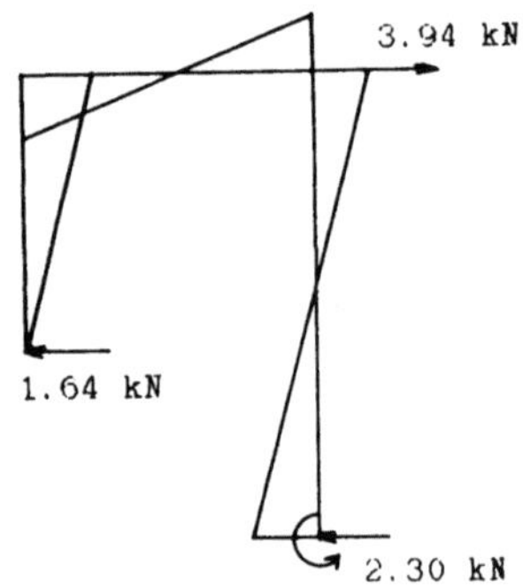

$F_{11} = 3.94$ kN

Equilibrium equation:

$$F_{10} + \frac{\Delta}{\Delta'}F_{11} = 0$$

$$\therefore \quad \frac{\Delta}{\Delta'} = \frac{3.10}{3.94}$$

$$= 0.79$$

Adjust arbitrary sway moments by 0.79

Calculation of sway displacement:

Since

$$\Delta' = \frac{160}{3EI}$$

$$\Delta = 0.79 \frac{160}{3EI}$$

$$= 2.1 \text{ mm}$$

5.4.3 Frames with Multi-sway Degrees of Freedom

In extending the moment distribution method to swaying frames of several stories, a restraining prop is introduced at each sway level to allow the no-sway moment distribution to proceed. This analysis results in the propping forces $F_{10}, F_{20}, \ldots, F_{n0}$, where n is the number of sway levels. Each level must then in turn be allowed to sway while the remaining levels are held at zero sway. The superposition of the results will then give the required frame behaviour. There is an equilibrium equation appropriate to each sway level but, since a permissible sway at any one level will induce further reactive forces in the remaining props, the equilibrium equations will be of the form

$$\begin{aligned} F_{10} + s_{11}\Delta_1 + s_{12}\Delta_2 + \cdots s_{1n}\Delta_n &= 0 \\ F_{20} + s_{21}\Delta_1 + s_{22}\Delta_2 + \cdots s_{2n}\Delta_n &= 0 \\ &\vdots \\ F_{n0} + s_{n1}\Delta_1 + s_{n2}\Delta_2 + \cdots s_{nn}\Delta_n &= 0 \end{aligned} \tag{5.18}$$

where s_{ij} is a sway stiffness coefficient expressing the force in the prop at level i due to a unit sway displacement at level j. Again, as a matter of convenience, arbitrary sways may be introduced at each level in order to calculate the sway stiffness coefficient. In a two-storey frame, for example, an arbitrary sway of Δ' at level 1 will require a force of F_{11} at level 1 and develop a reaction of F_{21} in the remaining prop, while for an arbitrary sway of Δ'' at level 2 the corresponding forces are F_{22} and F_{12}. Equation (5.18) then gives

$$F_{10}+F_{11}\frac{\Delta_1}{\Delta'}+F_{12}\frac{\Delta_2}{\Delta''}=0$$

$$F_{20}+F_{21}\frac{\Delta_1}{\Delta'}+F_{22}\frac{\Delta_2}{\Delta''}=0 \tag{5.19}$$

With the propping and sway forces known, equations (5.19) can be solved as a set of simultaneous equations for the ratios Δ_1/Δ' and Δ_2/Δ''. The assumed sway moments must then be adjusted by these ratios, as appropriate, before the sway moments are combined with the no-sway moments to give the final result.

Example 5.3 illustrates the use of equation (5.19) and introduces a number of other points relating to the moment distribution method. The first of these concerns the layout of the moment distribution table when the frame involves nodes with more than three elements framing into it. In example 5.3, recognizing the nature of the pinned bases, nodes 3, 4, 5 and 6 must be initially restrained and progressively relaxed. The carry-over moments from a distribution of moment at node 3 involve nodes 4 and 5 with end moments m_{53} and m_{43}. With the distribution of moment into three elements at node 3, requiring three columns to express the end moments there, the related end moments of '53' and '43' cannot both be adjacent to columns '35' and '34' of the table. Instead a symmetrical layout of the table has been nominated for the no-sway moment distribution, and the interaction of end moments '34' and '43' has been stressed by the arrow bar across the top of the table. Clearly other layouts can be devised and the order of relaxation of the nodes is not important, but it is important that all carry-over moments be recognized. With the end moments calculated, an equilibrium analysis is applied to the frame to determine the propping forces F_{10} and F_{20} as shown. The solution for the no-sway case converges rapidly because of the nature of the loads, and little bending moment is transferred to the second column line.

In carrying out the sway moment distribution, it is apparent that the symmetry of the frame can be used to advantage. Because of symmetry, the beams '34' and '56' will have a point of inflection at mid span and there will be no vertical deflection there under any sway displacements. Hence the frame can be modelled to reflect this, as shown in the sway moment distribution of example 5.3. The results can then be extended to the full frame to find the sway forces, again using equilibrium. It is helpful to note that a horizontal shear plane may be applied at each level and the resulting free body diagram must be in horizontal equilibrium. In this way the sway force and the propping restraints can be related to the shear in the columns calculated from the end moments there.

Example 5.3: Moment Distribution of a Multi-storey Frame

Given data:

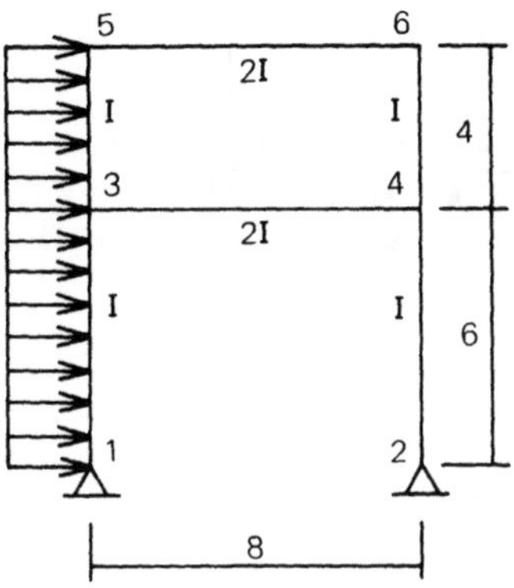

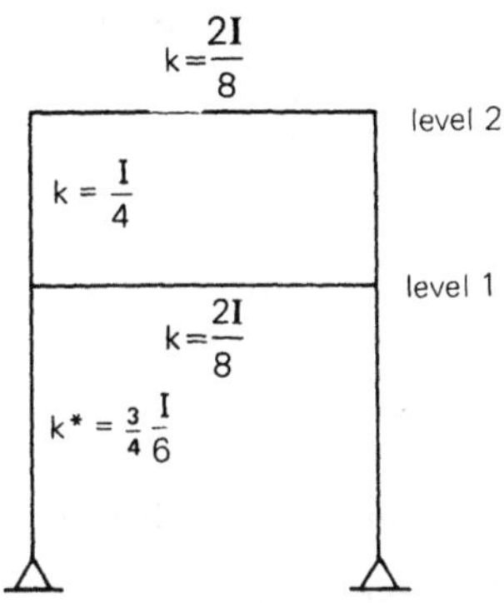

E constant: Relative I as shown

Part A—No-sway Moment Distribution (Moments kN m)

31	34	35	53	56	65	64	46	43	42
0.2	0.4	0.4	0.5	0.5	0.5	0.5	0.4	0.4	0.2
−5.40		1.60	−1.60						
0.76	1.52	1.52	0.76					0.76	
	−0.15	0.21	0.42	0.42	0.21	−0.15	−0.30	−0.30	−0.16
−0.01	−0.02	−0.02	−0.01	−0.01	−0.03	−0.03	−0.01	−0.01	
			0.01	0.01			0.01	0.01	. . .
−4.65	1.35	3.31	−0.42	0.42	0.18	−0.18	−0.30	0.46	−0.16

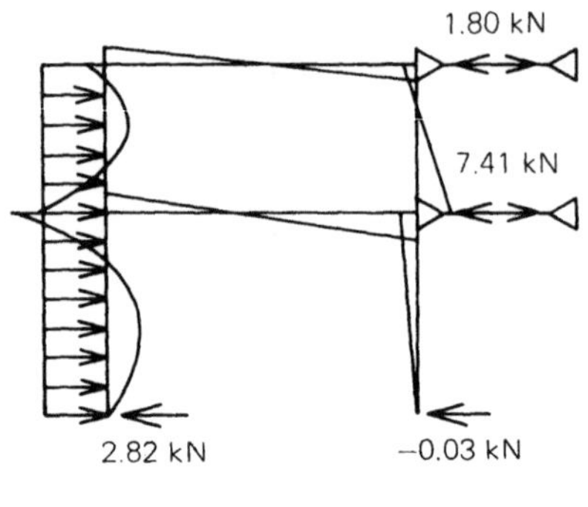

$F_{10} = -7.41$ kN

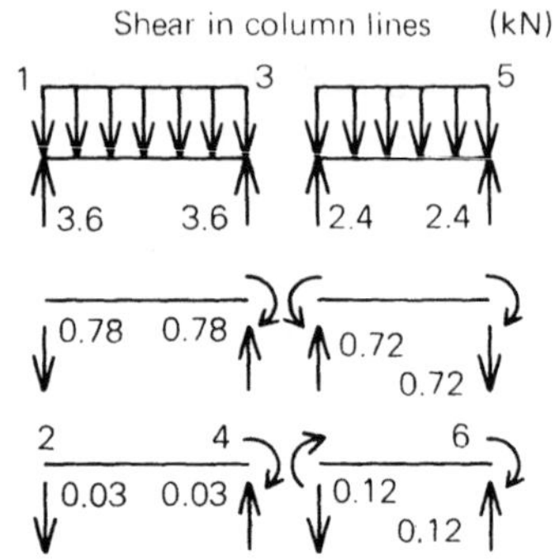

$F_{20} = -1.80$ kN

Part B—Sway Moment Distribution

Arbitrary sway forces

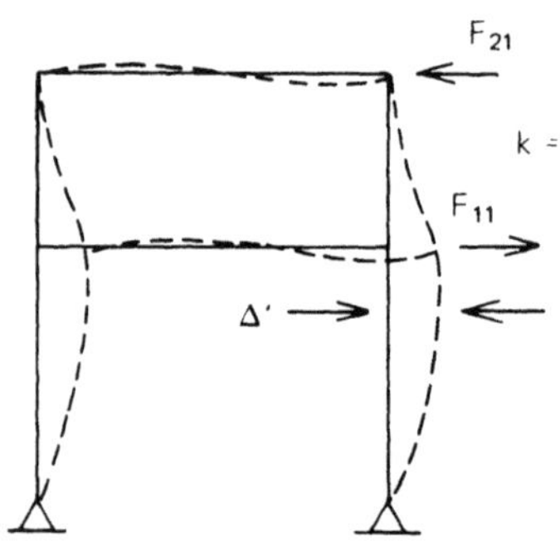

Level 1 sway

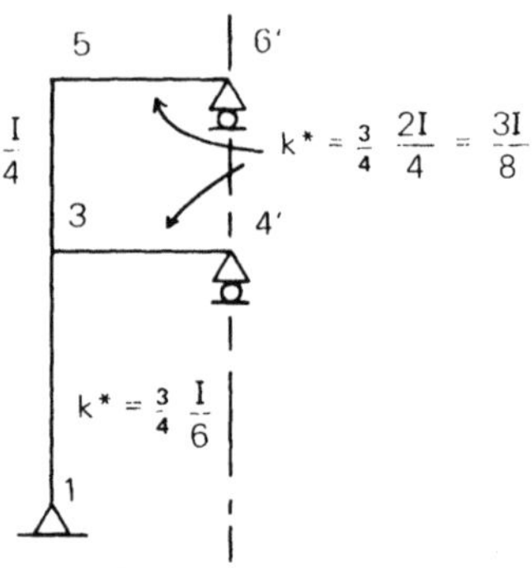

Sway Moment Distribution—Level 1 (Moments kN m)

31	34	35	53	56
$\frac{1}{6}$	$\frac{1}{2}$	$\frac{1}{3}$	$\frac{2}{5}$	$\frac{3}{5}$
10	—	−45	−45	
5.83	17.50	11.67	5.83	
		7.84	15.67	23.50
−1.31	−3.92	−2.61	−1.31	
		0.26	0.52	0.78
−0.04	−0.13	−0.09		
14.45	13.45	−27.94	−24.29	24.28

Sway Fixed End Moments:

$$m_{31} = \frac{3EI}{6^2} \cdot \Delta$$

$$m_{35} = m_{53} = -\frac{6EI}{4^2}\Delta$$

Assume $m_{31} = 10$

then $m_{35} = -45$

and $\Delta' = \frac{120}{EI}$

Shear in column line 1–3–5

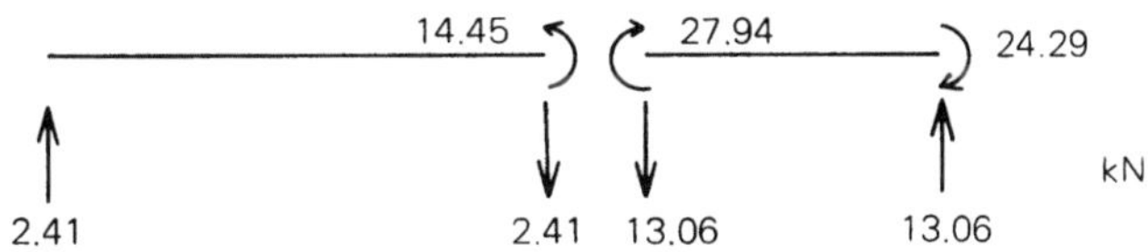

Column line 2–4–6 similar

$$\therefore \quad F_{11} = 30.94\ \text{kN}$$

$$\text{and} \quad F_{21} = -26.12\ \text{kN}$$

Arbitrary Sway Forces—Level 2 Sway

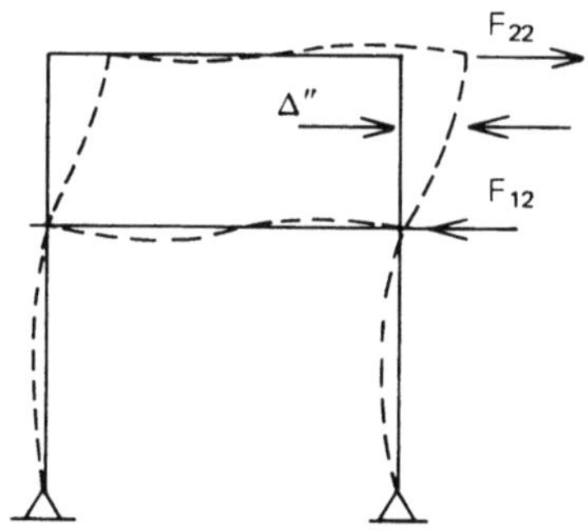

Sway Moment Distribution—Level 2 (Moments kN m)

31	34	35	53	56
$\frac{1}{6}$	$\frac{1}{2}$	$\frac{1}{3}$	$\frac{2}{5}$	$\frac{3}{5}$
		50	50	
−8.33	−25.00	−16.67	−8.34	
		−8.34	−16.67	−25.00
1.39	4.17	2.78	1.39	
		−0.28	−0.56	−0.83
0.05	0.14	0.09		
−6.89	−20.69	27.58	25.82	−25.83

Sway Fixed End Moments:

$$m_{35} = m_{53} = \frac{6EI}{4^2}\Delta$$

assume $m_{35} = 50$

then $\Delta'' = \dfrac{133.33}{EI}$

Shear in column line 1–3–5

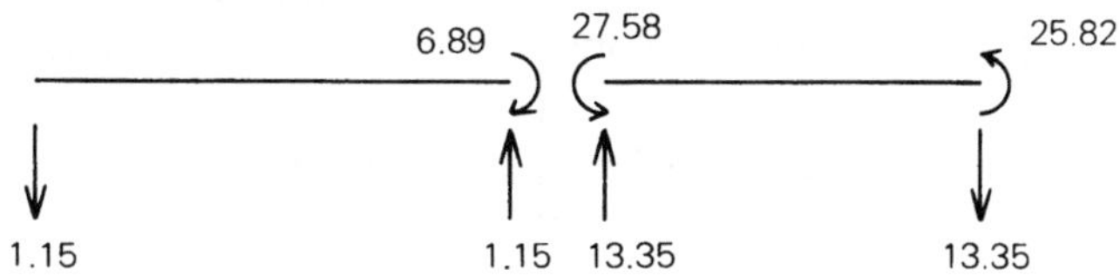

Column line 2–4–6 similar

$$\therefore \quad F_{12} = -29.00 \text{ kN}$$

and $$F_{22} = 26.70 \text{ kN}$$

Applying equation (5.19):

$$F_{10} + F_{11} \cdot \frac{\Delta_1}{\Delta'} + F_{12} \cdot \frac{\Delta_2}{\Delta''} = 0$$

$$F_{20} + F_{21} \cdot \frac{\Delta_1}{\Delta'} + F_{22} \cdot \frac{\Delta_2}{\Delta''} = 0$$

$$-7.41 + 30.94 \frac{\Delta_1}{\Delta'} - 29.00 \frac{\Delta_2}{\Delta''} = 0$$

$$-1.80 - 26.12 \frac{\Delta_1}{\Delta'} + 26.70 \frac{\Delta_2}{\Delta'} = 0$$

The solution is

$$\frac{\Delta_1}{\Delta'} = 3.64 \quad \text{and} \quad \frac{\Delta_2}{\Delta''} = 3.63$$

hence

$$\Delta_1 = \frac{437}{EI} \quad \text{and} \quad \Delta_2 = \frac{484}{EI}$$

The final moments are then given as follows (half the frame only shown):

31	34	35	53	56	
14.45	13.45	−27.94	−24.29	24.28	(a) Sway assumed; level 1
−6.89	−20.69	27.58	25.82	−25.83	(b) Sway assumed; level 2
52.66	49.01	−101.81	−88.51	88.48	(c) (a) × 3.64
−25.02	−75.15	100.17	93.78	−93.78	(d) (b) × 3.63
27.64	−26.14	−1.64	5.27	−5.27	(e) (c) + (d) Final sway moments
−4.65	1.35	3.31	−0.42	0.42	(f) No-sway moments
22.99	−24.79	1.67	4.85	−4.85	(g) (e) + (f)

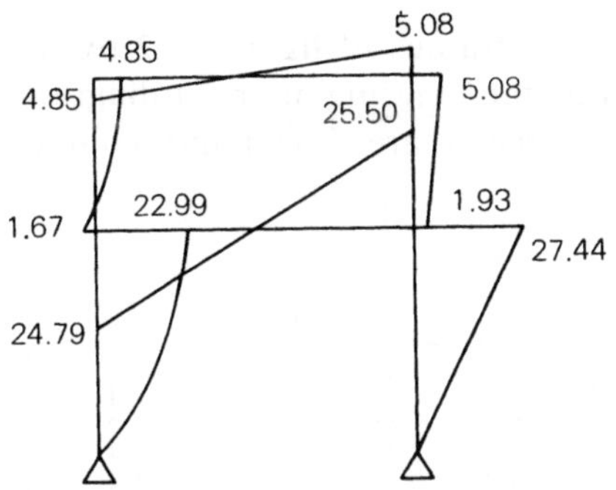

Frame BMD
(kN m)

5.5 PROBLEMS FOR SOLUTION

5.1 Analyse each of the continuous beams of figure P5.1 by the moment distribution method and draw the bending moment diagram. Note that the cantilever span of figure P5.1(b) effectively applies a moment directly to the beam at the end support that must be distributed entirely to the end moment of the adjacent span.

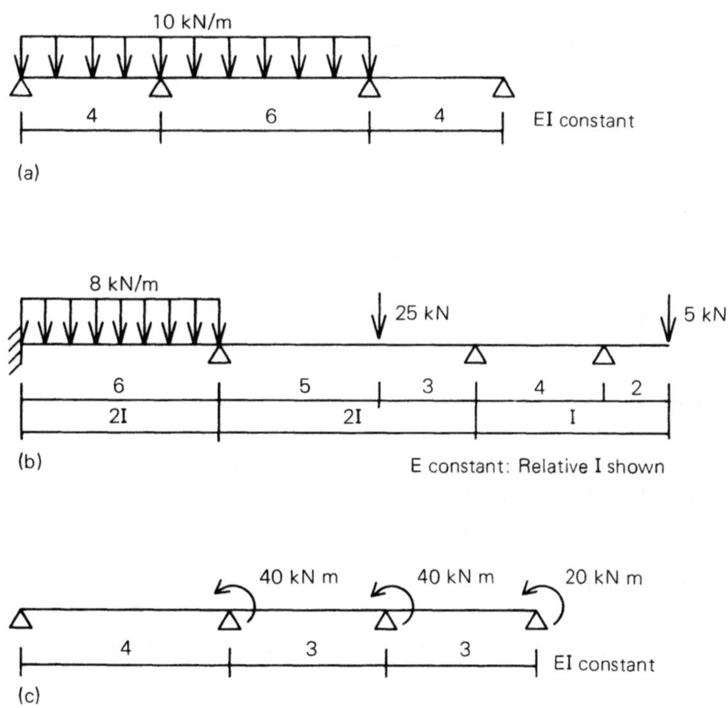

Figure P5.1

5.2 Analyse each of the rigid jointed plane frames of figure P5.2 by the moment distribution method and draw the bending moment diagram. Relative values of second moment of area, *I*, are shown and Young's Modulus, *E*, is constant throughout.

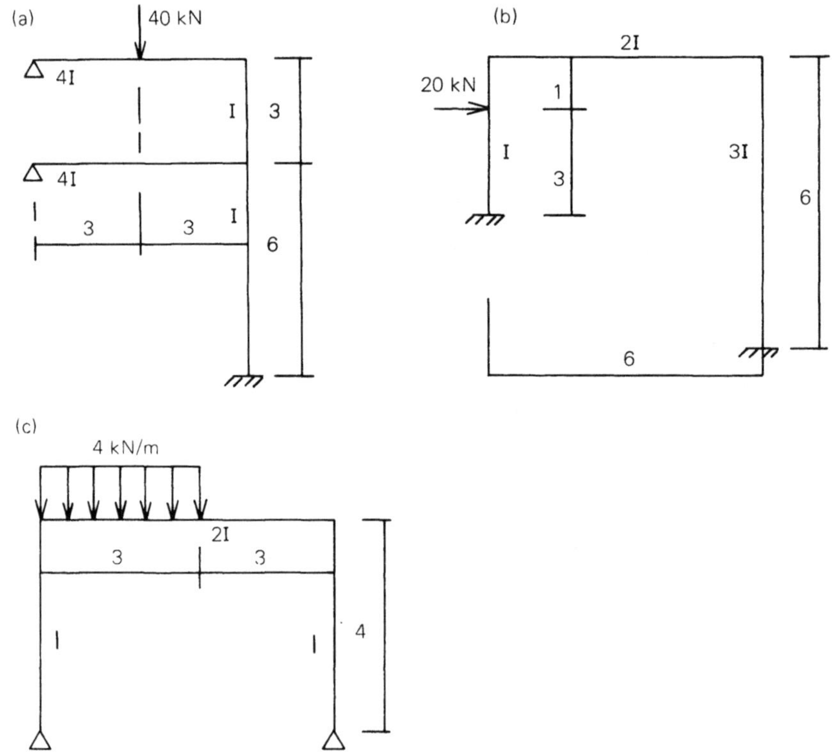

Figure P5.2

REFERENCE

Cross, H. (1936). 'Analysis of continuous frames by distributing fixed-end moments', *Trans. ASCE*, Vol. 96, No. 1793.

FURTHER READING

Gere, R. M., *Moment Distribution*, Van Nostrand, New York, 1963.

Chapter 6
The Matrix Stiffness Method—Part 2

In a final presentation of the matrix stiffness method of structural analysis, a general technique applicable to all classes of structure is outlined. The technique uses coordinate transformation and it is first necessary to discuss the axes of reference used to define the structure and its actions.

The geometry of a structure can be defined by reference to a suitably positioned set of axes. The axes can be described as global axes or system axes and are usually taken as a first quadrant right-hand cartesian set. An element of a structure may have any orientation with respect to the global axes, although the displacements of a structure are defined in terms of components in the directions of the global axes. On the other hand, element actions must be expressed in terms of local axes to provide meaningful results to the engineer. For example, the normal or transverse force at a

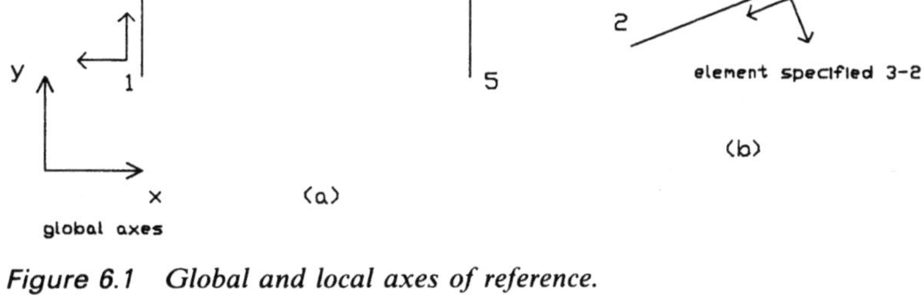

Figure 6.1 Global and local axes of reference.

section of a beam is recognized as a shear force, while the force in line with the element is recognized as an axial force. For one-dimensional elements it is usual and convenient to take the local x-axis in the direction of the element with the y-axis following as a consequence. Positive x is taken in the direction from end A to end B of the element according to how the element is specified in a reference list. The concepts of global and local axes are shown in figure 6.1(a), with the variation of local axes as a function of the element specification being shown in figure 6.1(b). It may now be appreciated that for continuous beams, the local axes for all of the elements specified in a logical sequence have the same orientation as the global axes. For this reason, continuous beam analysis does not involve coordinate transformation.

6.1 THE GENERAL ANALYSIS OF TRUSSES

A useful technique for the analysis of trusses has already been presented in chapter 3 with the formation of the structure stiffness matrix by matrix multiplication. A form of coordinate transformation was involved, in so far as the equilibrium equations were all expressed in terms of global axes, and the axial forces in the elements were resolved into components in the global x and y directions, before being introduced into the equilibrium equations. However that approach has a number of limitations and a more general approach is necessary.

6.1.1 The Plane Truss Element

Consider the behaviour of a single truss element under axial loads applied at either end as shown in figure 6.2(a). The element will move with displacements s_1 and s_2 and develop an internal force f_1 as shown in figure 6.2(b). The internal force is related to the element extension by the linear elastic stress–strain law, so that

$$f_1 = \frac{EA}{L}(s_2 - s_1) \tag{6.1}$$

For nodal equilibrium:

$$P_1 = -f_1 \quad \text{and} \quad P_2 = f_1$$

and substituting for f_1 from equation (6.1) gives a load–displacement relationship of the form

$$\begin{Bmatrix} P_1 \\ P_2 \end{Bmatrix} = \begin{bmatrix} \dfrac{EA}{L} & -\dfrac{EA}{L} \\ -\dfrac{EA}{L} & \dfrac{EA}{L} \end{bmatrix} \begin{Bmatrix} s_1 \\ s_2 \end{Bmatrix} \tag{6.2}$$

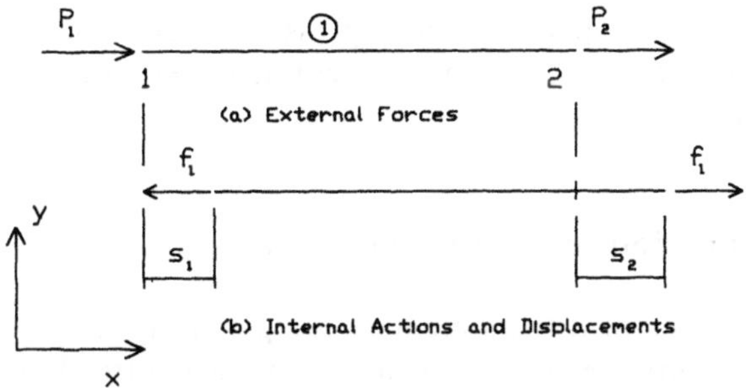

Figure 6.2 *Truss element under load.*

In more general terms, the nodal displacements at either end of an element must be described in both the x and y directions with respect to local element axes, as shown in figure 6.3. The end actions on the element, now recognized as internal actions, can also be more generally described with components in the local x and local y directions. However, the local y direction forces must be zero since any displacements normal to the element do not cause any element extension. The force-displacement relationship can now be written as

$$\begin{Bmatrix} p'_{12} \\ v'_{12} \\ \hline p'_{21} \\ v'_{21} \end{Bmatrix} = \left[\begin{array}{cc|cc} \dfrac{EA}{L} & 0 & -\dfrac{EA}{L} & 0 \\ 0 & 0 & 0 & 0 \\ \hline -\dfrac{EA}{L} & 0 & \dfrac{EA}{L} & 0 \\ 0 & 0 & 0 & 0 \end{array}\right] \begin{Bmatrix} d'_1 \\ d'_2 \\ \hline d'_3 \\ d'_4 \end{Bmatrix} \tag{6.3}$$

where the prime used in association with each of the terms of the action and displacement vectors indicates that those terms are defined with respect to the local axes.

The four by four matrix of equation (6.3) is the truss element stiffness matrix. The reason for the inclusion of the obviously zero shear actions, v'_{12} and v'_{21}, becomes apparent when the nodal equilibrium equations are

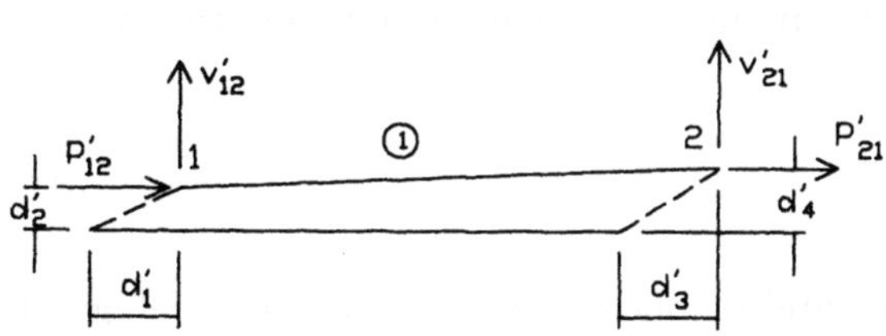

Figure 6.3 *General truss element actions.*

considered. The nodal equilibrium equations are expressed in terms of the global x and y axes, so that a coordinate transformation is necessary to adequately express equilibrium.

6.1.2 Coordinate Transformation for a Truss Element

As was mentioned, the local axes of a truss element are generally not the same as the global axes of reference set up to define the structure. It is necessary then to consider the influence of this on the actions and displacements defined with respect to the local axes. In essence, the transformation as indicated in figure 6.4 is required, both with regard to actions and displacements. Since the nett effect on both systems of figure 6.4 is the same then

$$p'_{ij} = p_{ij} \cos \alpha + v_{ij} \sin \alpha$$

$$v'_{ij} = -p_{ij} \sin \alpha + v_{ij} \cos \alpha$$

and

$$p'_{ji} = p_{ji} \cos \alpha + v_{ji} \sin \alpha$$

$$v'_{ji} = -p_{ji} \sin \alpha + v_{ji} \cos \alpha$$

The equations may be written in matrix form as

$$\left\{\begin{array}{c} p'_{ij} \\ v'_{ij} \\ \hline p'_{ji} \\ v'_{ji} \end{array}\right\} = \left[\begin{array}{cc|cc} \cos\alpha & \sin\alpha & 0 & 0 \\ -\sin\alpha & \cos\alpha & 0 & 0 \\ \hline 0 & 0 & \cos\alpha & \sin\alpha \\ 0 & 0 & -\sin\alpha & \cos\alpha \end{array}\right] \left\{\begin{array}{c} p_{ij} \\ v_{ij} \\ \hline p_{ji} \\ v_{ji} \end{array}\right\} \tag{6.4}$$

(local) (global)

where the four by four matrix is a coordinate transformation matrix that will be denoted by T. It is convenient to write equation (6.4) as

$$\left\{\begin{array}{c} f'_i \\ \hline f'_j \end{array}\right\} = [T] \left\{\begin{array}{c} f_i \\ \hline f_j \end{array}\right\} \tag{6.5}$$

and the displacements may be treated in a similar manner to give the equation

$$\left\{\begin{array}{c} d'_j \\ \hline d'_j \end{array}\right\} = [T] \left\{\begin{array}{c} d_i \\ \hline d_j \end{array}\right\} \tag{6.6}$$

where f_i and d_i are the actions and displacements at end i of the element, recalling that the prime used with the terms indicates a reference to local axes.

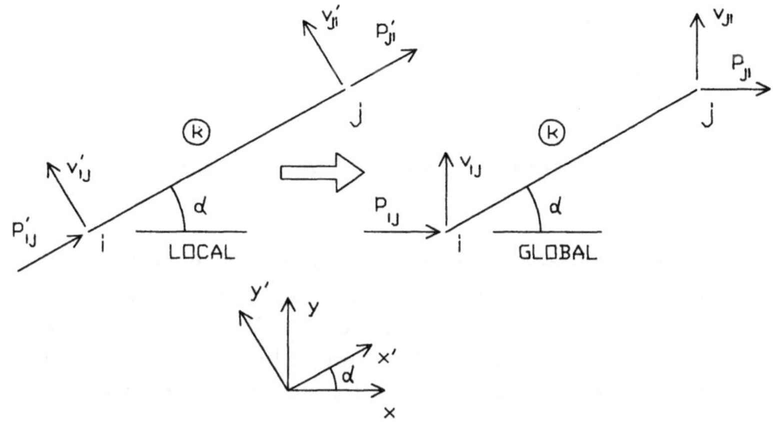

Figure 6.4 *Truss action coordinate transformation.*

Equation (6.3) can now be conveniently written as

$$\begin{Bmatrix} f'_i \\ -- \\ f'_j \end{Bmatrix} = \left[\begin{array}{cc|cc} \frac{EA}{L} & 0 & -\frac{EA}{L} & 0 \\ 0 & 0 & 0 & 0 \\ \hline -\frac{EA}{L} & 0 & \frac{EA}{L} & 0 \\ 0 & 0 & 0 & 0 \end{array}\right] \begin{Bmatrix} d'_i \\ -- \\ d'_j \end{Bmatrix} \tag{6.7}$$

Substituting for the local actions and displacements from equations (6.5) and (6.6) into equation (6.7) gives

$$[T] \begin{Bmatrix} f_i \\ -- \\ f_j \end{Bmatrix} = \left[\begin{array}{cc|cc} \frac{EA}{L} & 0 & -\frac{EA}{L} & 0 \\ 0 & 0 & 0 & 0 \\ \hline -\frac{EA}{L} & 0 & \frac{EA}{L} & 0 \\ 0 & 0 & 0 & 0 \end{array}\right] [T] \begin{Bmatrix} d_i \\ -- \\ d_j \end{Bmatrix}$$

Pre-multiplying both sides by T^{-1}, and noting that coordinate transformation matrices are necessarily orthogonal matrices which have the special property that the inverse is equal to the transpose, then

$$\begin{Bmatrix} f_i \\ -- \\ f_j \end{Bmatrix} = [T]^T \left[\begin{array}{cc|cc} \frac{EA}{L} & 0 & -\frac{EA}{L} & 0 \\ 0 & 0 & 0 & 0 \\ \hline -\frac{EA}{L} & 0 & \frac{EA}{L} & 0 \\ 0 & 0 & 0 & 0 \end{array}\right] [T] \begin{Bmatrix} d_i \\ -- \\ d_j \end{Bmatrix} \tag{6.8}$$

which can be conveniently written as

$$\left\{\frac{f_i}{f_j}\right\} = [\text{TESM}]\left\{\frac{d_i}{d_j}\right\} \tag{6.9}$$

6.1.3 Assembly of the Structure Stiffness Matrix using Truss Elements

With the truss element stiffness matrix now defined with respect to global axes as a transformed element stiffness matrix, it is possible to assemble a structure stiffness matrix. The assembly procedure follows on the basis of the nodal equilibrium equations and satisfies compatibility because common displacements are assigned to the ends of all elements terminating at the same node.

The procedure can best be explained with reference to an example. The truss of figure 3.2 and presented in a detailed study in section 3.5 will be used to illustrate the point. As a matter of convenience all elements will be addressed from low node number to high node number. This is not a general restriction and any automated process can readily cope with addressing the element in either order. The restriction simply assists in the explanation of the assembly process. The truss is shown again in figure 6.5 with the local axes defined for each element, and in particular, the angle α, defining the relationship between the global and local axes for a given element, is clearly stated.

The unrestrained structure has eight degrees of freedom and the structure stiffness matrix sought is therefore an eight by eight matrix. The matrix can be represented in a block diagram form as shown in figure 6.6 with an initial subdivision into four blocks each having four terms. That is to say

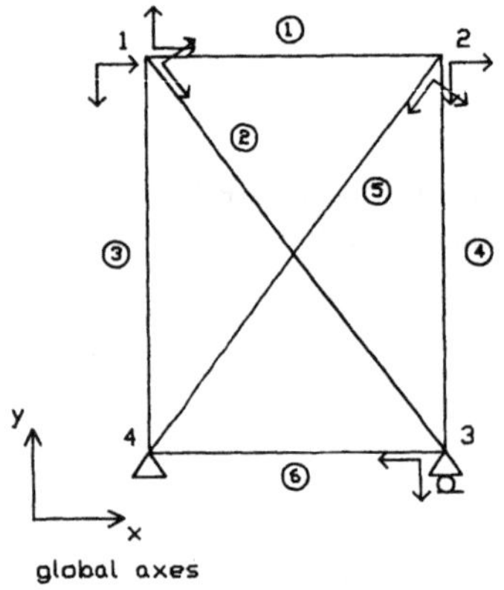

Element	Rotation
①	0
②	306.870
③	270
④	270
⑤	233.130
⑥	180

Figure 6.5 Truss analysis—relationship between global and local axes.

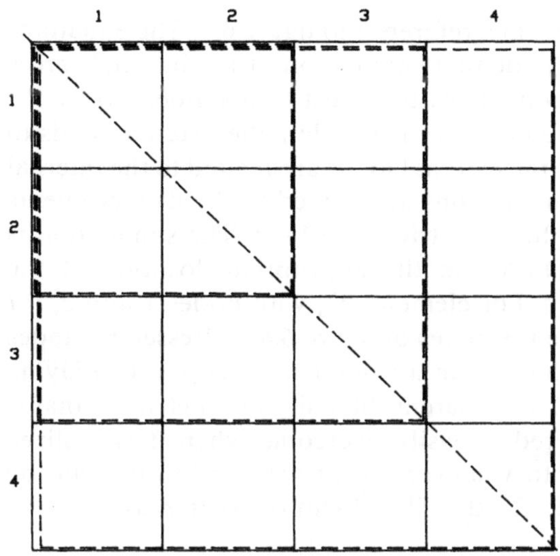

Figure 6.6 Block diagram of the structure stiffness matrix for the truss of figure 6.5.

each block represents a two by two submatrix and can be addressed by the node numbers arranged to define the row and column position of each.

Consider the equilibrium of node 1 for example, as shown in figure 6.7. The nodal equilibrium equations are given by

$$P_{1x} = \sum_{k=1}^{3} p_x, \quad \text{end 1 of } Ⓚ$$

and

$$P_{1y} = \sum_{k=1}^{3} p_y, \quad \text{end 1 of } Ⓚ$$

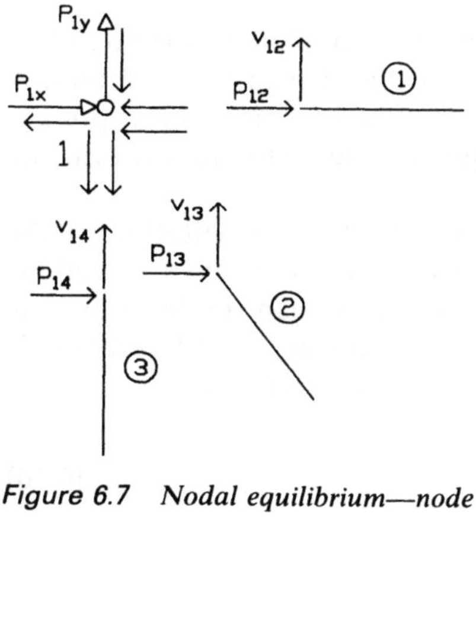

Figure 6.7 Nodal equilibrium—node 1.

where the notation is apparent from a reference to figure 6.7. The equations relate the external loads at the node to a summation of the internal forces resolved into appropriate components with respect to the global axes.

It should be recalled that the objective is to relate the external loads to the displacements. At this stage, the external loads are related to the internal actions, but of course the internal actions are related to the displacements by the transformed element stiffness matrices (TESM). The summation is achieved by assembling the TESMs in the appropriate location of the stiffness matrix for the structure. For element ①, with nodes 1 and 2, for instance, the TESM is located in the boxes of figure 6.6 addressed by nodes 1 and 2. All other TESMs are located in a similar manner and overlaying terms are added algebraically. The apparent difficulty of an element that is not consecutively node-numbered is easily overcome when it is realized that the TESM can be expanded to whatever size is necessary by the inclusion of zero terms. Thus for element ②, the TESM can be written as

$$\begin{array}{c} \\ 1 \\ 3 \end{array}\begin{array}{c} \begin{array}{cc} 1 & 3 \end{array} \\ \left[\begin{array}{c|c} T_{11} & T_{12} \\ \hline T_{21} & T_{22} \end{array}\right] \end{array} = \begin{array}{c} \\ 1 \\ 2 \\ 3 \end{array}\begin{array}{c} \begin{array}{ccc} 1 & 2 & 3 \end{array} \\ \left[\begin{array}{c|c|c} T_{11} & 0 & T_{12} \\ \hline 0 & 0 & 0 \\ \hline T_{21} & 0 & T_{22} \end{array}\right] \end{array}$$

and conveniently entered into the stiffness matrix.

In figure 6.6 the block diagram of the stiffness matrix is used to illustrate the assembly of three of the TESMs in this manner. The remainder follow in a similar way and have only been omitted in the interests of clarity. The structure stiffness matrix for the truss of figure 6.5 can now be confirmed on the basis of the above approach.

6.1.4 Solution for Element Actions

With the structure stiffness matrix and the load vector defined, the boundary conditions can be applied and a solution to the displacements found. A general procedure for the application of the boundary conditions is presented in section 6.3 and either this technique, or the techniques previously outlined, may be used.

The element actions can then be found by back-substitution of the displacements into equation (6.9). However this would result in the actions being expressed in terms of the global axes, rather than the local axes as required for meaningful interpretation. The necessary transformation is given in equation (6.5) so that equation (6.9) can be written as

$$\left\{\begin{array}{c} f_i' \\ \hline f_j' \end{array}\right\} = [T][\text{TESM}]\left\{\begin{array}{c} d_i \\ \hline d_j \end{array}\right\} \tag{6.10}$$

6.2 THE GENERAL ANALYSIS OF PLANE FRAMES

The analysis of rigid-jointed plane frames has already been introduced in chapter 4 with a technique for the analysis of rectangular frames, ignoring axial deformation. It is now appropriate to consider a more general method, applicable to any plane frame. The method will be seen as a logical extension to the general stiffness method for the analysis of trusses. Indeed the method is even broader than that and is in fact the basis of the finite element method, where a wide range of different element types can be considered.

Restricting the elements to one-dimensional types for skeletal frames, the range of structures that can be analysed in this way includes continuous beams, plane trusses, plane frames, plane grids, space trusses and space frames. With both an element stiffness matrix and a coordinate transformation matrix defined for each element, the procedure for the assembly of the structure stiffness matrix and the subsequent solution for displacements and back-substitution to determine the element actions is the same for all classes of structure. Differences do arise though in the treatment of the general loads on the structure. The matrix operations of the stiffness method are defined for the case of nodal loads only. The load vector is simply defined as the set of loads that may be applied at the nodes, according to the degrees of freedom admitted. For the analysis of a truss this has not presented any problem, since the assumption is usually made that the loads can only be applied at the nodes.

The apparent restriction is readily overcome and the appropriate technique has already been used in chapter 4. However, it is an important concept and worthy of restatement here. Any general set of loads on a linear elastic structure can be considered through a two-part approach to the problem. From the structure whose analysis is required, a restrained primary structure can be introduced. The imposed restraints correspond to the degrees of freedom at the nodes. The actions in the restrained primary structure under the loads are well known as fixed end actions that can be readily calculated as moments and forces. Therefore the actions in the imposed restraints are identified. These actions can be negated in a second solution where the structure is loaded at the nodes. Clearly the superposition of the two solutions then gives the solution to the problem. The procedure is shown in principle in the diagrams of figure 6.8 using a simple portal frame.

6.2.1 The General Plane Frame Element

The general plane frame element is subjected to the internal actions of axial force, shear force and moment. These actions can be readily defined by a combination of the continuous beam element and the truss element, using the principle of superposition. In effect, additional independent equations relating to the axial force–axial displacement relationship are introduced

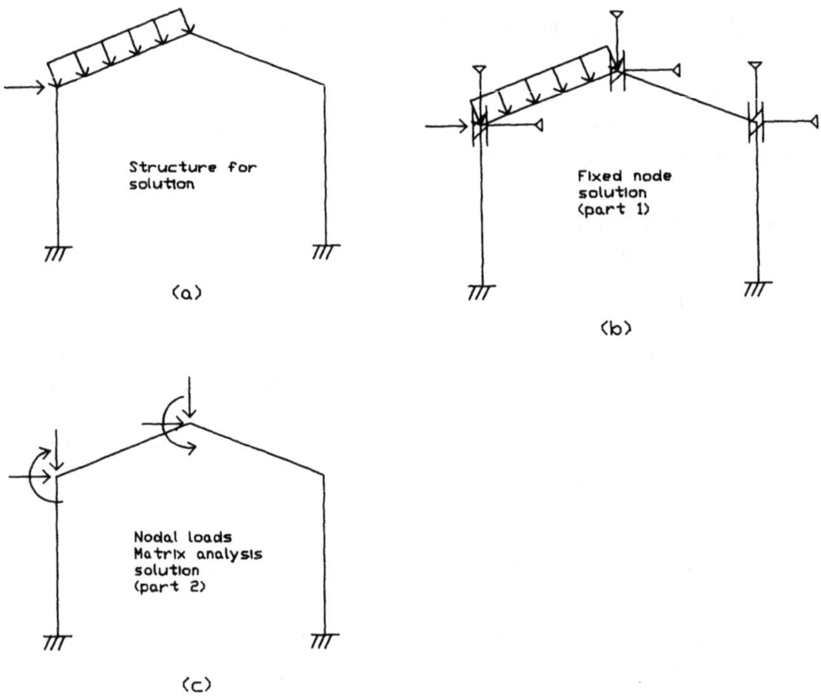

Figure 6.8 *Nodal loads on a plane frame.*

into the continuous beam action–displacement relationship. The continuous beam element stiffness matrix is expanded to a six by six matrix to accommodate the three degrees of freedom per node. The result is given in equation (6.11) which defines the required element stiffness matrix:

$$\begin{Bmatrix} p_{ij} \\ v_{ij} \\ m_{ij} \\ p_{ji} \\ v_{ji} \\ m_{ji} \end{Bmatrix} = \left[\begin{array}{ccc|ccc} \frac{EA}{L} & 0 & 0 & -\frac{EA}{L} & 0 & 0 \\ 0 & \frac{12EI}{L^3} & \frac{6EI}{L^2} & 0 & -\frac{12EI}{L^3} & \frac{6EI}{L^2} \\ 0 & \frac{6EI}{L^2} & \frac{4EI}{L} & 0 & -\frac{6EI}{L^2} & \frac{2EI}{L} \\ \hline -\frac{EA}{L} & 0 & 0 & \frac{EA}{L} & 0 & 0 \\ 0 & -\frac{12EI}{L^3} & -\frac{6EI}{L^2} & 0 & \frac{12EI}{L^3} & -\frac{6EI}{L^2} \\ 0 & \frac{6EI}{L^2} & \frac{2EI}{L} & 0 & -\frac{6EI}{L^2} & \frac{4EI}{L} \end{array}\right] \begin{Bmatrix} s_i \\ d_i \\ \theta_i \\ s_j \\ d_j \\ \theta_j \end{Bmatrix} \tag{6.11}$$

The actions and displacements of equation (6.11) are defined in the figure shown in section B2.7 of appendix B.

6.2.2 Coordinate Transformation for a Plane Frame Element

It is apparent from figure 6.1(a) that the general analysis of a plane frame involves coordinate transformation. Further, the process can be seen as an extension of the procedure involved in the analysis of a truss. Equation (6.11) introduces the additional degree of freedom of rotation at each end of the element, accompanied by the corresponding internal moment. The rotation and the internal moment are defined with respect to the z-axis which is the axis normal to the plane of the structure, and this axis is unaffected by coordinate transformation of the type considered here. In effect, the z-axis is common to both the local and global axes. For this reason the coordinate transformation matrix appropriate to the plane frame element is simply the extension of the matrix given in equation (6.4), so that for the plane frame element

$$[T] = \left[\begin{array}{ccc|ccc} \cos\alpha & \sin\alpha & 0 & 0 & 0 & 0 \\ -\sin\alpha & \cos\alpha & 0 & 0 & 0 & 0 \\ 0 & 0 & 1 & 0 & 0 & 0 \\ \hline 0 & 0 & 0 & \cos\alpha & \sin\alpha & 0 \\ 0 & 0 & 0 & -\sin\alpha & \cos\alpha & 0 \\ 0 & 0 & 0 & 0 & 0 & 1 \end{array}\right]$$

The general nature of the transformed element stiffness matrix can now be seen as

$$[\mathrm{TESM}] = [T]^{\mathrm{T}}[\mathrm{ESM}][T] \tag{6.12}$$

where both the element stiffness matrix, ESM, and the coordinate transformation matrix, T, are appropriate to the nature of the structure. As has already been indicated, the procedure for the assembly of the structure stiffness matrix for the plane frame follows along exactly the same lines as that used in the general truss analysis.

6.2.3 Application of Boundary Conditions—Solution for Displacements

In the preceding sections the transformed element stiffness matrices have been used to assemble the structure stiffness matrix without regard to the boundary conditions. This is a general approach and has the advantage of expressing the structure stiffness characteristics in such a way that alternative boundary conditions could be readily applied. Mathematically, the general load-displacement characteristics for the structure are expressed by

$$P = K \cdot d \tag{6.13}$$

as has already been seen.

The solution technique used in chapters 3 and 4, based on the partitioning of an ordered form of equation (6.13), could be used. However, in keeping with the more general line presented in this chapter, a general technique for proceeding to the solution of equation (6.13) will be given. The technique was first presented by Zienkiewicz (1971).

The general form of equation (6.13) is

$$\begin{Bmatrix} P_1 \\ P_2 \\ P_3 \\ \vdots \\ P_{n-2} \\ P_{n-1} \\ P_n \end{Bmatrix} = \begin{bmatrix} k_{11} & k_{12} & k_{13} & \dots & & k_{1n} \\ k_{21} & k_{22} & \dots & & & \\ k_{31} & \dots & & & & \\ & & & & & \\ & & & & & \\ & & & & & \\ k_{n1} & \dots & & & & k_{nn} \end{bmatrix} \begin{Bmatrix} d_1 \\ d_2 \\ d_3 \\ \vdots \\ d_{n-2} \\ d_{n-1} \\ d_n \end{Bmatrix} \tag{6.14}$$

Suppose the prescribed displacements, that is, the boundary restraints, are at d_1, d_2, d_{n-2} and d_{n-1}, and that the values are

$$d_1 = \alpha_1$$

$$d_2 = \alpha_2$$

$$d_{n-2} = \alpha_3$$

and

$$d_{n-1} = \alpha_4$$

If each of the leading diagonal terms of the stiffness matrix, corresponding to a prescribed displacement, is multiplied by a weighting factor of say 10^{20}, and the corresponding term in the load vector is replaced with the weighted diagonal term times the prescribed displacement, then equation (6.14) becomes

$$\begin{Bmatrix} \alpha_1 k_{11} 10^{20} \\ \alpha_2 k_{22} 10^{20} \\ P_3 \\ \vdots \\ \alpha_3 k_{n-2\,n-2} 10^{20} \\ \alpha_4 k_{n-1\,n-1} 10^{20} \\ P_n \end{Bmatrix} = \begin{bmatrix} k_{11}10^{20} & k_{12} & k_{13} & \dots & & & k_{1n} \\ k_{21} & k_{22}10^{20} & \dots & & & & \\ k_{31} & \dots & & & & & \\ & & & & & & \\ & & & & k_{n-2\,n-2}10^{20} & & \\ & & & & & k_{n-1\,n-1}10^{20} & \\ k_{n1} & \dots & & & & & k_{nn} \end{bmatrix} \begin{Bmatrix} d_1 \\ d_2 \\ d_3 \\ \vdots \\ d_{n-2} \\ d_{n-1} \\ d_n \end{Bmatrix} \tag{6.15}$$

The first equation represented in the matrix relationship of equation (6.15) is

$$\alpha_1 \cdot k_{11} \cdot 10^{20} = k_{11} \cdot 10^{20} \cdot d_1 + k_{12} \cdot d_2 + k_{13} \cdot d_3 + \cdots + k_{1n} \cdot d_n$$

Solving for d_1 gives

$$d_1 = [\alpha_1 \cdot k_{11} \cdot 10^{20} - (k_{12} \cdot d_2 + k_{13} \cdot d_3 + \cdots + k_{1n} \cdot d_n)]/(k_{11} \cdot 10^{20})$$

that is

$$d_1 \simeq \alpha_1 \text{ (to a very good approximation)}$$

A similar result is obtained by considering the remaining equations directly related to the restraints and, of course, the results are equally valid when α_i equals zero. The standard solution to equation (6.15) will return the complete displacement vector.

In considering what the technique does, it is relevant to return to the general relationship given in equation (6.13), and to recall that, prior to the consideration of the boundary conditions, a solution is not possible. This is because P is not a vector of completely known terms and d is not a vector of completely unknown terms. The general load vector contains the unknown reactions, and the general displacement vector contains the specified or prescribed displacements at the restraints. What the above technique does is to specify a term in the load vector to replace an unknown reaction, in such a way as to ensure that the corresponding displacement has its prescribed value on solution.

A simplified version of the technique, valid only when the prescribed displacements are zero, serves to illustrate the point further. In this case, suppose the required leading diagonal terms were replaced by a unit value, with the remaining terms in both the corresponding rows and columns set to zero. The result is that equation (6.14) would be written as

$$\begin{Bmatrix} 0 \\ 0 \\ P_3 \\ \vdots \\ 0 \\ 0 \\ P_n \end{Bmatrix} = \begin{bmatrix} 1 & 0 & 0 & \cdots & & & & \\ 0 & 1 & 0 & \cdots & & & & \\ 0 & 0 & k_{33} & k_{34} & \cdots & & & \\ & & & & & & & \\ 0 & 0 & 0 & 0 & \cdots & 1 & 0 & 0 \\ 0 & 0 & 0 & 0 & \cdots & 0 & 1 & 0 \\ 0 & 0 & k_{n3} & k_{n4} & \cdots & 0 & 0 & k_{nn} \end{bmatrix} \begin{Bmatrix} d_1 \\ d_2 \\ d_3 \\ \vdots \\ d_{n-2} \\ d_{n-1} \\ d_n \end{Bmatrix} \tag{6.16}$$

Effectively the equations represented in equation (6.16) are

$$P_F = K_F \cdot d_F$$

plus a set of equations

$$\{O\} = [I] \cdot \{d_R\}$$

where $[I]$ is the identity matrix. However the equations are all distributed throughout the matrix rather than being arranged in a compact form. The solution of equation (6.16) will return the required displacements for all the nodes of the structure.

The general procedure for handling the boundary conditions is demonstrated in examples 6.1 and 6.2. Both examples use the MATOP command, MODDG, which has been specifically designed to handle the boundary conditions by permitting a modification to the diagonal elements of a matrix. In addition, the SELECT command facilitates the back-substitution process to recover the element actions. Displacement terms appropriate to each element can be selected from the displacement vector for structures where the elements have consecutive node numbering.

Example 6.1: Analysis of a Continuous Beam

Given data:

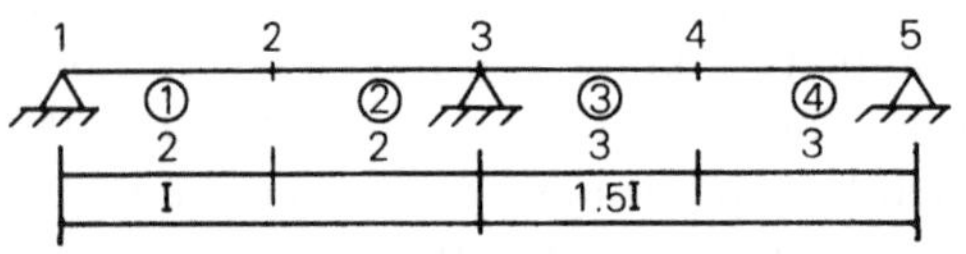

$E = 200\ \text{kN/mm}^2$
$I = 50 \times 10^6\ \text{mm}^4$

Load Case 1:

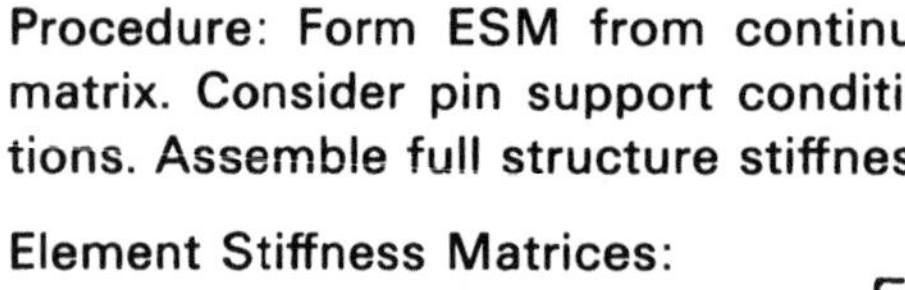

Load Case 2: Support 3 settles vertically by 10 mm

Procedure: Form ESM from continuous beam element stiffness matrix. Consider pin support condition with the boundary conditions. Assemble full structure stiffness matrix.

Element Stiffness Matrices:

Elements ① and ②

$$\text{ESM} = EI\left[\begin{array}{cc|cc} 1.5 & 1.5 & -1.5 & 1.5 \\ 1.5 & 2 & -1.5 & 1 \\ \hline -1.5 & -1.5 & 1.5 & -1.5 \\ 1.5 & 1 & -1.5 & 2 \end{array}\right]$$

Elements ③ and ④

$$\text{ESM} = EI\left[\begin{array}{cc|cc} 0.66\dot{6} & 1 & -0.66\dot{6} & 1 \\ 1 & 2 & -1 & 1 \\ \hline -0.66\dot{6} & -1 & 0.66\dot{6} & -1 \\ 1 & 1 & -1 & 2 \end{array}\right]$$

(kN m units)

Structure Stiffness Matrix:

$$K = EI\begin{bmatrix}
1.5 & 1.5 & -1.5 & 1.5 & & & & & & \\
1.5 & 2 & -1.5 & 1 & & & & & & \\
-1.5 & -1.5 & 3 & 0 & -1.5 & 1.5 & & & & \\
1.5 & 1 & 0 & 4 & -1.5 & 1 & & & & \\
 & & -1.5 & -1.5 & 2.16\dot{6} & -0.5 & -0.66\dot{6} & 1 & & \\
 & & 1.5 & 1 & -0.5 & 4 & -1 & 1 & & \\
 & & & & -0.66\dot{6} & -1 & 1.33\dot{3} & 0 & -0.66\dot{6} & 1 \\
 & & & & 1 & 1 & 0 & 4 & -1 & 1 \\
 & & & & & & -0.66\dot{6} & -1 & 0.66\dot{6} & -1 \\
 & & & & & & 1 & 1 & -1 & 2
\end{bmatrix}$$

Load Vector: Load case 1—Fixed End Actions

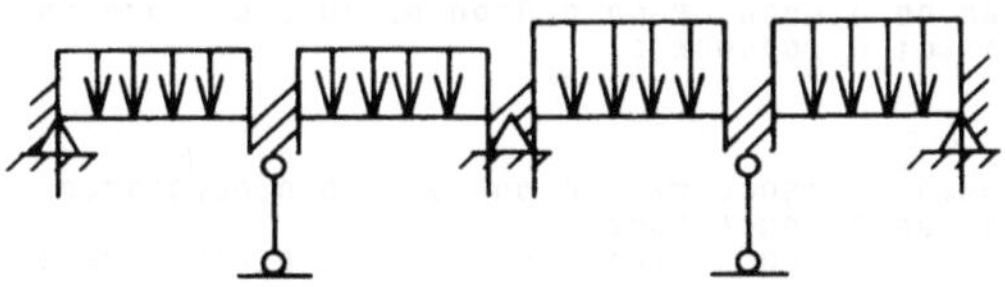

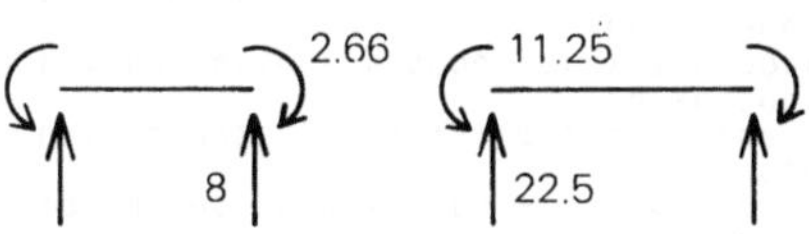

Typical fixed end actions

Hence nodal loads are:

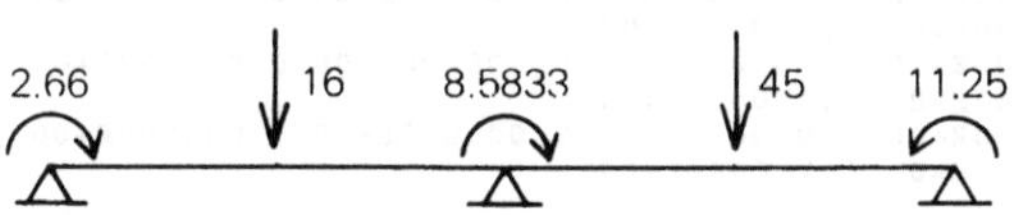

Load case 2—$d_3 = -0.010$m

Hence

$$P = \begin{Bmatrix}
R_1 & R_1 \\
-2.66\dot{6} & 0 \\
-16 & 0 \\
0 & 0 \\
R_3 & R_3 \\
-8.583\dot{3} & 0 \\
-45 & 0 \\
0 & 0 \\
R_5 & R_5 \\
11.25 & 0
\end{Bmatrix} \text{ and } P(\text{modified}) = \begin{Bmatrix}
0 & 0 \\
-2.66\dot{6} & 0 \\
-16 & 0 \\
0 & 0 \\
0 & -0.01 \cdot k_{11} \cdot 10^{20} \\
-8.5833 & 0 \\
-45 & 0 \\
0 & 0 \\
0 & 0 \\
11.25 & 0
\end{Bmatrix}$$

Solution: Using MATOP, and the commands MODDG and SELECT in particular, the solution can now be found as shown in the following output.

```
LOAD.K.10.10
PRINT.K  Structure Stiffness Matrix (scaled)
0.150000E+01 0.150000E+01 -.150000E+01 0.150000E+01 0.000000E+00 0.000000E+00
0.000000E+00 0.000000E+00 0.000000E+00 0.000000E+00
0.150000E+01 0.200000E+01 -.150000E+01 0.100000E+01 0.000000E+00 0.000000E+00
0.000000E+00 0.000000E+00 0.000000E+00 0.000000E+00
-.150000E+01 -.150000E+01 0.300000E+01 0.000000E+00 -.150000E+01 0.150000E+01
0.000000E+00 0.000000E+00 0.000000E+00 0.000000E+00
0.150000E+01 0.100000E+01 0.000000E+00 0.400000E+01 -.150000E+01 0.100000E+01
0.000000E+00 0.000000E+00 0.000000E+00 0.000000E+00
0.000000E+00 0.000000E+00 -.150000E+01 -.150000E+01 0.216667E+01 -.500000E+00
-.666666E+00 0.100000E+01 0.000000E+00 0.000000E+00
0.000000E+00 0.000000E+00 0.150000E+01 0.100000E+01 -.500000E+00 0.400000E+01
-.100000E+01 0.100000E+01 0.000000E+00 0.000000E+00
0.000000E+00 0.000000E+00 0.000000E+00 0.000000E+00 -.666666E+00 -.100000E+01
0.133333E+01 0.000000E+00 -.666666E+00 0.100000E+01
0.000000E+00 0.000000E+00 0.000000E+00 0.000000E+00 0.100000E+01 0.100000E+01
0.000000E+00 0.400000E+01 -.100000E+01 0.100000E+01
0.000000E+00 0.000000E+00 0.000000E+00 0.000000E+00 0.000000E+00 0.000000E+00
-.666666E+00 -.100000E+01 0.666666E+00 -.100000E+01
0.000000E+00 0.000000E+00 0.000000E+00 0.000000E+00 0.000000E+00 0.000000E+00
0.100000E+01 0.100000E+01 -.100000E+01 0.200000E+01
LOAD.P.10.2
MODDG.K
PRINT.K  Now modified for solution
0.150000E+21 0.150000E+01 -.150000E+01 0.150000E+01 0.000000E+00 0.000000E+00
0.000000E+00 0.000000E+00 0.000000E+00 0.000000E+00
0.150000E+01 0.200000E+01 -.150000E+01 0.100000E+01 0.000000E+00 0.000000E+00
0.000000E+00 0.000000E+00 0.000000E+00 0.000000E+00
-.150000E+01 -.150000E+01 0.300000E+01 0.000000E+00 -.150000E+01 0.150000E+01
0.000000E+00 0.000000E+00 0.000000E+00 0.000000E+00
0.150000E+01 0.100000E+01 0.000000E+00 0.400000E+01 -.150000E+01 0.100000E+01
0.000000E+00 0.000000E+00 0.000000E+00 0.000000E+00
0.000000E+00 0.000000E+00 -.150000E+01 -.150000E+01 0.216667E+21 -.500000E+00
-.666666E+00 0.100000E+01 0.000000E+00 0.000000E+00
0.000000E+00 0.000000E+00 0.150000E+01 0.100000E+01 -.500000E+00 0.400000E+01
-.100000E+01 0.100000E+01 0.000000E+00 0.000000E+00
0.000000E+00 0.000000E+00 0.000000E+00 0.000000E+00 -.666666E+00 -.100000E+01
0.133333E+01 0.000000E+00 -.666666E+00 0.100000E+01
0.000000E+00 0.000000E+00 0.000000E+00 0.000000E+00 0.100000E+01 0.100000E+01
0.000000E+00 0.400000E+01 -.100000E+01 0.100000E+01
0.000000E+00 0.000000E+00 0.000000E+00 0.000000E+00 0.000000E+00 0.000000E+00
-.666666E+00 -.100000E+01 0.666667E+20 -.100000E+01
0.000000E+00 0.000000E+00 0.000000E+00 0.000000E+00 0.000000E+00 0.000000E+00
0.100000E+01 0.100000E+01 -.100000E+01 0.200000E+01
SCALE.K.10000
SOLVE.K.P
PRINT.P  The Displacement Vector
0.162500E-23 -.260417E-23
0.650000E-03 -.354167E-02
0.150833E-02 -.656250E-02
0.695833E-03 -.276042E-02
-.221057E-22 -.100000E-01
-.343333E-02 -.416663E-03
-.106125E-01 -.734374E-02
-.695834E-03 0.192708E-02
-.233124E-22 -.390615E-23
0.621667E-02 0.270833E-02
REMARK. Calculating Element Actions
LOAD.E.4.4  (Elements 1 and 2)
SCALE.E.10000
SELECT.D.P.4.2.1.1
MULT.E.D.F
PRINT.F  Element 1 Actions
-.243750E+01 0.390626E+01
-.266667E+01 -.393540E-05
0.243750E+01 -.390626E+01
-.220833E+01 0.781251E+01
DELETE.D
DELETE.F
SELECT.D.P.4.2.3.1
MULT.E.D.F
```

```
PRINT.F  Element 2 Actions
-.184375E+02 0.390626E+01
0.220833E+01 -.781251E+01
0.184375E+02 -.390626E+01
-.390833E+02 0.156250E+02
DELETE.D
DELETE.F
DELETE.E
LOAD.E.4.4  (Elements 3 and 4)
SCALE.E.10000
SELECT.D.P.4.2.5.1
MULT.E.D.F
PRINT.F  Element 3 Actions
0.294583E+02 -.260417E+01
0.305000E+02 -.156250E+02
-.294583E+02 0.260417E+01
0.578750E+02 0.781245E+01
DELETE.D
DELETE.F
SELECT.D.P.4.2.7.1
MULT.E.D.F
PRINT.F  Element 4 Actions
-.155416E+02 -.260410E+01
-.578750E+02 -.781245E+01
0.155416E+02 0.260410E+01
0.112500E+02 -.211227E-05
End of File
```

Final Actions:
Load Case 1:

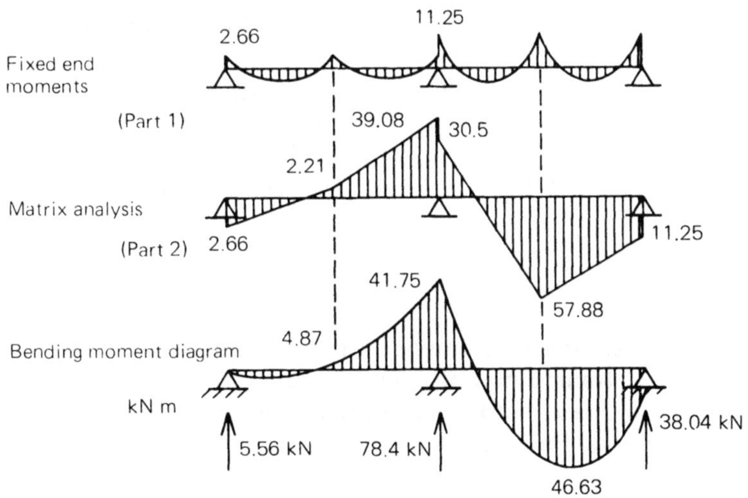

Load Case 2: Since there are no moments in the fixed end case here, the results are given directly by the matrix analysis.

Bending Moment Diagram

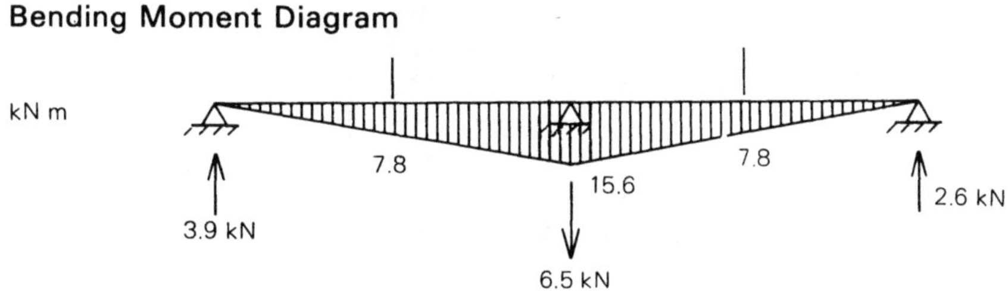

Deflections:

	y translation (mm)	
Node	Load case 1	Load case 2
2	1.5	−6.6
3	0	−10
4	−10.6	−7.3

Example 6.2: Analysis of a Plane Frame

Given data:

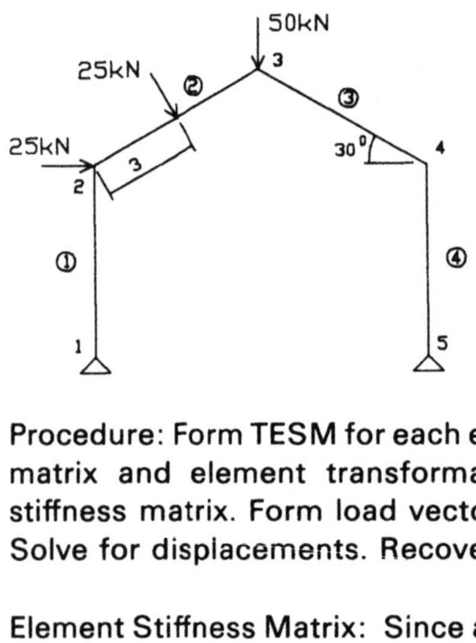

Properties for all elements:

$$E = 2 \times 10^5 \text{ MPa}$$
$$l = 6000 \text{ mm}$$
$$A_x = 7500 \text{ mm}^2$$
$$I_x = 400 \times 10^6 \text{ mm}^4$$

Procedure: Form TESM for each element based on element stiffness matrix and element transformation matrix. Assemble structure stiffness matrix. Form load vector and apply boundary conditions. Solve for displacements. Recover element actions.

Element Stiffness Matrix: Since all elements have the same properties only one ESM is evaluated

ESM (all elements) = (Units kN) mm)

$$\begin{bmatrix} 250 & 0 & 0 \\ 0 & 4.4\dot{4} & 13\,333.3\dot{3} \\ 0 & 13\,333.3\dot{3} & 53\,333\,333.\dot{3} \\ \hline -250 & 0 & 0 \\ 0 & -4.4\dot{4} & -13\,333.3\dot{3} \\ 0 & 13\,333.3\dot{3} & 26\,666\,666.\dot{6} \end{bmatrix}$$

$$
\left.\begin{array}{|ccc}
-250 & 0 & 0 \\
0 & -4.4\dot{4} & 13\,333.3\dot{3} \\
0 & -13\,333.3\dot{3} & 26\,666\,666.\dot{6} \\
\hline
250 & 0 & 0 \\
0 & 4.4\dot{4} & -13\,333.3\dot{3} \\
0 & -13\,333.3\dot{3} & 53\,333\,333.\dot{3}
\end{array}\right]
$$

Transformation Matrices: Based on

Element	α
①	90°
②	30°
③	330°
④	270°

$$
T=\left[\begin{array}{ccc|ccc}
\cos\alpha & \sin\alpha & 0 & 0 & 0 & 0 \\
-\sin\alpha & \cos\alpha & 0 & 0 & 0 & 0 \\
0 & 0 & 1 & 0 & 0 & 0 \\
\hline
0 & 0 & 0 & \cos\alpha & \sin\alpha & 0 \\
0 & 0 & 0 & -\sin\alpha & \cos\alpha & 0 \\
0 & 0 & 0 & 0 & 0 & 1
\end{array}\right]
$$

Transformed Element Stiffness Matrices: The appropriate TESM can now be calculated from

$$
\text{TESM}=[T]^T\cdot[\text{ESM}]\cdot[T]
$$

Using MATOP the following results are obtained:

```
LOAD.E.6.6
LOAD.T.6.6   Matrix T for Element 1
TRANS.T.R
MULT.R.E.S
MULT.S.T.K
PRINT.K   TESM for Element 1
0.444400E+01 0.000000E+00 -.133333E+05 -.444400E+01 0.000000E+00 -.133333E+05
0.000000E+00 0.250000E+03 0.000000E+00 0.000000E+00 -.250000E+03 0.000000E+00
-.133333E+05 0.000000E+00 0.533333E+08 0.133333E+05 0.000000E+00 0.266667E+08
-.444400E+01 0.000000E+00 0.133333E+05 0.444400E+01 0.000000E+00 0.133333E+05
0.000000E+00 -.250000E+03 0.000000E+00 0.000000E+00 0.250000E+03 0.000000E+00
-.133333E+05 0.000000E+00 0.266667E+08 0.133333E+05 0.000000E+00 0.533333E+08
DELETE.T
DELETE.R
DELETE.S
DELETE.K
LOAD.T.6.6   Matrix T for Element 2
TRANS.T.R
MULT.R.E.S
MULT.S.T.K
PRINT.K   TESM for Element 2
0.188600E+03 0.106326E+03 -.666667E+04 -.188600E+03 -.106326E+03 -.666667E+04
0.106326E+03 0.658328E+02 0.115467E+05 -.106326E+03 -.658328E+02 0.115467E+05
-.666667E+04 0.115467E+05 0.533333E+08 0.666667E+04 -.115467E+05 0.266667E+08
-.188600E+03 -.106326E+03 0.666667E+04 0.188600E+03 0.106326E+03 0.666667E+04
-.106326E+03 -.658328E+02 -.115467E+05 0.106326E+03 0.658328E+02 -.115467E+05
-.666667E+04 0.115467E+05 0.266667E+08 0.666667E+04 -.115467E+05 0.533333E+08
DELETE.T
DELETE.R
DELETE.S
DELETE.K
LOAD.T.6.6   Matrix T for Element 3
TRANS.T.R
MULT.R.E.S
MULT.S.T.K
```

```
PRINT.K  TESM for Element 3
0.188600E+03 -.106326E+03 0.666667E+04 -.188600E+03 0.106326E+03 0.666667E+04
-.106326E+03 0.658328E+02 0.115467E+05 0.106326E+03 -.658328E+02 0.115467E+05
0.666667E+04 0.115467E+05 0.533333E+08 -.666667E+04 -.115467E+05 0.266667E+08
-.188600E+03 0.106326E+03 -.666667E+04 0.188600E+03 -.106326E+03 -.666667E+04
0.106326E+03 -.658328E+02 -.115467E+05 -.106326E+03 0.658328E+02 -.115467E+05
0.666667E+04 0.115467E+05 0.266667E+08 -.666667E+04 -.115467E+05 0.533333E+08
DELETE.T
DELETE.R
DELETE.S
DELETE.K
LOAD.T.6.6  Matrix T for Element 4
TRANS.T.R
MULT.R.E.S
MULT.S.T.K
PRINT.K  TESM for Element 4
0.444400E+01 0.000000E+00 0.133333E+05 -.444400E+01 0.000000E+00 0.133333E+05
0.000000E+00 0.250000E+03 0.000000E+00 0.000000E+00 -.250000E+03 0.000000E+00
0.133333E+05 0.000000E+00 0.533333E+08 -.133333E+05 0.000000E+00 0.266667E+08
-.444400E+01 0.000000E+00 -.133333E+05 0.444400E+01 0.000000E+00 -.133333E+05
0.000000E+00 -.250000E+03 0.000000E+00 0.000000E+00 0.250000E+03 0.000000E+00
0.133333E+05 0.000000E+00 0.266667E+08 -.133333E+05 0.000000E+00 0.533333E+08
End of File
```

Structure Stiffness Matrix: Since MATOP does not provide for the assembly of a structure stiffness matrix, this operation must be carried out by hand. The block diagram of the stiffness matrix is

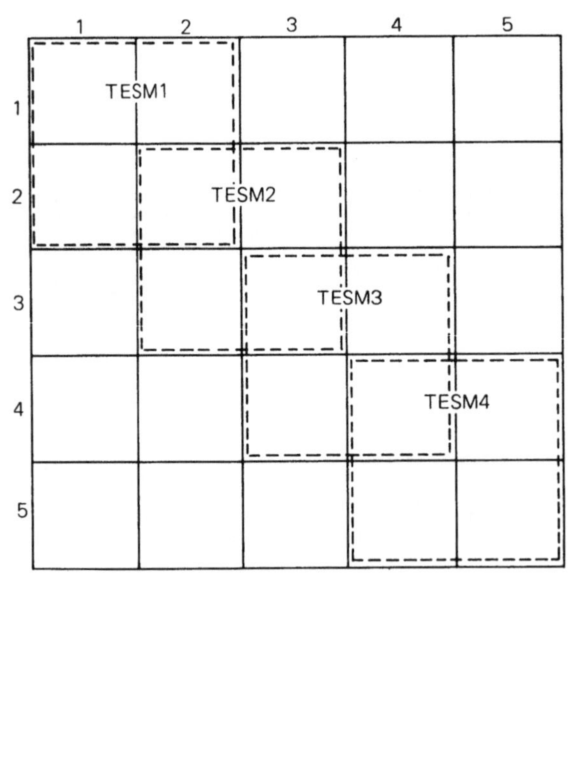

and the resulting matrix is

```
4.444 0 -13333.333 -4.444 0 -13333.333 0 0 0 0 0 0 0 0 0
0 250 0 0 -250 0 0 0 0 0 0 0 0 0 0
-13333.333 0 53333333.333 13333.333 0 26666666.666 0 0 0 0 0 0 0 0 0
-4.444 0 13333.333 193.044 106.325 6666.666 -188.6 -106.325 -6666.666 0 0 0 0 0 0
0 250 0 106.325 315.833 11546.666 -106.325 -65.833 11546.666 0 0 0 0 0 0
-13333.333 0 26666666.666 6666.666 11546.666 106666666.666 6666.666 -11546.666 26666666.666 0 0 0 0 0 0
0 0 0 -188.6 -106.325 6666.666 377.2 0 13333.333 -188.6 106.325 6666.666 0 0 0
0 0 0 -106.325 -65.833 -11546.666 0 131.6 0 106.325 -65.833 11546.666 0 0 0
0 0 0 -6666.666 11546.666 26666666.666 13333.333 0 106666666.666 -6666.666 -11546.666 26666666.666 0 0 0
0 0 0 0 0 0 -188.6 106.325 -6666.666 193.044 -106.32 6666.666 -4.444 0 13333.33
0 0 0 0 0 0 106.325 -65.833 -11546.666 -106.32 315.83 -11546.666 0 -250 0
0 0 0 0 0 0 6666.666 11546.666 26666666.666 6666.666 -11546.6 106666666.666 -13333.333 0 26666666.666
0 0 0 0 0 0 0 0 0 -4.444 0 -13333.333 4.444 0 -13333.333
0 0 0 0 0 0 0 0 0 0 -250 0 0 250 0
0 0 0 0 0 0 0 0 0 13333.333 0 26666666.666 -13333.333 0 53333333.333
```

Load Vector: Fixed End Actions

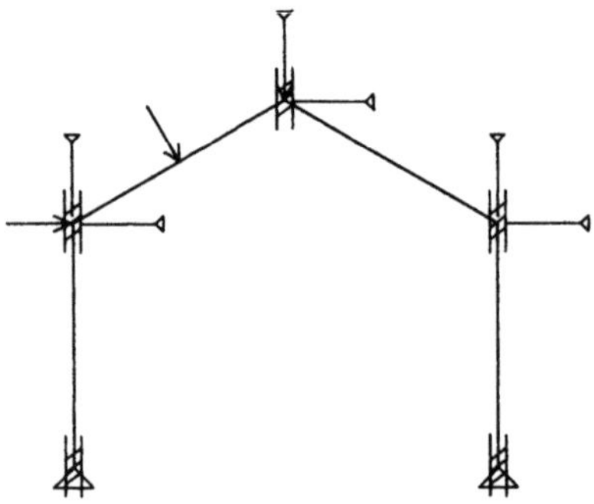

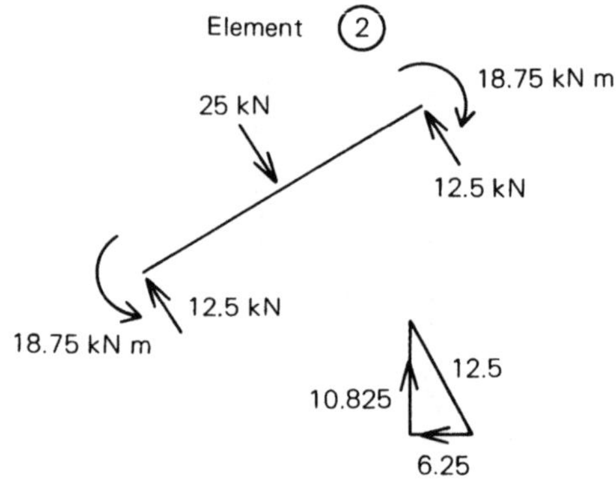

Nodal Loads

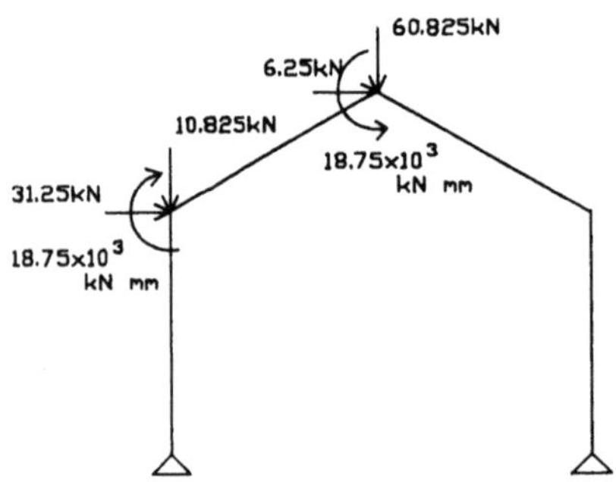


```
LOAD.K.15.15  Structure Stiffness Matrix
MODDG.K
LOAD.P.15.1
PRINT.P  The Modified Load Vector
0.000000E+00
0.000000E+00
0.000000E+00
0.312500E+02
-.108250E+02
-.187500E+05
0.625000E+01
-.608250E+02
0.187500E+05
0.000000E+00
0.000000E+00
0.000000E+00
0.000000E+00
0.000000E+00
0.000000E+00
SOLVE.K.P
PRINT.P  The Displacement Vector
0.253468E-19
-.717048E-21
-.641482E-02
0.334134E+02
-.717048E-01
-.387706E-02
0.361150E+02
-.491770E+01
0.173951E-02
0.385992E+C2
-.216968E+00
-.249510E-02
0.590390E-19
-.216968E-20
-.840226E-02
```

Solution: The results as shown were obtained from MATOP

Displacements

Node	x (mm)	y (mm)	Rot. (radian)
1	0	0	−0.0064
2	33.55	−0.07	−0.0039
3	36.25	−4.93	0.0017
4	38.75	−0.22	−0.0025
5	0	0	−0.0084

Element Actions: In a final series of operations using MATOP, the element actions were recovered as shown:

```
LOAD.E.6.6   Element Stiffness Matrix
LOAD.T.6.6   Matrix T for Element 1
TRANS.T.R
MULT.R.E.S
MULT.S.T.M
SELECT.D.P.6.1.1.1
MULT.T.M.U
MULT.U.D.F
PRINT.F  Element 1 Actions
0.179262E+02
0.112641E+02
0.756298E-03
-.179262E+02
-.112641E+02
0.676738E+05
DELETE.T
DELETE.R
DELETE.S
DELETE.M
DELETE.U
DELETE.D
LOAD.T.6.6  Matrix T for Element 2
TRANS.T.R
MULT.R.E.S
MULT.S.T.M
SELECT.D.P.6.1.4.1
MULT.T.M.U
MULT.U.D.G
PRINT.G  Element 2 Actions
0.208598E+02
-.384770E+01
-.864238E+05
-.208598E+02
0.384770E+01
0.633514E+05
DELETE.T
DELETE.R
DELETE.S
DELETE.M
DELETE.U
DELETE.D
DELETE.F
DELETE.G
```

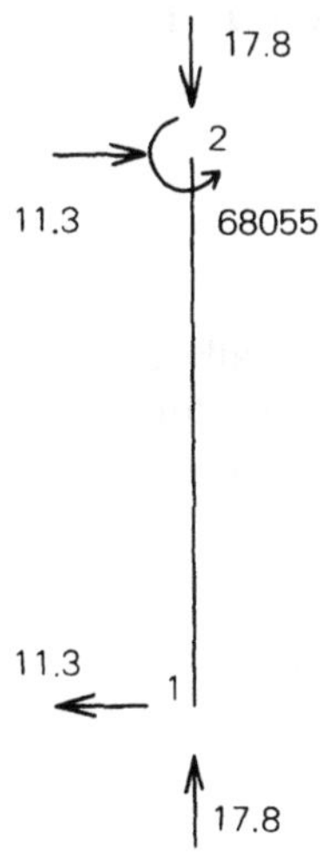

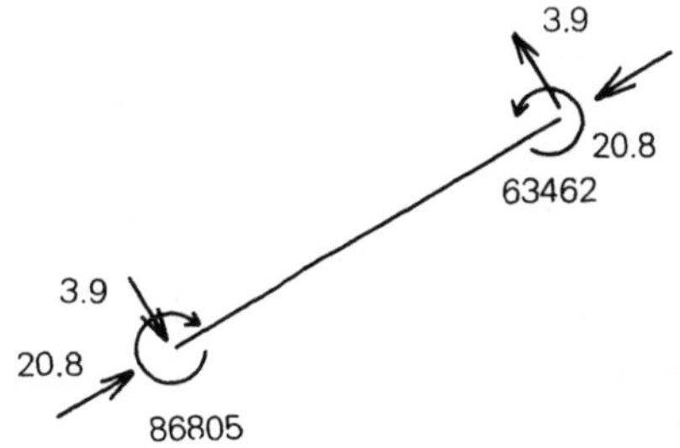

```
LOAD.T.6.6  Matrix T for Element 3
TRANS.T.R
MULT.R.E.S
MULT.S.T.M
SELECT.D.P.6.1.7.1
MULT.T.M.U
MULT.U.D.F
PRINT.F  Element 3 Actions
0.497460E+02
-.336838E+02
-.446015E+05
-.497460E+02
0.336838E+02
-.157524E+06
DELETE.T
DELETE.R
DELETE.S
DELETE.M
DELETE.U
DELETE.D
LOAD.T.6.6  Matrix T for Element 4
TRANS.T.R
MULT.R.E.S
MULT.S.T.M
SELECT.D.P.6.1.10.1
MULT.T.M.U
MULT.U.D.G
PRINT.G  Element 4 Actions
0.542420E+02
0.262369E+02
0.157524E+06
-.542420E+02
-.262369E+02
0.126848E-01
End of File
```

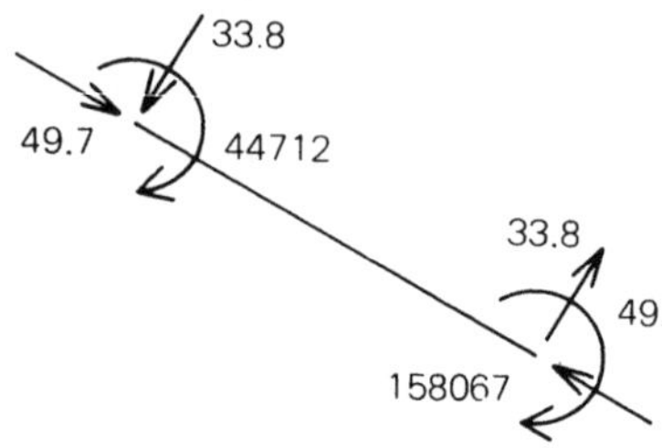

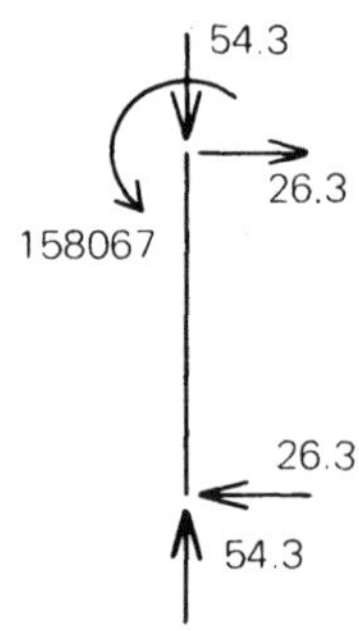

Final Actions: Finally the Frame BMD is given by a combination of the element actions under nodal loads and the fixed end actions as shown:

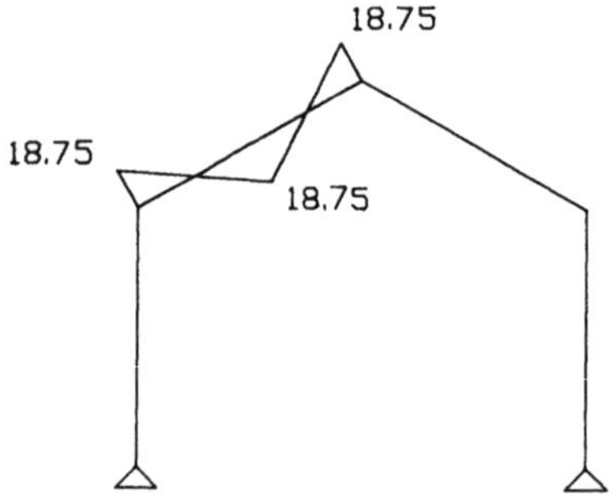

Fixed End Moments
(Part 1 Solution)

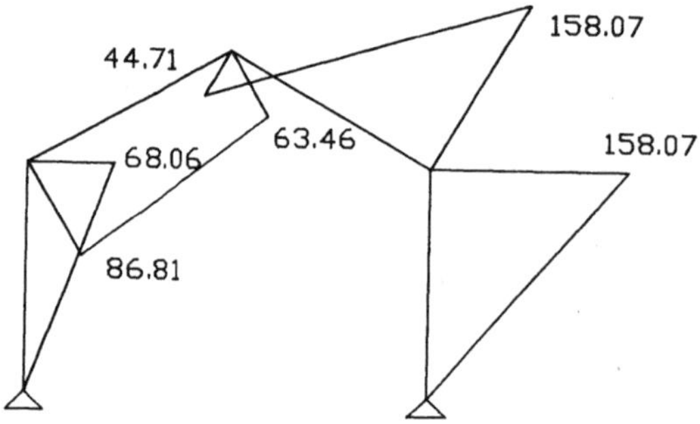

Element Moments
due to Nodal Loads
(Part 2 Solution)

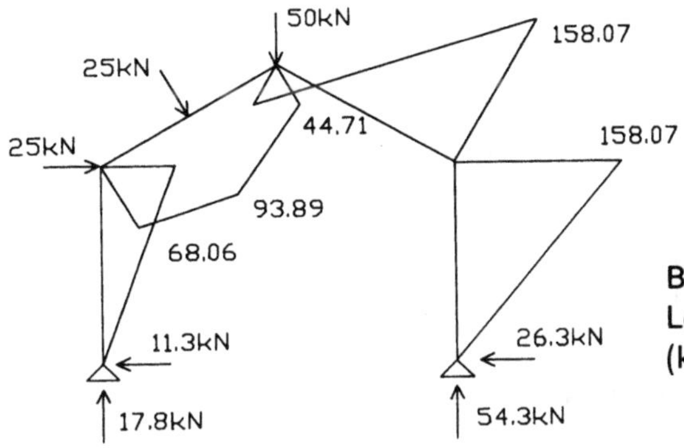

Bending Moment Diagram,
Loads and Reactions
(kN m) (kN)

Equilibrium Check:

$$\Sigma V=0 \qquad 17.8+54.3-21.65-50=0.45$$

$$\Sigma H=0 \qquad 25+12.5-11.3-26.3=-0.10$$

(small errors due to rounding off)

6.2.4 The Bandwidth of the Stiffness Matrix

As has been previously noted, a structure stiffness matrix has several characteristics. The matrix is generally sparse, and is both symmetrical and banded. That is to say, it is possible to define a band within the matrix, parallel to the leading diagonal, such that the band encloses all non-zero terms of the matrix. This is illustrated in figure 6.9 which demonstrates the assembly of a structure stiffness matrix in a similar manner to that presented in section 6.1.3 for a truss. Figure 6.9(a) shows a two-bay gabled frame of a type often used in light industrial buildings. The unrestrained structure has 24 degrees of freedom and the stiffness matrix can be represented by the block diagram shown. In this case each block represents a three by three sub-matrix corresponding to the three degrees of freedom per node. The structure stiffness matrix is assembled by locating the TESMs in the locations addressed through the element node numbers. Each of the elements is conveniently consecutively node-numbered with the exception of element number five. The transformed element stiffness matrix for that element extends, with zero terms, over the locations four to six and this defines the band width in this case. The band width is clearly a function of the maximum difference between the node numbers of any element and it may be seen that it is given by

$$\text{bandwidth} = \text{number of DOF/node (max. diff in node numbers} + 1)$$

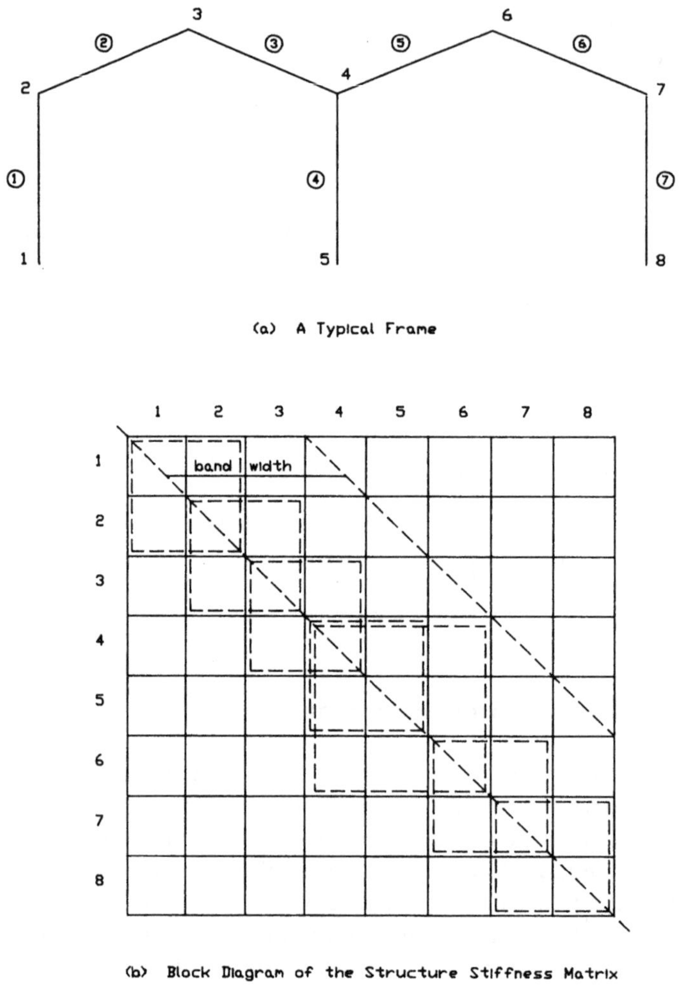

Figure 6.9 *Band width of a stiffness matrix.*

The significance of the band width may be appreciated when it is realised that it is only necessary to store the terms within the bandwidth, for any computer-based solution to the problem. Most solution routines used in structural analysis programs take advantage of the banded and symmetric nature of the structure stiffness matrix in the interests of computational efficiency. It is therefore desirable to number the nodes in such a way that the band width is kept to a minimum. The solution routine in the program MATOP is a general one however, applicable to both symmetric and nonsymmetric equations, and it does not use the special features of a structure stiffness matrix.

6.2.5 Frame Elements with End Moment Releases

In the general analysis of plane frames based on forming the structure stiffness matrix, it is a relatively simple matter to include elements with special features. One example is the frame element with end moment release. In chapter 4, details of the modification for the continuous beam element stiffness matrix for end moment release were presented. The resulting matrices for both a left-hand and a right-hand end-pinned element are summarised in appendix B.

Just as the general frame element was seen to be the result of the superposition of the general continuous beam element and the truss element, so the frame element with end moment release can be derived by combining the corresponding beam element stiffness matrix with that for a truss element. It is important to note that the matrix is retained as a six by six matrix, in spite of the presence of a complete row and column of zero terms. The reason for this is that, while the element end rotation is unspecified, and the element end moment is zero at the release end, that does not necessarily apply to other elements terminating at that node. Overall the frame retains its degrees of freedom.

The element stiffness matrix, now modified for the end moment release, can be transformed and assembled into the structure stiffness matrix in exactly the same way as before. Whether the element is considered to be left-hand end or right-end end-pinned, is a function of the element specification since this determines the orientation of the local axes. A comment needs to be made with respect to a situation where the node is fully released from all moment. In this case the procedure will require the pin to be nominated in all elements terminating at that node except one. This is necessary to retain a rotational displacement. If only two elements are involved, the pin can be considered either just to the right or just to the left of the node, so that additional displacement data (rotation) is obtained from the analysis.

Figure 6.10 illustrates two examples of frames involving end moment release. In figure 6.10(a) for instance, element 6 would be formed as a left-hand end-pinned element stiffness matrix, provided the element is specified as 5–7. In figure 6.10(b) the moment release can be exclusively assigned to either the right-hand end of element 2 (element specified 2–3) or the left-hand end of element 3 (element specified 3–4).

Finally, it should be appreciated that while the end moment release concept can be applied to an element to meet the requirements of the boundary conditions, this is not a necessary condition. Rather it is a function of the overall approach to the problem. The structure stiffness matrix can always be assembled without regard to the boundary conditions, and these can then be accommodated by the technique described in section 6.3. This is actually the more general approach and it is the basis of many analysis programs.

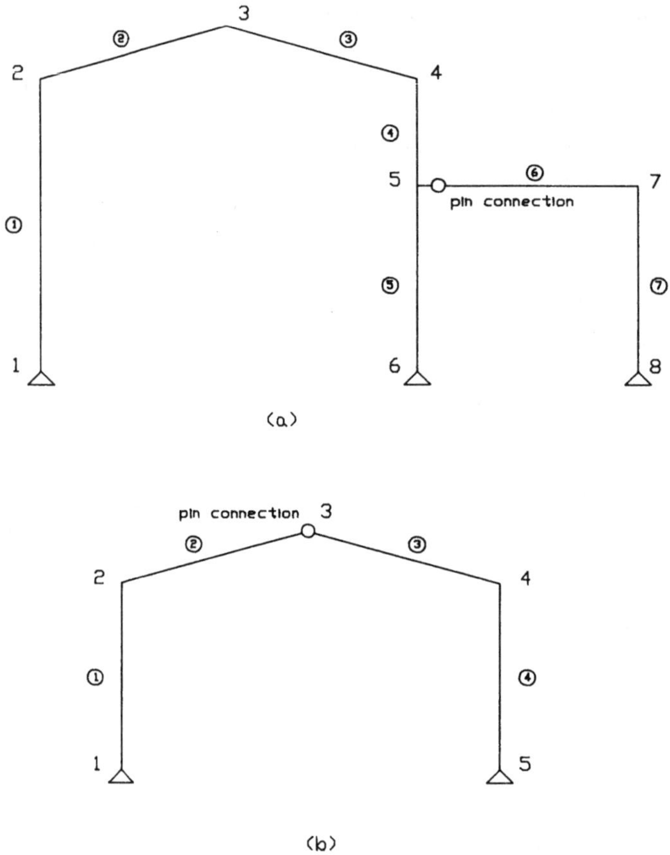

Figure 6.10 *Frames with end moment release.*

6.3 COMPOSITE STRUCTURES—TRUSS AND FRAME ELEMENTS

The mathematical model of any structure is an idealisation of the actual behaviour. For the most part, plane frames can be modelled essentially as a collection of rigid-jointed flexural elements, with trusses as a collection of pin-jointed axial elements. In the case of frames, many joints may not be fully rigid and it is certainly true that truss joints are seldom pin-connected in a physical sense. However, the approximations are reasonable and, although the modelling of structures can be extended to include flexible joints, this is not necessary in routine analysis. It is appropriate though to model some structures as combinations of truss elements and frame or flexural elements, and it is in this context that the structures are referred to as composite structures.

A classical problem that falls into the category is the case of a beam on an elastic foundation. If a strip footing is supported on a sub-grade that is considered to be elastic, then it can be modelled as a beam supported on a foundation which will develop a continuous reaction proportional to its deflection. The problem was first studied by E. Winkler in 1867 and such a beam is sometimes referred to as a Winkler Beam. Various results based on the closed form solution to the governing differential equation have been published over the years. The problem as defined in figure 6.11(a) may be modelled by the beam and spring system indicated in figure 6.11(b). In this case the springs are assigned a stiffness corresponding to the sub-grade modulus of elasticity. The spring element is equivalent to a truss element with the spring stiffness, k, equal to the axial rigidity, EA/L.

Some discretion must be used in the selection of the number of elements since the beam is modelled, in figure 6.11(b), with support at finite intervals, rather than the continuous support provided by the sub-grade. The model of figure 6.11(b) is shown as a typical arrangement, and a sensitivity analysis would have to be carried out in a given problem to determine an adequate interval for the springs. The model is of course only valid for transverse loads on the beam. At this stage, interest is in the assembly of the structure stiffness matrix once the model has been defined.

Consider the behaviour of a typical spring element as shown in figure 6.11(c). The element action-displacement relationship is given by the equation

$$\begin{Bmatrix} F_i \\ \hline F_j \end{Bmatrix} \begin{bmatrix} k & \vert & -k \\ \hline -k & \vert & k \end{bmatrix} \begin{Bmatrix} d_i \\ \hline d_j \end{Bmatrix} \tag{6.17}$$

Equation (6.17) has the same form as equation (6.2) used in the development of the truss element stiffness matrix. Details of the nodal loads and displacements at nodes 2 and 6 of the model of figure 6.11(b) are shown in figure 6.11(d), and this suggests the relationship between the terms of both the continuous beam element stiffness matrices, representing elements 1 to 3, and the spring element stiffness matrices. The full assembly of the structure stiffness matrix is given in the block diagram of figure 6.11(e). It may be noted from that figure that, for a typical spring element, say element 5, the element stiffness matrix is expanded in the relationship

$$\begin{array}{c} \\ 2 \\ 6 \end{array}\overset{\begin{array}{cc} 2 & 6 \end{array}}{\begin{bmatrix} k & -k \\ -k & k \end{bmatrix}} \qquad \begin{array}{c} \\ 2 \\ \\ \\ 6 \end{array}\overset{\begin{array}{ccccc} 2 & & & & 6 \end{array}}{\left[\begin{array}{cc|c|c} k & 0 & 0 & -k \\ 0 & 0 & & 0 \\ \hline \multicolumn{2}{c|}{0} & 0 & 0 \\ \hline -k & 0 & 0 & k \end{array}\right]}$$

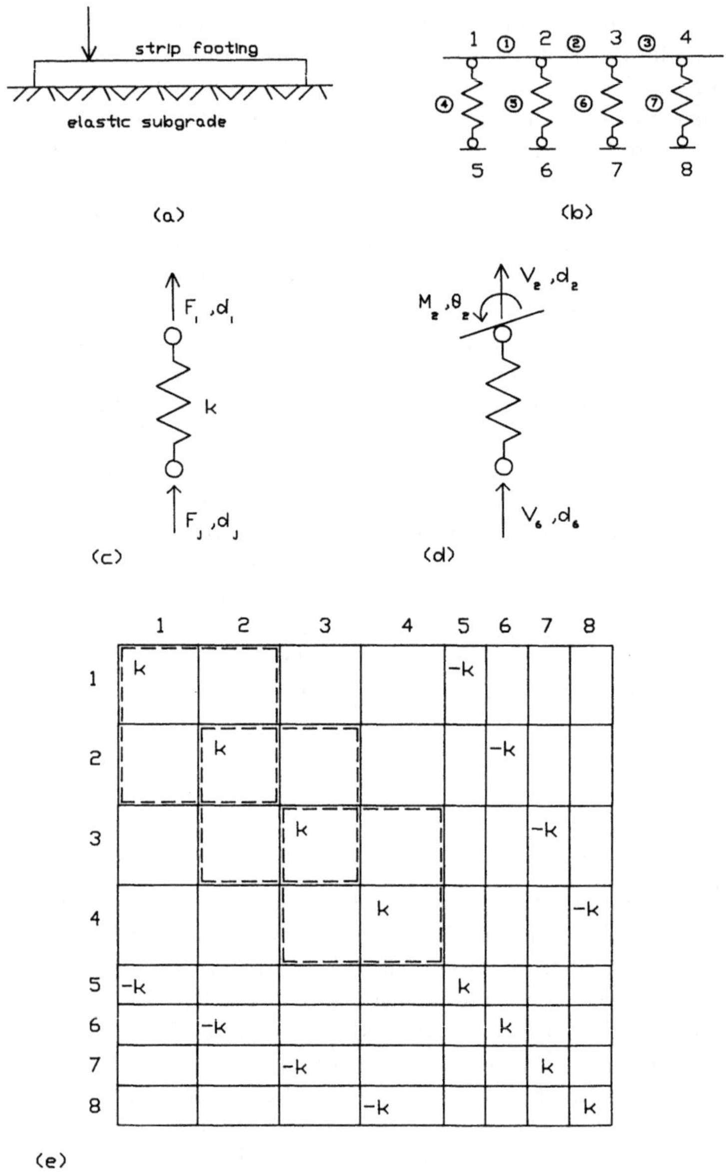

Figure 6.11 *Beam on elastic foundation.*

to facilitate its assembly into the structure stiffness matrix. Of course the boundary conditions are that the displacements, d_5 to d_8, are zero, and interest centres on the structure stiffness matrix, K_F, related to the unrestrained degrees of freedom from which the solution can proceed. A detailed study of a beam on an elastic foundation is presented in example 6.3.

Example 6.3: Beam on an Elastic Foundation

Given data:

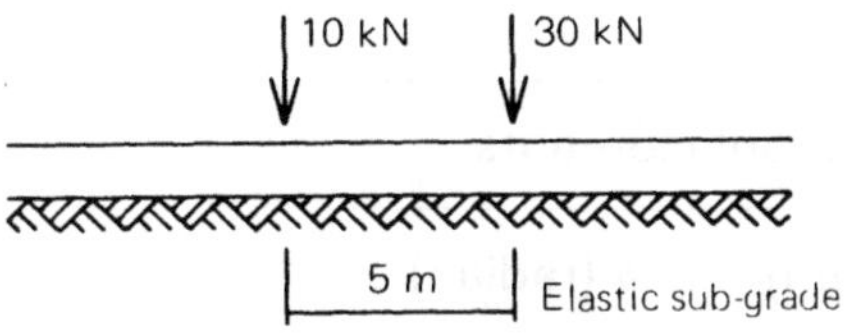

Model as:

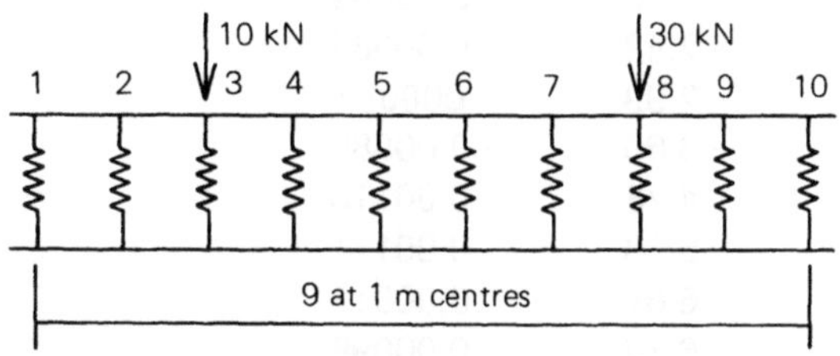

Beam $EI = 10\text{MN m}^2$
Spring $k = 1$ MN/m

Procedure: Form beam element stiffness matrix; assemble to give k_F with beam elements and spring stiffness. Solve for displacements, recover element actions.
Beam Element Stiffness Matrices:
All elements:

$$\text{ESM} = \begin{bmatrix} 120 & 60 & -120 & 60 \\ 60 & 40 & -60 & 20 \\ -120 & -60 & 120 & -60 \\ 60 & 20 & -60 & 40 \end{bmatrix} \quad \text{(Units: MN and m)}$$

Structure Stiffness Matrix:

$$\begin{bmatrix}
121 & 60 & -120 & 60 & & & & & & & & & & & & & & & & \\
60 & 40 & -60 & 20 & & & & & & & & & & & & & & & & \\
-120 & -60 & 241 & 0 & -120 & 60 & & & & & & & & & & & & & & \\
60 & 20 & 0 & 80 & -60 & 20 & & & & & & & & & & & & & & \\
 & & -120 & -60 & 241 & 0 & -120 & 60 & & & & & & & & & & & & \\
 & & 60 & 20 & 0 & 80 & -60 & 20 & & & & & & & & & & & & \\
 & & & & -120 & -60 & 241 & 0 & -120 & 60 & & & & & & & & & & \\
 & & & & 60 & 20 & 0 & 80 & -60 & 20 & & & & & & & & & & \\
 & & & & & & -120 & -60 & 241 & 0 & -120 & 60 & & & & & & & & \\
 & & & & & & 60 & 20 & 0 & 80 & -60 & 20 & & & & & & & & \\
 & & & & & & & & -120 & -60 & 241 & 0 & -120 & 60 & & & & & & \\
 & & & & & & & & 60 & 20 & 0 & 80 & -60 & 20 & & & & & & \\
 & & & & & & & & & & -120 & -60 & 241 & 0 & -120 & 60 & & & & \\
 & & & & & & & & & & 60 & 20 & 0 & 80 & -60 & 20 & & & & \\
 & & & & & & & & & & & & -120 & -60 & 241 & 0 & -120 & 60 & & \\
 & & & & & & & & & & & & 60 & 20 & 0 & 80 & -60 & 20 & & \\
 & & & & & & & & & & & & & & -120 & -60 & 241 & 0 & -120 & 60 \\
 & & & & & & & & & & & & & & 60 & 20 & 0 & 80 & -60 & 20 \\
 & & & & & & & & & & & & & & & & -120 & -60 & 121 & -60 \\
 & & & & & & & & & & & & & & & & 60 & 20 & -60 & 40
\end{bmatrix}$$

Solution for displacements: The structure stiffness matrix has been formed as K_F. The applied loads are immediately nodal loads and the solution follows from MATOP as:

```
LOAD.K.20.20
LOAD.P.20.1
SOLVE.K.P
PRINT.P
-.794184E+00
-.884034E+00
-.166498E+01
-.844325E+00
-.242861E+01
-.641948E+00
-.299309E+01
-.572224E+00
-.369358E+01
-.864069E+00
-.473031E+01
-.118315E+01
-.592856E+01
-.110826E+01
-.663787E+01
-,106478E+00
-.612310E+01
0.950534E+00
-.500571E+01
0.120082E+01
End of File
```

Node	Displacements	
	d (mm)	θ (radian)
1	−0.79	−0.00088
2	−1.66	−0.00084
3	−2.43	−0.00064
4	−2.99	−000057
5	−3.69	−0.00086
6	−4.73	−0.00118
7	−5.93	−0.00111
8	−6.64	−0.00011
9	−6.12	0.00095
10	−5.01	0.00012

Note that particular care must be taken to ensure that consistent units are used.

Element Actions: A typical solution is that for element ⑦ (node 7–8)

$$\begin{Bmatrix} v_{78} \\ m_{78} \\ v_{87} \\ m_{87} \end{Bmatrix} = \begin{bmatrix} 120 & 60 & -120 & 60 \\ 60 & 40 & -60 & 20 \\ -120 & -60 & 120 & -60 \\ 60 & 20 & -60 & 40 \end{bmatrix} \begin{Bmatrix} -5.928\times10^{-3} \\ -1.10826\times10^{-3} \\ -6.6378\times10^{-3} \\ -0.10647\times10^{-3} \end{Bmatrix}$$

$$= \begin{Bmatrix} 12.23\times10^{-3} \\ -3.90\times10^{-3} \\ -12.23\times10^{-3} \\ 16.13\times10^{-3} \end{Bmatrix}$$

(Shear MN; moment MN m)

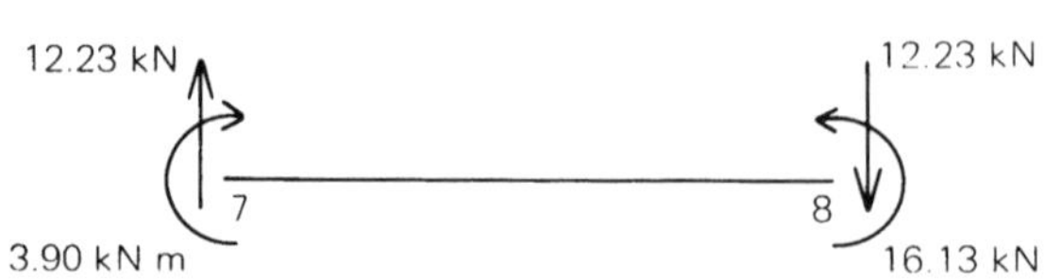

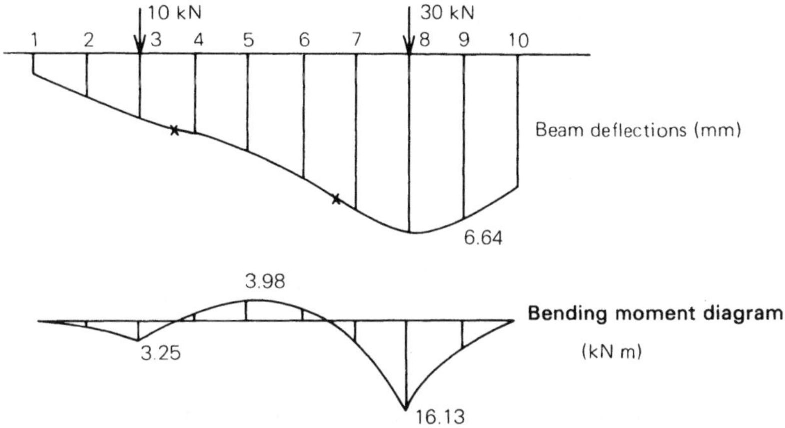

6.4 PROBLEMS FOR SOLUTION

6.1 Using the general stiffness method, form the structure stiffness matrix for the truss of figure P6.1 and hence analyse the structure to find the displacements at each node.

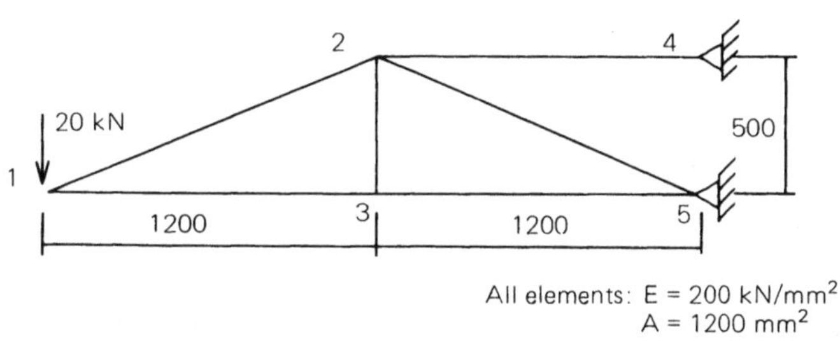

Figure P6.1.

6.2 Modify the structure of figure P6.1 by the addition of an element 3–4 with the same properties as those given. Compare the resulting structure stiffness matrix with that of problem **6.1**, and analyse the truss to find the displacements and the internal forces.

6.3 Figure P6.2 shows a cranked beam modelled with three elements. Using the general plane frame element, form the transformed element stiffness matrix for each element and assemble the structure stiffness matrix. Calculate the displacements at nodes 2 and 3 and determine the beam bending moment diagram.

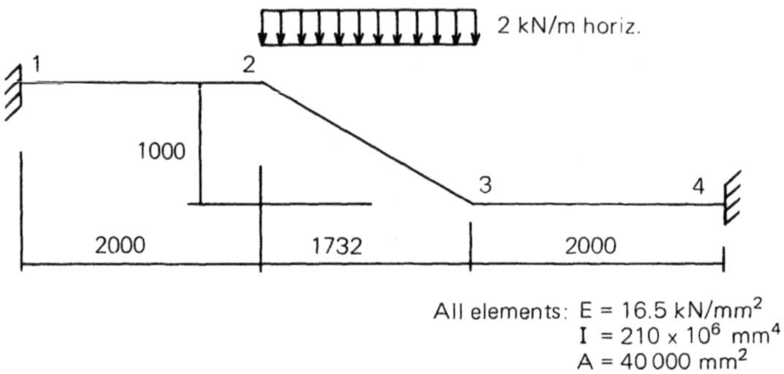

Figure P6.2.

6.4 Modify the problem presented as example 6.2 by introducing fixed bases at nodes 1 and 5 and a horizontal restraint at node 4, and analyse the resulting structure.

6.5 Analyse the two hinged gable roof structure shown in figure P6.3 by the general stiffness method to determine the reactions and the bending moment diagram.

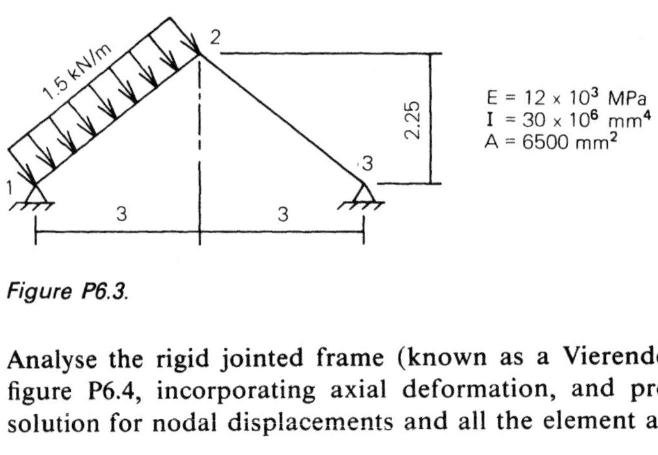

Figure P6.3.

6.6 Analyse the rigid jointed frame (known as a Vierendeel Truss) of figure P6.4, incorporating axial deformation, and proceed to the solution for nodal displacements and all the element actions.

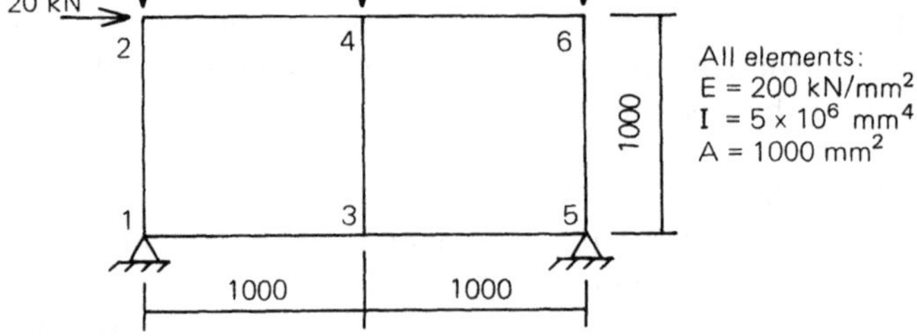

Figure P6.4.

6.7 (a) Calculate the nodal loads that would be used in a load vector for the matrix analysis of the frame of figure P6.4, if the frame carried a uniformly distributed load of 15 kN/m along the beams '2–4' and '4–6' and 7.5 kN/m along the beams '1–3' and '3–5', in addition to the loads shown.

(b) Some of the results of a matrix analysis using the load vector of part (a) are shown in table P6.1. Using these results, draw the bending moment diagram for the frame. The results have not been combined with any fixed end moments and the given end moments are anticlockwise positive.

Table P6.1

Element		Moment (kN m)	
A	B	End A	End B
1	3	12.52	18.94
3	5	−23.35	−18.17
2	4	11.77	18.02
4	6	−22.38	−17.35

6.8 The beam of figure P6.5 is fully restrained at node 1 and supported at node 2 by one linear spring of stiffness k_1, and one rotational spring of stiffness k_2.

(a) Form the structure stiffness matrix clearly, showing the influence of the spring stiffnesses, k_1 and k_2.

(b) Given that $k_1 = 3000$ kN/m and that $k_2 = 2000$ kN m/radian, determine the beam bending moment diagram and actions in the springs.

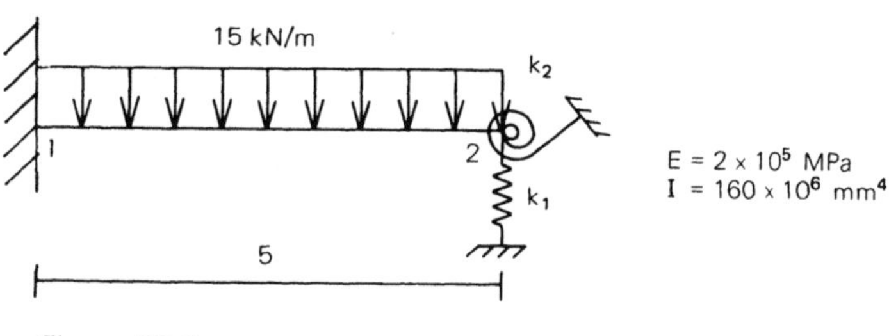

Figure P6.5.

6.9 The two cantilever beams of figure P6.6 are connected by a linear spring between nodes 2 and 3. Analyse the system to determine the bending moment diagram for each beam and the force in the spring.

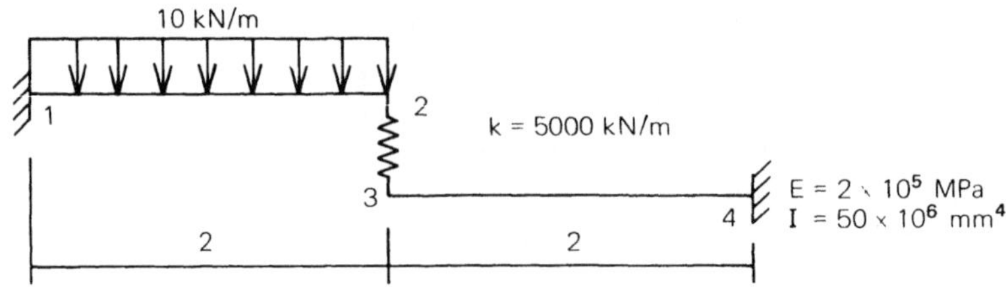

Figure P6.6.

6.10 Modify the structure of figure P6.4 by introducing diagonal bracing in both directions in both panels. Assume that the bracing is pin-connected to the frame elements with properties $E = 200\ \text{kN/mm}^2$ and $A = 500\ \text{mm}^2$, and that the bracing elements are not connected where they cross. Analyse the resulting structure by modifying the structure stiffness matrix of problem **6.6** and compare the results.

REFERENCE

Zienkiewicz, O. C. (1971). *The Finite Element Method in Engineering Science*, McGraw-Hill, London.

FURTHER READING

Jenkins, W. M., *Structural Mechanics and Analysis*, Van Nostrand Reinhold, New York, 1982.

Willens, N. and Lucas, W. M., *Structural Analysis for Engineers*, McGraw-Hill, New York, 1978.

Chapter 7
The Principle of Virtual Work

The methods of structural analysis presented in the previous chapters have been based on relationships between internal actions and displacements of the elements of a structure, and on relationships between the external loads and the internal actions. The relationships have been based on two fundamental requirements in the analysis of static structural systems: firstly that the structure must be in equilibrium, and secondly that the deformation of the structure must be compatible and meet the boundary conditions. Equilibrium has been satisfied by applying the equations of statics, while compatibility has been based on the geometry of the elastic curve or the geometry of the deformed element. The resulting techniques have been general enough to cover a wide range of structural analysis problems for both statically determinate and statically indeterminate systems.

However there is an alternative approach to deriving the fundamental equations of structural behaviour, based on a consideration of work done by forces acting through displacements. Such a consideration leads into the principle of virtual work and its two parts: the principle of virtual displacements and the principle of virtual forces. The principle of virtual work is perhaps one of the most fundamental principles in mechanics. It is applicable to both non-linear and inelastic systems, and, as stated by Malvern (1969), it is not dependent on the conservation of energy. The principle has been widely used in studies of elastic structural analysis and in this context, where the principle of conservation of energy applies, the principle of virtual work can be seen to be underlying what are collectively known as the energy theorems in structural analysis. While the major emphasis of this chapter is on the application of the principle of virtual forces to the deflection of linear elastic structures, the general nature of the principle should not be overlooked.

7.1 WORK CONCEPTS

Work is equal to the product of a force and the displacement of its point of application in the direction of the force. More precisely, the differential work dW, done by a force F, in moving through a differential displacement ds, in the direction of F, is defined as

$$dW = F\,ds$$

If the displacement occurs from point A to point B then the work done, W, is

$$W = \int_A^B F\,ds \tag{7.1}$$

Further, if the force is constant during a displacement Δ in the direction of the force, then

$$W = F\Delta \tag{7.2}$$

Although equation (7.2) is simply a special case of equation (7.1), it is important to distinguish between the two expressions. In many applications in structural mechanics, the displacement is as a result of the force acting. For instance, in applying a gravity load to the end of a deformable rod arranged vertically as a pin-ended bar, the bar will extend as the load is applied and continue to do so until the elongation of the bar develops an internal force to balance the external load and the system is in equilibrium. In such a case the force is simply not constant throughout the displacement Δ. The effect is illustrated for the elastic rod of figure 7.1(a) with the

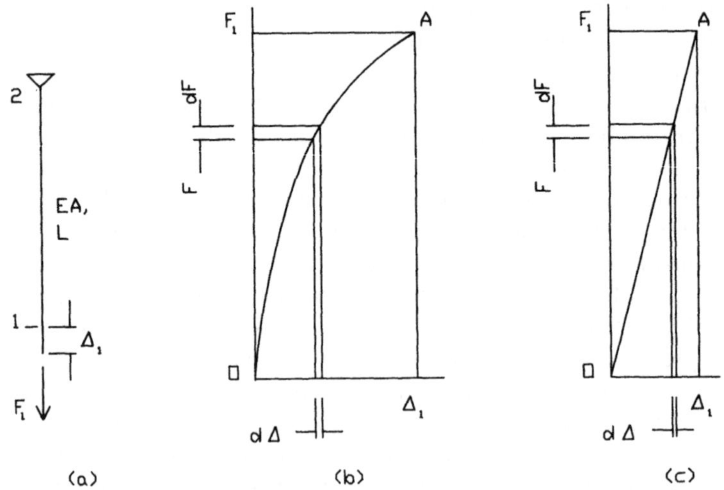

Figure 7.1 *Force–extension relationship in an elastic rod.*

force–extension relationship shown for both the non-linear and linear case in figures 7.1(b) and 7.1(c) respectively. In both cases the work done is equal to the area under the curve OA.

7.2. THE PRINCIPLE OF VIRTUAL DISPLACEMENTS

An important abstract extension to the work principle was introduced by the early 18th Century mathematician John Bernoulli, when he put forward the principle of virtual displacements. Prior to that time, work principles were used as the basis for the calculations of static equilibrium and were applied, for example, to the mechanics of a simple lever. Bernoulli's contribution was his emphasis on the independence of the force and the displacement systems in considering work; hence the term virtual work. The word 'virtual' is widely understood to mean 'in effect but not in fact', and in this context virtual displacements refer to a set of imaginary displacements caused by some agency independent of the forces acting on the system.

7.2.1 The Principle of Virtual Displacements applied to a Rigid Body

The principle will be introduced as it applies to a rigid body before going on to the application to deformable bodies. Consider a rigid body in three-dimensional space in equilibrium under the action of a set of generalised forces designated by Q as shown in figure 7.2. Suppose the body is then subjected to a small virtual displacement during which the forces remain acting in their original directions and equilibrium is maintained. The virtual displacement of the body which is a rigid body movement may be defined by the set of displacements Δ, with each term corresponding to a displacement in the direction of a force Q. Since the forces are constant during the virtual displacement, the total amount of virtual work done is given by

$$W_{\mathrm{T}} = \sum Q_i \Delta_i \tag{7.3}$$

Each generalized force, Q_i, on the body may be resolved into three force and three moment components with respect to the coordinate axes taken through an origin O, and designated as

$$F_{ix}, F_{iy}, F_{iz}, M_{ix}, M_{iy} \text{ and } M_{iz}$$

These components may then be transformed into an equivalent set of actions acting through the origin of coordinates and described as

$$\sum F_{ix}, \sum F_{iy}, \sum F_{iz}, \sum M'_{ix}, \sum M'_{iy} \text{ and } \sum M'_{iz}$$

The prime notation has been introduced since, although the moments

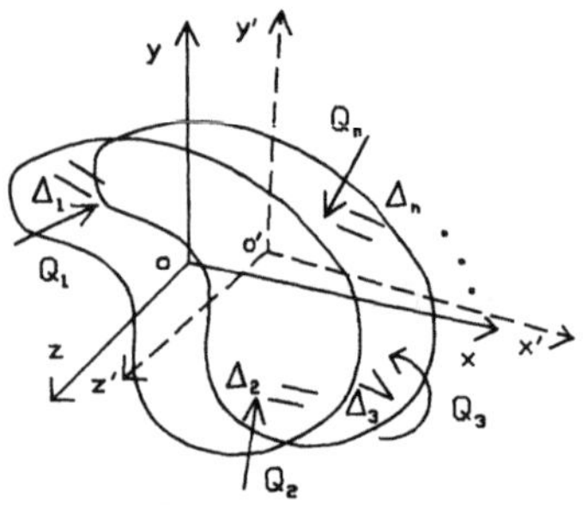

Figure 7.2 Virtual work done on a rigid body.

transform directly, in general the force components transform with a force and a moment component.

On the other hand, any virtual displacement can be defined by the displacement of the origin of the coordinate axes system, expressed by its six components

$$\delta_{ox}, \delta_{oy}, \delta_{oz}, \alpha_{ox}, \alpha_{oy} \text{ and } \alpha_{oz}$$

It should be noted that any virtual displacement term Δ_i is necessarily geometrically related to the defined displacement of the origin. The total virtual work done is now seen as the sum of six expressions and is given by

$$\begin{aligned} W_T = &(\sum F_{ix})\delta_{ox} + (\sum F_{iy})\delta_{oy} + (\sum F_{iz})\delta_{oz} \\ &+ (\sum M'_{ix})\alpha_{ox} + (\sum M'_{iy})\alpha_{oy} + (\sum M'_{iz})\alpha_{oz} \end{aligned} \tag{7.4}$$

However, in order to satisfy the equations of static equilibrium, each of the summations in equation (7.4) must be zero, so that the total virtual work done is zero.

The principle of virtual displacements as applied to a rigid body may be formally stated as:

> *If a rigid body, in equilibrium under the action of a system of forces Q, is subjected to a virtual displacement, then the virtual work done by the Q force system is zero*

The converse of the principle of virtual displacements as stated, may be used to establish equilibrium conditions as an alternative to using the equations of statics. If the virtual work done on a rigid body is equated to zero, the resulting expression must be an equilibrium condition if the body is to be in equilibrium. Adequate structural systems are constrained and at first sight it appears difficult to see how any virtual displacement can be introduced to a constrained rigid body. However each constraint can be replaced with the reactive force associated with it so that a kinematic boundary condition (that is, a prescribed displacement) is replaced with a static boundary condition, namely the reactive force on the boundary. Such action effectively releases the system and it is possible to be selective about the number of releases so as to establish data about specific reactions. For two-dimensional systems, three independent rigid body movements can be introduced as virtual displacements corresponding to the three equations of equilibrium available from statics. As an example, consider the beam of Figure 7.3(a). The kinematic boundary condition of zero vertical displacement at node 2 can be replaced with the reaction Y_2 and a small virtual displacement Δ can be introduced in the direction of Y_2. The resulting virtual displacement diagram is shown in figure 7.3(b). From the principle of virtual displacements it is then necessary that

$$-20(0.5\Delta) + Y_2(\Delta) - 12(0.75\Delta) = 0$$

that is

$$Y_2 = 19 \text{ kN}$$

Other constraint releases can be introduced, with corresponding virtual displacement patterns, to establish the remaining reactions.

It is important to note that in all cases, all other displacements throughout the system are necessarily related to the specified virtual displacement through geometry. Further, although compatibility appears to have been violated in the virtual displacements, this is not the case. In figure 7.3(b), a kinematic boundary condition has been replaced with a static boundary condition and all kinematic and static boundary conditions have been met. Virtual displacement patterns established under these conditions can be described as geometrically compatible displacements.

Equilibrium equations involving internal actions can be established in a similar manner. In this case, the constraint may be regarded as the

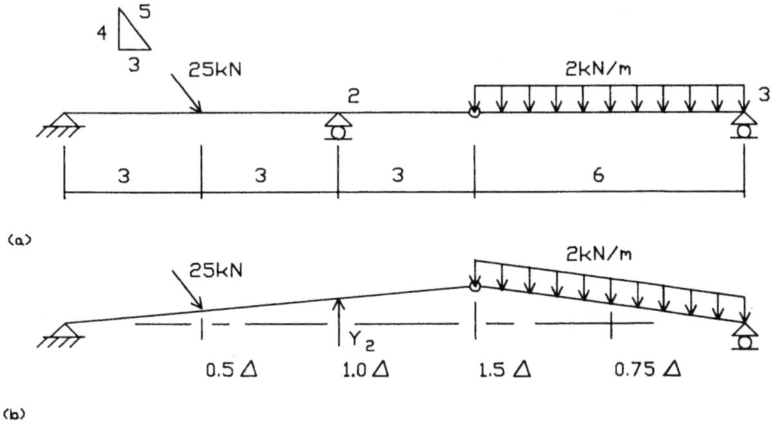

Figure 7.3 Reactions from principle of virtual displacements.

compatibility of the element associated with the type of action. The structure can be released from this constraint, which is then replaced with the appropriate internal action acting as an external static condition on the introduced boundary. For axial force, the element can be allowed to slip in the axial direction; for shear force, the element can slip in the transverse direction; and for moment, the element can be allowed to rotate. The release of internal actions and some corresponding displacement patterns are shown in figure 7.4 for a simply supported beam. Each of the patterns are geometrically compatible since all of the virtual displacements throughout the structure can be expressed in terms of the specified virtual displacement, Δ. With the release of the shearing constraint only, moment continuity must be preserved and this is met by having the same slope on either side of the shear release. The virtual displacements of figures 7.4(b) and (c) both satisfy this requirement, although a more useful work equation can be derived from figure 7.4(c) since only the shear force V_X and the Q load do virtual work. In figure 7.4(b), the reaction Y_1 is also involved in the work equation. A similar situation occurs with the virtual displacements of figures 7.4(d) and 7.4(e) where the latter figure generally proves more useful.

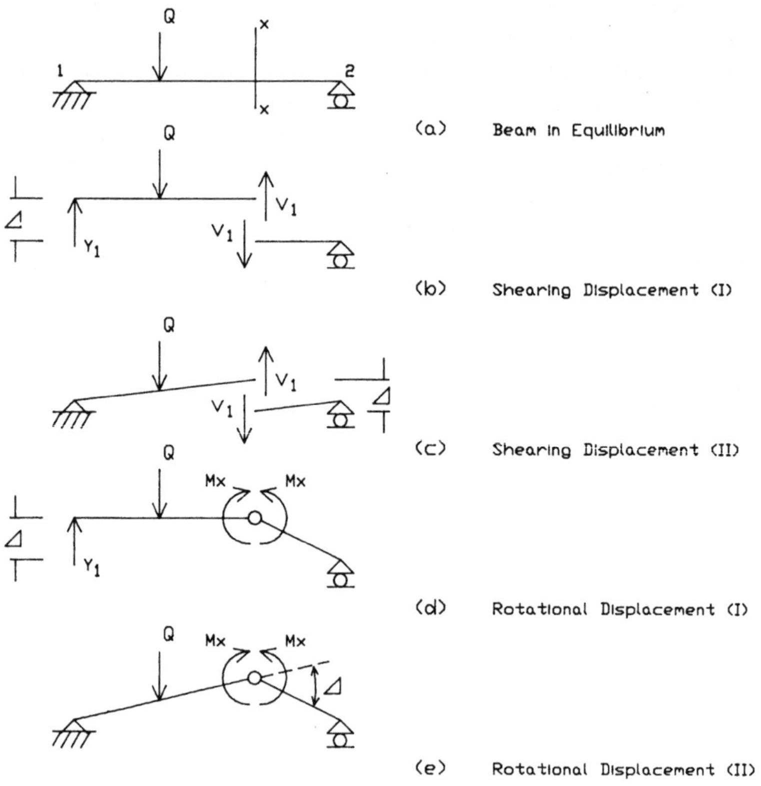

Figure 7.4 Virtual displacement patterns for a beam.

The application of the principle of virtual displacements is not restricted to statically determinate systems. In fact the principle can be used to advantage with statically indeterminate systems where several internal actions are known and reactions are required. Such a case arises in the analysis of frames by the moment distribution method, as shown by example 5.2 of chapter 5. In the initial part of the analysis, the frame is prevented from swaying by the introduced reaction at node 3 that must be calculated. The problem is solved in example 7.1 using some of the results from example 5.2.

Example 7.1: Reactions in an Indeterminate Frame

Given data:

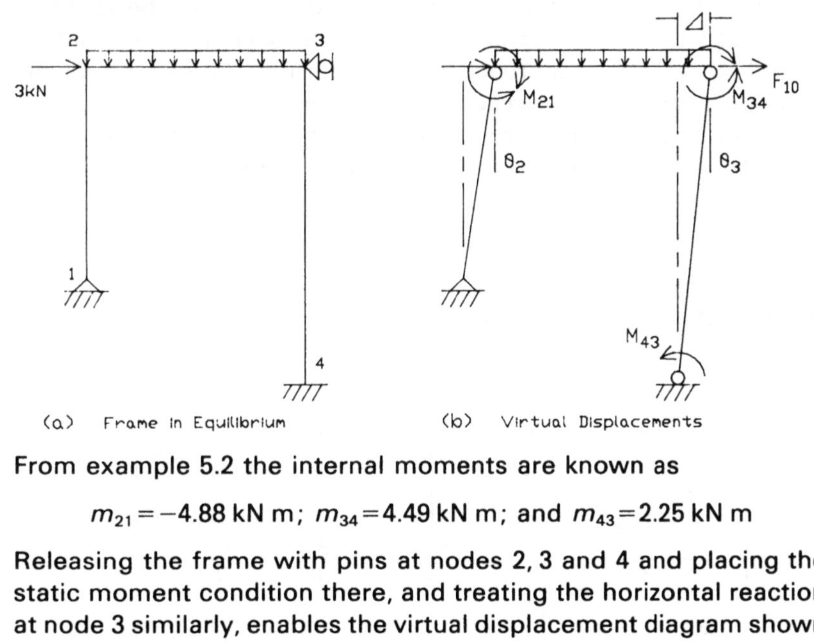

From example 5.2 the internal moments are known as

$$m_{21}=-4.88 \text{ kN m};\ m_{34}=4.49 \text{ kN m};\ \text{and}\ m_{43}=2.25 \text{ kN m}$$

Releasing the frame with pins at nodes 2, 3 and 4 and placing the static moment condition there, and treating the horizontal reaction at node 3 similarly, enables the virtual displacement diagram shown to be introduced.

By the principle of virtual displacements, the virtual work done must be zero, hence

$$3\Delta + F_{10}\Delta - m_{21}\theta_2 - m_{34}\theta_3 - m_{43}\theta_4 = 0$$

but

$\theta_2=\Delta/4$ and $\theta_3=\theta_4=\Delta/6$ for small Δ, so that

$$3\Delta + F_{10}\Delta + 4.88\Delta/4 - 4.49\Delta/6 - 2.25\Delta/6 = 0$$

$$\therefore\quad F_{10} = -3.10 \text{ kN}$$

For a statically indeterminate system, it is of course necessary to introduce a sufficient number of releases, either internal or external, to allow the resulting assemblage of rigid elements to act as a mechanism.

Example 7.2 demonstrates an application in the analysis of trusses. For any statically indeterminate truss, any internal force can be expressed in terms of the external loads and the reactions. The procedure involves releasing the nominated element by a cut, permitting longitudinal slip, and replacing the continuity by the static condition of the internal force. As with all such releases, the static condition must act on both cut faces. Of course the value of example 7.2 can be questioned since Y_{13} must be known before F_{35} can be calculated. However, the example has been introduced in order to demonstrate principles more than anything else. A solution for the force is immediately available for the statically determinate version of the truss where the support at node 13 is removed. It can then be seen that virtual work equation of example 7.2 achieves the same type of result as would a method of sections approach from equilibrium. As an approximate analysis for the given truss, the reaction Y_{13} can be estimated by assuming that the structure acts like a propped cantilever beam. In this case, Y_{13} is readily calculated as 12.5 kN giving the approximate value of F_{35} as 17.5 kN.

Example 7.2: Forces in an Indeterminate Truss

Given data:

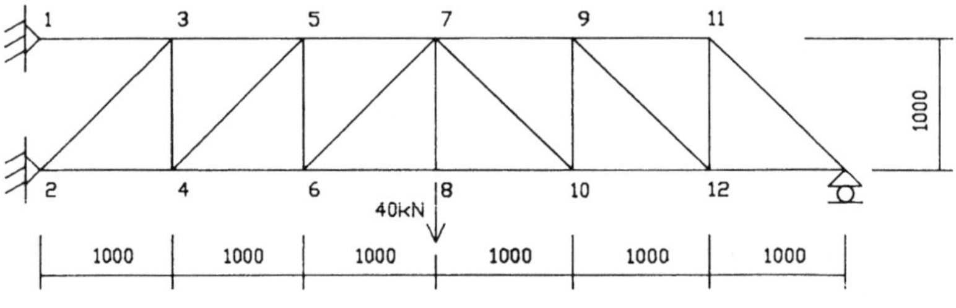

with reaction at node 13 known as 13.5 kN.

To find the force in element 3–5, release the force by a cut and place the static force F_{35} there. Similarly replace the kinematic boundary condition at node 13 with the reactive force Y_{13}. Develop the following virtual displacement diagram by introducing a small displacement of say 1 mm at the cut, causing a rotation of the rest of the structure about node 4. From the geometry of the structure, nodes 8 and 13 drop by 2 and 5 mm respectively.

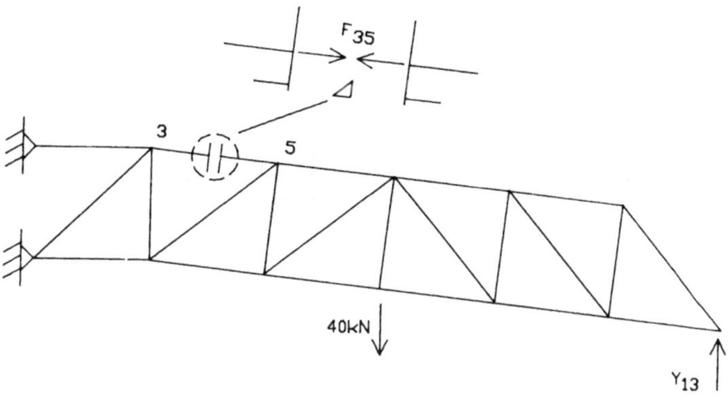

By the principle of virtual displacements, the virtual work done must be zero, hence

$$-F_{35}(1)+40(2)-Y_{13}(5)=0$$

$$\therefore F_{35}=40(2)-Y_{13}(5)$$

$$=12.5\text{ kN}$$

7.2.2 The Principle of Virtual Displacements applied to a Deformable Body

In the preceding section, it has been assumed that all of the structural systems have behaved as rigid bodies during the virtual displacements, that is, the virtual displacements have been rigid body displacements. In fact, the systems may well be deformable but the deformation of the systems as a result of the loads applied has not been a consideration. The rigid body assumption is simply consistent with the application of the equations of statics where changes of geometry under the loads are ignored. Nevertheless, any compatible displacement pattern of a deformable body in equilibrium can be used as a virtual displacement pattern. When such a virtual displacement is used then the question of virtual work of the internal stresses in the body arises. This leads to the principle of virtual displacements for a deformable body, which applies to any structural system although it is developed here for a planar structure for simplicity.

Consider now a planar deformable body in equilibrium under the action of a set of generalized forces, Q, as shown in figure 7.5(a). Over any interface, the internal stresses may be described as the set q typically illustrated by figure 7.5(b), where an element from the body has been isolated as a free body diagram. Suppose the body is now subjected to a virtual displacement which, because the body is a deformable one, involves both

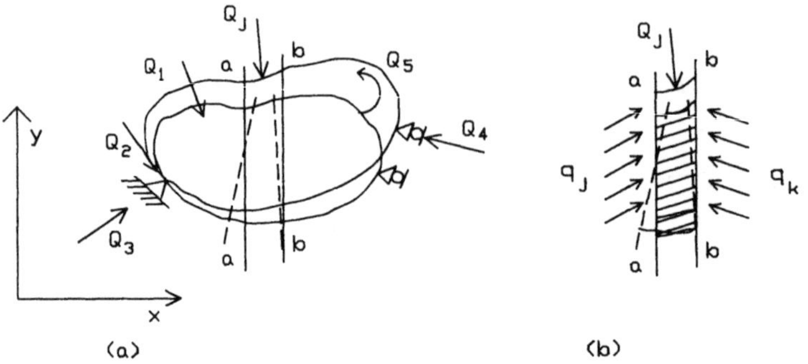

Figure 7.5 *Virtual work done on a deformable body.*

external displacements, say Δ, and internal displacements, δ. Both the external forces and the internal stresses (as stress resultants) do virtual work.

Any displacement of a deformable body can be described in two parts. The first of these can be defined as a rigid body movement to be followed by deformation, involving the internal stresses, to take up the final form. This is illustrated in figure 7.5(b) for the element from figure 7.5(a). Similarly, the virtual work of a deformable body can be described in two parts, namely that done during rigid body movement and that done during deformation. The total amount of virtual work done by the system can be represented as W_T. Although both internal stresses and external loads are involved, the summation of work done by Q and q to find W_T will only involve the external loads Q. This is because, for every interface in the body where the stresses q are defined, equal and opposite stresses acting on adjacent boundaries necessarily undergo the same displacements. Thus the total virtual work done by the internal stresses cancels out, leaving the total amount of virtual work done by the system as

$$W_T = \sum Q_i \Delta_i \tag{7.5}$$

If the virtual work done by the system during deformation is denoted as W_E, then the virtual work done by the body during rigid body movement must be given by $W_T - W_E$. However, by the principle of virtual displacements as applied to a rigid body, this work must be zero, so that

$$W_T = W_E \tag{7.6}$$

The virtual work done during deformation is done by the internal stresses and it may be written as

$$W_E = \sum q_i \delta_i \tag{7.7}$$

Thus, equation (7.6) becomes

$$\sum Q_i \Delta_i = \sum q_i \delta_i \tag{7.8}$$

which is an expression of the principle of virtual displacements as it applies to a deformable body. The principle may be formally stated as:

> *If a deformable body, in equilibrium under the action of a system of forces Q, is subjected to a virtual displacement, then the external virtual work done by the Q force system is equal to the internal virtual work done by the internal stresses q*

Table 7.1 Internal actions on various elements.

Type of deformation	Internal forces	Internal displacements
Axial	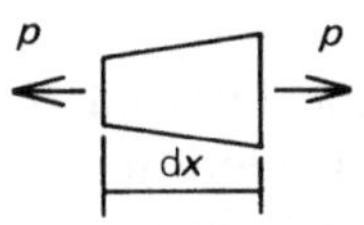 $q = p$	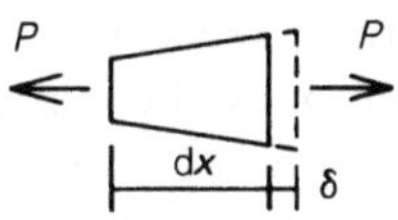 $\delta = \varepsilon\, dx = \frac{\sigma}{E} dx = \frac{P}{EA} dx$
Bending	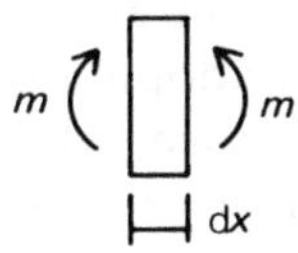 $q = m$	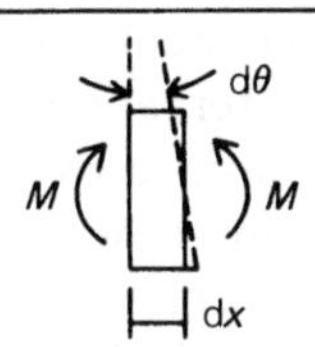 $\delta = d\theta = \frac{M}{EI} dx$
Shearing	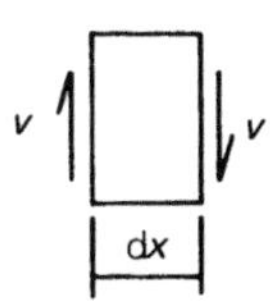 $q = v$	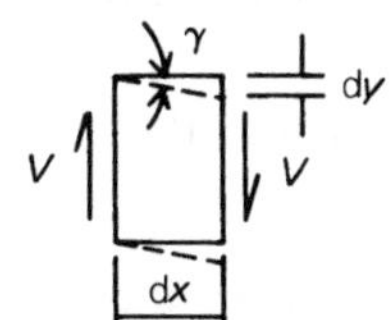 $\delta = dy = \gamma dx = \frac{\tau}{G} dx$ $= \kappa \frac{V}{GA} dx$
Torsional	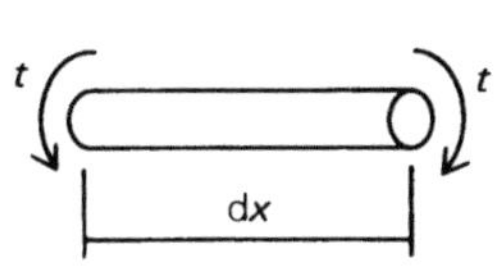 $q = t$	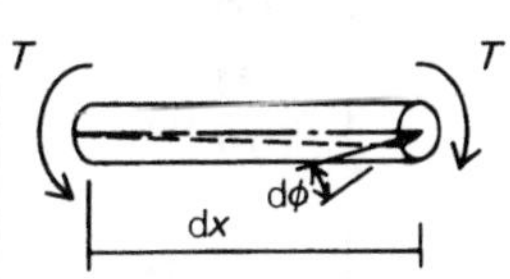 $\delta = d\phi = \frac{T}{GJ} dx$ (circular section)

It remains now for suitable expressions to be found for internal virtual work. While the principle of virtual displacements is quite general, further consideration will only be given to its application to linear elastic systems. Once the deformation behaviour of the system is specified, the internal deformations follow from the nature of the external deformation. For one-dimensional structural elements, four types of deformation are involved: namely, axial, flexural, shearing and torsional deformation. In order to calculate the internal virtual work, each of these deformations must be linked with the corresponding, but independent, internal stress due to axial force, bending moment, shear force and torsional moment. Since the resultant stress effects under such actions are well known, it is more convenient to look at the internal force and the nett effect of the internal deformation acting on a section.

Table 7.1 summarizes the nature of the internal force and internal displacement for the four actions under consideration. In table 7.1, each of the internal displacements has been also linked to an internal action associated with an applied load. While this is a useful relationship that will be used extensively, it is not a necessary one. Internal deformations do not have to be caused by applied loads; they may be due to temperature or lack of fit of specified elements, as will be seen. The independence of the internal forces and the internal displacements, central to the whole concept of virtual work, is emphasized by the notation adopted in table 7.1, where one set of internal actions is written in lower case, while the other set is written in upper case. Under specific circumstances, the required expressions for internal virtual work can be found from the data of table 7.1 by integrating over the length of the elements of the structure. For example, provided the virtual displacement is assumed to be associated with some bending moment, M, and it is applied to a flexural system already under the action of another

Table 7.2 Internal virtual work expressions

Type of deformation	*Internal virtual work*
Axial	$\int p \frac{P}{EA} \mathrm{d}x$
Bending	$\int m \frac{M}{EI} \mathrm{d}x$
Shearing	$\kappa \int v \frac{V}{GA} \mathrm{d}x$
Torsional	$\int t \frac{T}{GJ} \mathrm{d}x$

bending moment, m, then the internal virtual work due to bending is given by

$$\sum q\delta = \int m \frac{M}{EI} \, dx \tag{7.9}$$

The integral of equation (7.9) is to be interpreted as the sum of all such integrals taken over the length of each element making up the structure. Other integral expressions for internal virtual work, applicable for similar conditions, are summarized in table 7.2.

7.2.3 A Mathematical Illustration of the Principle of Virtual Displacements

The validity of the principle of virtual displacements may be demonstrated in the behaviour of a simply supported linear elastic beam carrying a uniformly distributed load of intensity q over a span L. Denoting the deflection at any point along the beam as v, and introducing a small virtual displacement δv there, so that both v and δv follow elastic curves and meet the compatibility requirements, then equation (7.8) may be written as

$$\int_0^L q(\delta v) \, dx - \int_0^L m \frac{M}{EI} \, dx = 0 \tag{7.10}$$

The first integral of equation (7.10) represents the external virtual work while the second is the internal virtual work due to bending. The second integral expression may also be written in terms of the displacements using equation (3.10), so that

$$\int_0^L m \frac{M}{EI} \, dx = \int_0^L EI \frac{d^2 v}{dx^2} \left(\frac{d^2 \delta v}{dx^2} \right) dx$$

Integrating by parts results in

$$\begin{aligned} &\int_0^L EI \frac{d^2 v}{dx^2} \left(\frac{d^2 \delta v}{dx^2} \right) dx \\ &\quad = \left[EI \frac{d^2 v}{dx^2} \left(\frac{d\delta v}{dx} \right) \right]_0^L - \int_0^L \left(\frac{d\delta v}{dx} \right) EI \frac{d^3 v}{dx^3} \, dx \\ &\quad = - \int_0^L \left(\frac{d\delta v}{dx} \right) EI \frac{d^3 v}{dx^3} \, dx \end{aligned}$$

where the static boundary condition of zero moment at either end of the span has reduced the term in the square brackets to zero. Integrating again by parts gives

$$-\int_0^L \left(\frac{\mathrm{d}\delta v}{\mathrm{d}x}\right) EI\frac{\mathrm{d}^3 v}{\mathrm{d}x^3}\,\mathrm{d}x$$

$$= -\left[EI\frac{\mathrm{d}^3 v}{\mathrm{d}x^3}\delta v\right]_0^L + \int_0^L \delta v EI\frac{\mathrm{d}^4 v}{\mathrm{d}x^4}\,\mathrm{d}x$$

$$= \int_0^L \delta v EI\frac{\mathrm{d}^4 v}{\mathrm{d}x^4}\,\mathrm{d}x$$

where now the kinematic boundary condition of zero virtual displacement at either end of the span has reduced the term in the square brackets to zero. From simple beam theory

$$\frac{\mathrm{d}^4 v}{\mathrm{d}x^4} = \frac{q(x)}{EI}$$

so that

$$\int_0^L EI\frac{\mathrm{d}^2 v}{\mathrm{d}x^2}\left(\frac{\mathrm{d}^2\delta v}{\mathrm{d}x^2}\right)\mathrm{d}x = \int_0^L q(\delta v)\,\mathrm{d}x$$

which confirms equation (7.10). A rigorous mathematical approach to the principle of virtual work along these lines is given in chapter 8 of Oden and Ripperger (1981).

7.3. THE PRINCIPLE OF VIRTUAL FORCES

Since work involves both forces and displacements, it is reasonable to suppose that some advantage could be gained by considering virtual work on the basis of a virtual force system and the true displacements. The principle of virtual forces arises from such a consideration as an important concept that is complementary to the principle of virtual displacements. The principle may be formally stated as:

> *If a deformable body, in equilibrium under the action of a system of virtual forces Q, is subjected to a set of compatible displacements consistent with the constraints, then the external virtual work done is equal to the internal virtual work*

In practical terms, the principle can be seen as an alternative interpretation of equation (7.8) where the set of loads, Q, and the internal stresses, q, are now regarded as virtual, while the set of external displacements, Δ, and the internal displacements, δ, are the true displacements. The data of tables 7.1 and 7.2 can now be seen as relevant expressions for internal virtual work for both of the virtual work principles.

Considering the expression for external virtual work, which is the left-hand side of equation (7.8), an expansion of the summation involved gives

$$\sum Q_i\Delta_i = Q_1\Delta_1 + Q_2\Delta_2 + Q_3\Delta_3 + \cdots + Q_n\Delta_n \tag{7.11}$$

In the principle of virtual forces, the set of displacements are those caused by the actual loading condition on the structure in question. As has already been mentioned, such displacements may be caused by applied loads but they may also be caused by other effects such as temperature and lack of fit. The set of forces Q are of course the virtual forces which are an imaginary set of forces necessarily in equilibrium on the structure. By a careful selection of the virtual force system, the principle of virtual forces may be used to calculate the deflections at nominated points throughout a structure. For example, consider the deflection of a cantilever beam carrying a uniformly distributed load as shown in figure 7.6(a). The beam will deflect along its length and the vertical deflection at node 2 may be nominated as Δ_1. Suppose now that a virtual force system, comprising a single unit load applied vertically at node 2 and the consequent equilibrating reactive forces Q_2 and Q_3, is applied to the same structure as shown in figure 7.6(b).

Because of the selection of the Q force system, equation (7.11) is simply $1(\Delta_1)$. Internal virtual work is done by the virtual stresses of bending and shear in the single element 1–2 of the structure. Applying equation (7.8) gives

$$\sum Q_i\Delta_i = \sum q\delta$$

which is

$$1(\Delta_1) = \int_0^L m\frac{M}{EI}\,dx + \kappa\int_0^L v\frac{V}{GA}\,dx$$

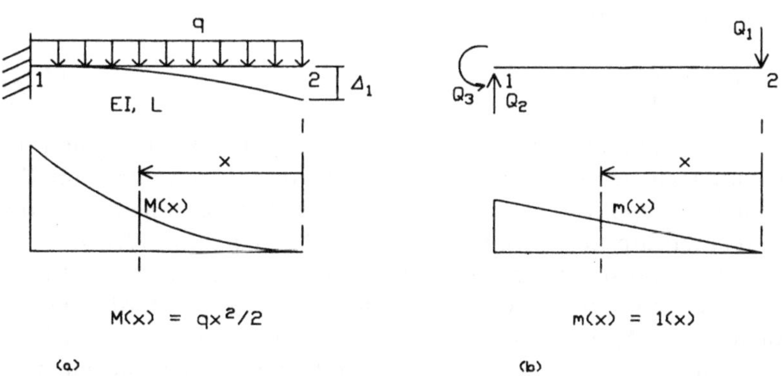

Figure 7.6 Deflection of a cantilever by the principle of virtual forces.

The contribution to the beam deflection due to shear deformation is generally regarded as a second-order effect and may be neglected, so that

$$1(\Delta_1) = \int_0^L m \frac{M}{EI} \mathrm{d}x$$

$$= \int_0^L 1x \frac{qx^2}{2EI} \mathrm{d}x$$

that is

$$\Delta_1 = \frac{qL^4}{8EI}$$

which is the required result.

If the deflection is required at any other point along the beam, it is only necessary to introduce another Q force system involving a unit load in a position corresponding to the deflection sought.

7.3.1 General Application to the Deflection of Frames

Example 7.3 serves to illustrate the general application of the principle of virtual forces to the deflection of frames. In addition, the example introduces the concept of the standard integrals which are summarised in table B1.5 of appendix B. The example shows a light building frame which is statically determinate. The horizontal deflection of the frame at nodes 1, 5 and 6 is to be calculated using the principle of virtual forces. The principal action in the frame is bending, and although there is shearing action and axial deformation, these will not contribute significantly to the deflection. The internal virtual work will thus be considered to be carried out by the virtual bending stresses, in association with the real internal bending deformation throughout the elements. The appropriate integral expression for internal virtual work due to bending is given as

$$\sum q\delta = \int_0^L m \frac{M}{EI} \mathrm{d}x \tag{7.12}$$

where the integral implies the sum of all such integrals over the length of all participating elements of the structure. Interest then centres on the moment diagrams applicable under various loads, since these give the functional relationship to be used in the integral of equation (7.12).

There are several significant points to be made in connection with the use of the principle of virtual forces and the use of the standard integrals in the example 7.3. Appropriate care must be taken with the sign of the bending moment functions, since the result of integrating along any element may represent either positive or negative internal virtual work. The standard integrals are quite general and negative values may be substituted into the

Example 7.3: Deflections in a Statically Determinate Frame

Given data:

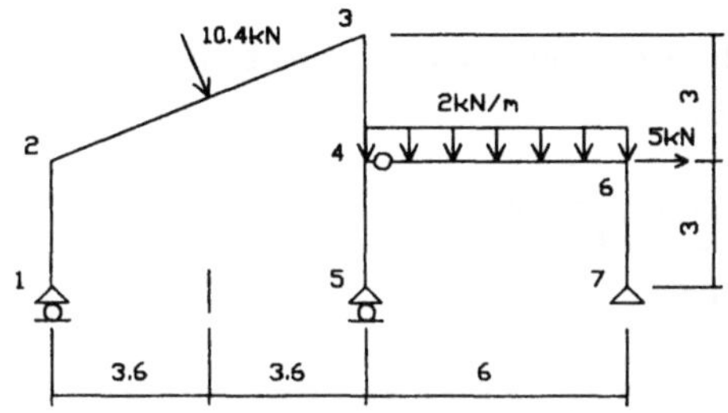

$E = 200\ kN/mm^2$
$I = 85 \times 10^6\ mm^4$

The horizontal deflection at nodes 1, 5 and 6 is required. The frame reactions and bending moment diagram are

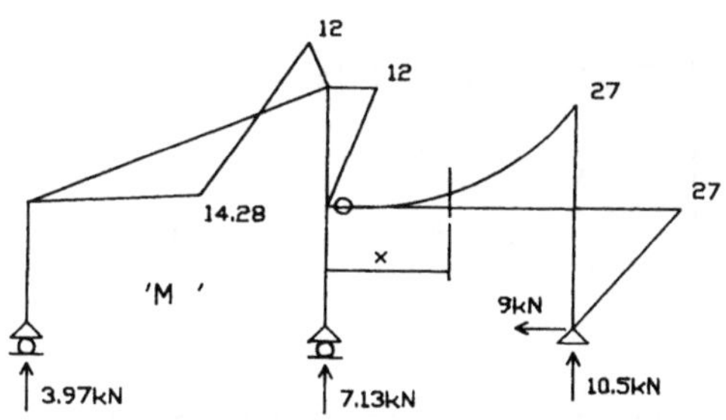

Apply unit horizontal force in turn to nodes 1, 5 and 6 to give the following bending moment diagrams:

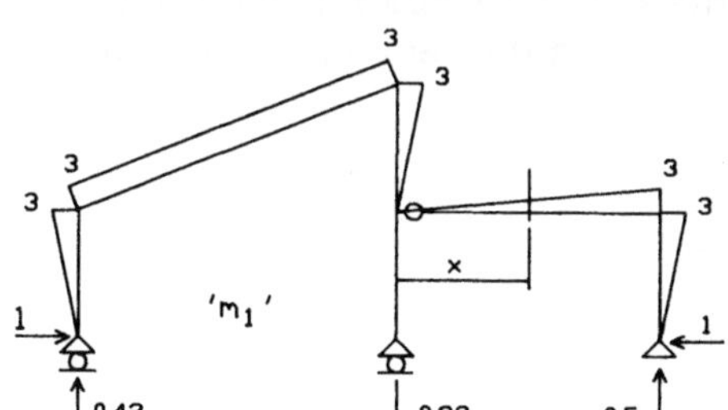

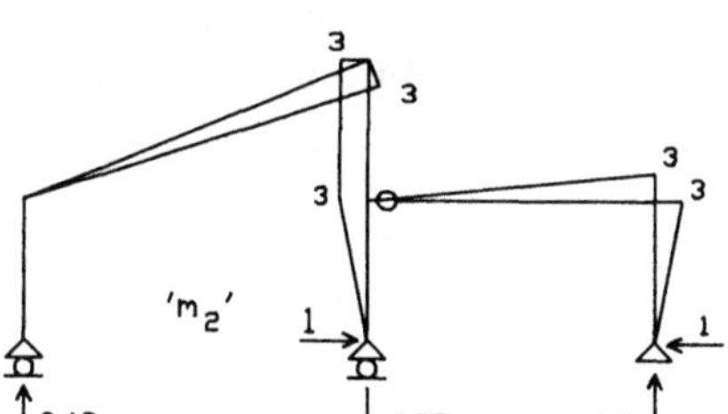

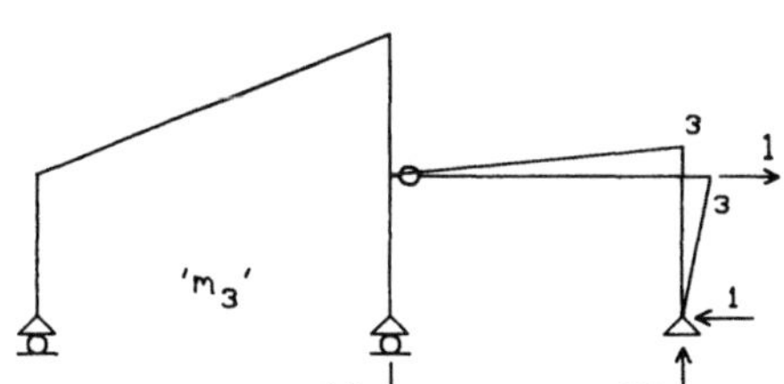

Ignoring axial and shear deformation as second-order effects, the internal virtual work expression is

$$\sum q \cdot \delta = \int m \frac{M}{EI} \mathrm{d}x$$

Considering each unit force application as a virtual force system, the principle of virtual forces gives

$$1(\Delta_i) = \int m_i \frac{M}{EI} \mathrm{d}x$$

(a) Horizontal Deflection at Node 1
Integration over element 4–6:

$$M(x) = 1.5x - 2\frac{x^2}{2} \qquad 0 \leqslant x \leqslant 6$$

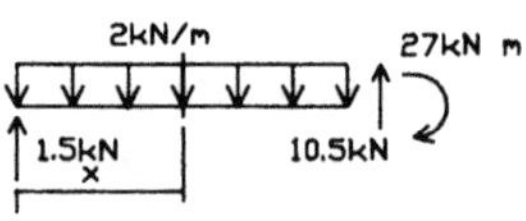

$$m_1(x) = -0.5x \qquad 0 \leqslant x \leqslant 6$$

$$\therefore \quad \int_0^6 m_1(x) \frac{M(x)}{EI} \mathrm{d}x = \int_0^6 -(0.75x^2 - 0.5x^3)/EI \, \mathrm{d}x$$

$$= -\frac{1}{EI}\left[0.75\frac{x^3}{3} - 0.5\frac{x^4}{4}\right]_0^6$$

$$= \frac{108}{EI}$$

The remaining integrals can be evaluated in a similar way, but the standard integrals of table B1.5 in appendix B can be used to advantage. These are based on standard functions to describe $m(x)$ and $M(x)$. Integrating over elements 2–3 (in two parts), 3–4 and 6–7 gives the following tabulated results, with the remaining elements not contributing in this case:

M (x)	m_1 (x)	$\int m_1 M \, dx$
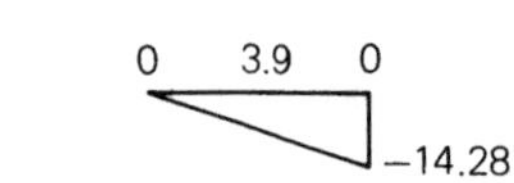	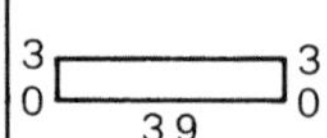	$\frac{L}{2} M_2 m_1 = -83.54$
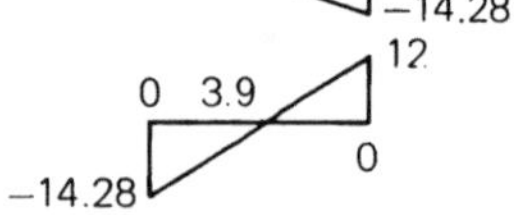	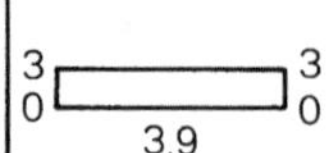	$\frac{L}{2}[m_1(M_1 + M_2)] = -13.34$
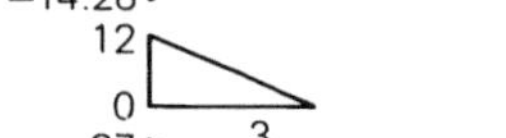	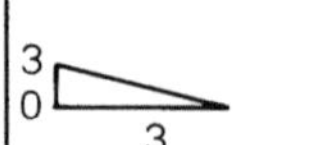	$\frac{L}{3} M_1 m_1 = 36.0$
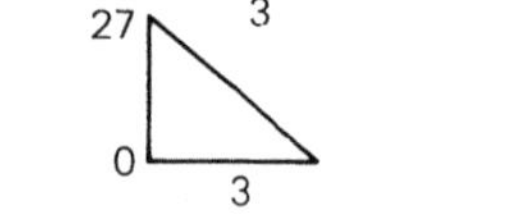	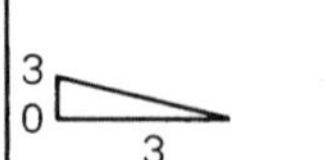	$\frac{L}{3} M_1 m_1 = 81.0$

Summing all of the integral expressions gives

$$\Delta_1 = \frac{128.12}{EI}$$

$$= 0.007 \text{ m}$$

(b) Horizontal Deflection at Nodes 5 and 6
Using the remaining two moment diagrams, m_2 and m_3, the horizontal deflections at nodes 5 and 6 can be found as

$$\Delta_5 = \frac{178.04}{EI} = 0.010 \text{ m}$$

$$\Delta_6 = \frac{189.00}{EI} = 0.011 \text{ m}$$

and from a more general analysis, the deflected shape is known to have the following form:

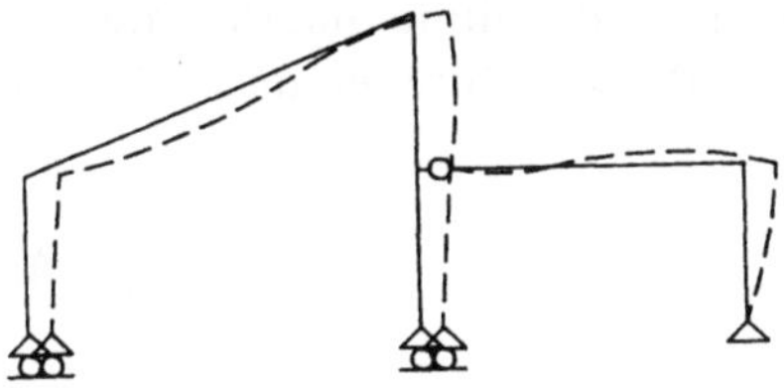

---

expressions as required. It is only necessary consistently to declare a local sign convention to determine positive bending for a given element. Since the sense of any displacement may not be known at the outset, it is quite feasible to obtain a negative result for a deflection. This simply means that the actual deflection has the opposite sense to the applied unit load in the Q force system. While it is possible to derive more complex standard integrals, they become increasingly difficult to interpret. It is considered that it is better to use the limited set given in appendix B, and complete the work by carrying out any other necessary integration from first principles.

It should be appreciated that the Q force system is a completely general one. As such, it may include a unit moment to be used in conjunction with a rotation, as part of the displacements sought on a given structure. With this in mind it is appropriate to comment on the consistency of the units involved. It is of no consequence whether the unit action of the Q force system is regarded as having force dimensions or not. Equation (7.8) will always be dimensionally correct, since it should be realized that the final result for the displacement is found by dividing both sides by the unit action. Provided consistent units are used throughout, the correct values will be obtained.

7.3.2 General Application to the Deflection of Trusses

In a similar manner to that presented in the previous section, an example will be considered in some detail to illustrate the application of the principle of virtual forces to the deflection of trusses.

Example 7.4 shows a simple pin-jointed truss subjected to lateral loads. The horizontal and vertical deflection at node 2, and the relative deflection between nodes 1 and 4, in the direction of a line joining them, are to be calculated using the principle of virtual forces. The truss carries the load through the action of axial force only. The internal virtual work is thus only that associated with axial effects. From table 7.2, the expression for internal virtual work due to axial force is given as

$$\sum q\delta = \int p\frac{P}{EA}\,\mathrm{d}x \tag{7.13}$$

For prismatic elements, the axial forces are constant along the element and the integral expression becomes a simple product. Recalling that the integral implies a summation over all the elements of the structure, equation (7.13) becomes

$$\sum q\delta = \sum p\frac{PL}{EA} \tag{7.14}$$

Example 7.4: Deflections in a Statically Determinate Truss

Given data:

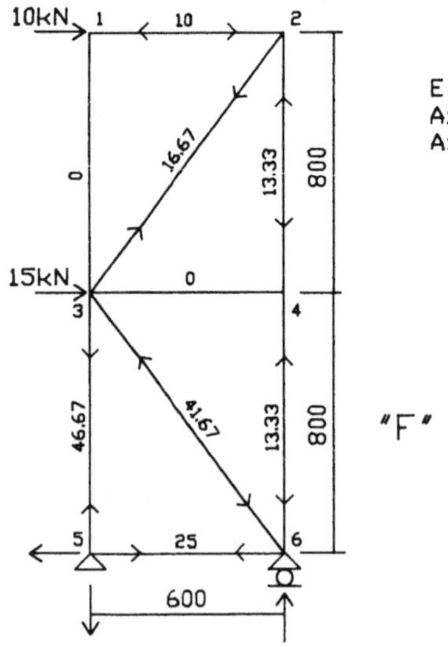

E = 200 kN/mm²
Ax = 800 mm² (diagonals)
Ax = 1000 mm² (others)

The horizontal and vertical deflection at node 2 and the relative deflection of nodes 1 and 4, along a line joining them, is required. The forces in the truss under the loads, designated as the set 'F', are shown as part of the data. Apply unit forces as shown to give the sets of internal force designated as 'f_1', 'f_2' and 'f_3'.

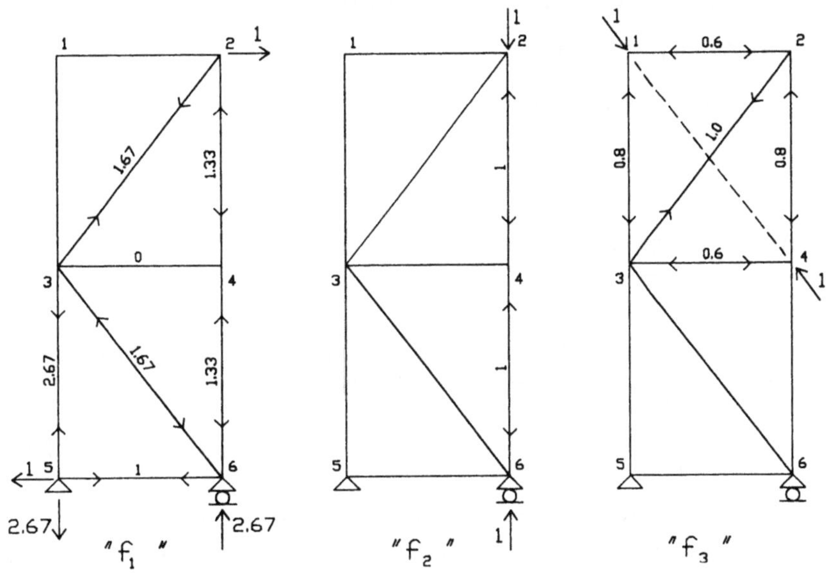

Equation (7.14) can then be seen as

$$1 \cdot \Delta_i = \sum f_i \frac{FL}{AE}$$

and the summation is best evaluated in tabular form as follows:

Element	$L/EA \times 10^3$	F	f_1	f_2	f_3	$Ff_1 \frac{L}{EA}$	$Ff_2 \frac{L}{EA}$	$Ff_3 \frac{L}{EA}$
1–2	3	−10			−0.6			0.02
1–3	4	0			−0.8			
2–3	6.25	16.67	1.67		1.0	0.17		0.10
2–4	4	−13.33	−1.33	−1.0	−0.8	0.07	0.05	0.04
3–4	3	0			−0.6			
3–5	4	46.67	2.67			0.50		
3–6	6.25	−41.67	−1.67			0.43		
4–6	4	−13.33	−1.33	−1.0		0.07	0.05	
5–6	3	25	1.00			0.08		
					Σ	1.32	0.10	0.16

That is, the virtual force systems have been combined with the real displacement system to give

(a) Horizontal deflection at node 2, $\Delta_{2H}=1.32$ mm

(b) Vertical deflection at node 2, $\Delta_{2V}=0.10$ mm

and

(c) Denoting the displacements at nodes 1 and 4 along the line 1–4 as Δ_{1D} and $\Delta 4_D$ respectively, equation (7.14) becomes

$$1 \cdot \Delta_{1D} + 1 \cdot \Delta_{4D} = 0.16$$

$$\therefore 1(\Delta_{1D} + \Delta_{4D}) = 0.16$$

which is

$$\Delta_{1-4} = 0.16 \text{ mm}$$

where Δ_{1-4} is the required relative deflection.

It follows, then, that equation (7.8), as an expression of the principle of virtual forces applied to the deflection of trusses, is

$$\sum Q\Delta = \sum q \frac{PL}{EA} \tag{7.14}$$

where P represents the set of internal forces due to the loads causing the deflection, and q represents the set of internal virtual forces due to a suitably defined virtual force system.

7.3.3 Deflections due to Temperature, Lack of Fit and Support Movements

The principle of virtual forces has been shown to be an effective method of calculating the deflections due to applied loads. Deflections may also occur because of temperature variation in the elements, discrepancies between the length of the elements and their design length or specified movements at the supports. For statically determinate systems, these effects result in deflections without stresses and the deflections can be readily calculated using the principle of virtual forces. In the first two cases the internal displacement of the element is specified by the effect. For example, the total internal deformation of a truss element of length L, subjected to a temperature rise of ΔT, is given as

$$\delta = \alpha \Delta T L$$

where α is the coefficient of linear expansion of the material.

On the other hand, any variation in the length of a truss element from the design length is necessarily given directly by measurement. Deflections in a truss due to any of these effects can be calculated directly from equation (7.8) as an expression of virtual work in the form

$$\sum Q\Delta = \sum q\delta$$

For temperature loads the relevant expression is

$$\sum Q\Delta = \sum q(\alpha \Delta TL)$$

while for lack of fit the expression becomes

$$\sum Q\Delta = \sum q(\Delta L)$$

Consistent with the 'tension positive' sign convention adopted for axial effects, a temperature rise will produce a positive internal deformation, as will an excessive length of an element. In either case, the virtual Q force system is selected to suit the displacement required, as previously explained. A useful application of virtual work in this context arises with the need to camber trusses to offset deflections under load. In this case an external deflection can be specified as a suitable camber and the necessary variation, ΔL, in selected elements can be calculated. For instance, each element in the bottom chord of a truss can be shortened to camber the truss.

A further aspect of temperature effects occurs with a temperature gradient across an element. The internal deformation is now the rotation of a plane section through the element, resulting in curvature. The effect is illustrated in figure 7.7 where it can be noted that the factors influencing the curvature are the temperature variation about the mean value and the depth of the section. The mean temperature, if it is a variation from an accepted normal value for the structure, is responsible for longitudinal expansion or contraction and does not effect curvature. Deflections due to

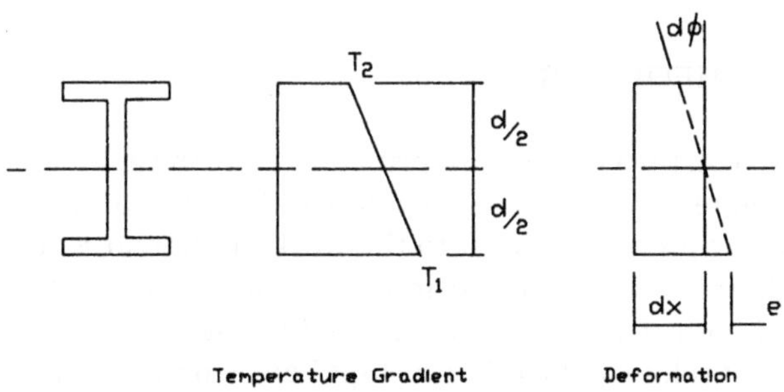

$$e = \alpha \Delta T\, \mathrm{d}x$$

where α = coefficient of linear expansion

and $\Delta T = \dfrac{(T_1 - T_2)}{2}$ so that $\mathrm{d}\phi = \dfrac{2\alpha \Delta T}{d}\mathrm{d}x$

Figure 7.7 Effect of temperature gradient on an element.

the temperature gradient may then be calculated as required from the virtual work expression:

$$\sum QA = \int m \, d\phi$$

A statically determinate structure subjected to support movements will simply move as a rigid body. Although the deflections throughout the rest of the structure can usually be deduced from the geometry of the system, the principle of virtual forces can be used to advantage here also. Since there are no internal deformations, the external virtual work must be zero. With the external deflections at the supports specified, the deflections at other points in the structure can be calculated by using a virtual Q force system in the usual way. In this case the virtual work expression is

$$\sum Q\Delta = 0$$

which is

$$Q_1\Delta_1 + Q_2\Delta_2 + \cdots + Q_nA_n = 0$$

The expansion of the summation does not now reduce to a single term, since the Q system necessarily includes the forces at the reactive supports where the movement has taken place. Instead, the expansion includes the known virtual work done by the reactive Q forces moving through the support movements, and the unknown virtual work term associated with the required displacement which can now be determined.

The question of deflections of statically determinate systems due to effects other than applied loads is again taken up in chapter 8 as part of a study of the behaviour of statically indeterminate structures.

7.4 THE RECIPROCAL THEOREMS

Two related theorems based on the virtual work principles, one the generalisation of the other, are often referred to as the reciprocal theorems and they will be presented here since they are relevant to the work of chapter 8. The theorem was first put forward in a specific form by the nineteenth century British engineer, Maxwell, and it was later generalised by Betti.

The reciprocal theorems relate to the relationships between the independent force and displacement systems acting on the same structure under virtual work concepts. The general form of the reciprocal theorem, or Betti's Law, may be stated as follows:

> *For a linear elastic structure, the external work done by a set of Q forces acting through the displacements Δ' produced by a set of Q' forces, is equal to the external work done by the set of Q' forces acting through the displacements Δ produced by the set of Q forces*

The theorem is a direct result of the application of the principle of virtual work and it may be demonstrated in an application to a simple planar truss. The more general application of the theorem should be readily appreciated as a result of the following study.

A suitable notation to describe the loads (external forces) and displacements of a truss was introduced in chapter 3 with the concepts of load and displacement vectors. The displacements may also be identified with the degrees of freedom of the structure nominated, in part, by the selection of the nodes as discussed in chapter 2. For a truss, the nodes are identified at the joints with two degrees of freedom at each node. The complete set of forces acting on the structure is given by a vector of $2n$ terms, where n is the number of nodes. At the ith node, the forces can be described as Q_{2i-1} and Q_{2i}, in the x and y direction respectively. The displacements at the ith node may be described in a similar manner as Δ_{2i-1} and Δ_{2i}. The forces and displacement systems acting on the truss of figure 7.8 can be readily described with this notation.

In figure 7.8(a) for example, a Q force system with internal forces q is shown, with the forces Q_1 and Q_5 producing the set of displacements, Δ, as shown in figure 7.8(b). Similarly, figure 7.8(c) represents another force system Q', with the forces Q_4' and Q_7' producing the set of displacements Δ' (as shown in figure 7.8(d)) and the internal forces q'.

In line with example 7.4, the Q' force system may be regarded as a virtual force system to be considered in conjunction with the displacement system Δ. Equation (7.8) is then

$$\sum Q'\Delta = \sum q'\delta \tag{7.15}$$

where δ represents the internal deformation of the element due to the Q force system; hence

$$\delta_i = \frac{q_i L_i}{EA}$$

Equation (7.15) is then

$$\sum Q'\Delta = \sum q' \frac{qL}{EA} \tag{7.16}$$

Similarly, the Q force system may be regarded as a virtual force system to be considered in conjunction with the displacement system Δ'. Equation (7.8) now gives

$$\sum Q\Delta' = \sum q\delta' \tag{7.17}$$

but now δ' represents the internal deformation given by

$$\delta_i' = \frac{q_i' L_i}{EA}$$

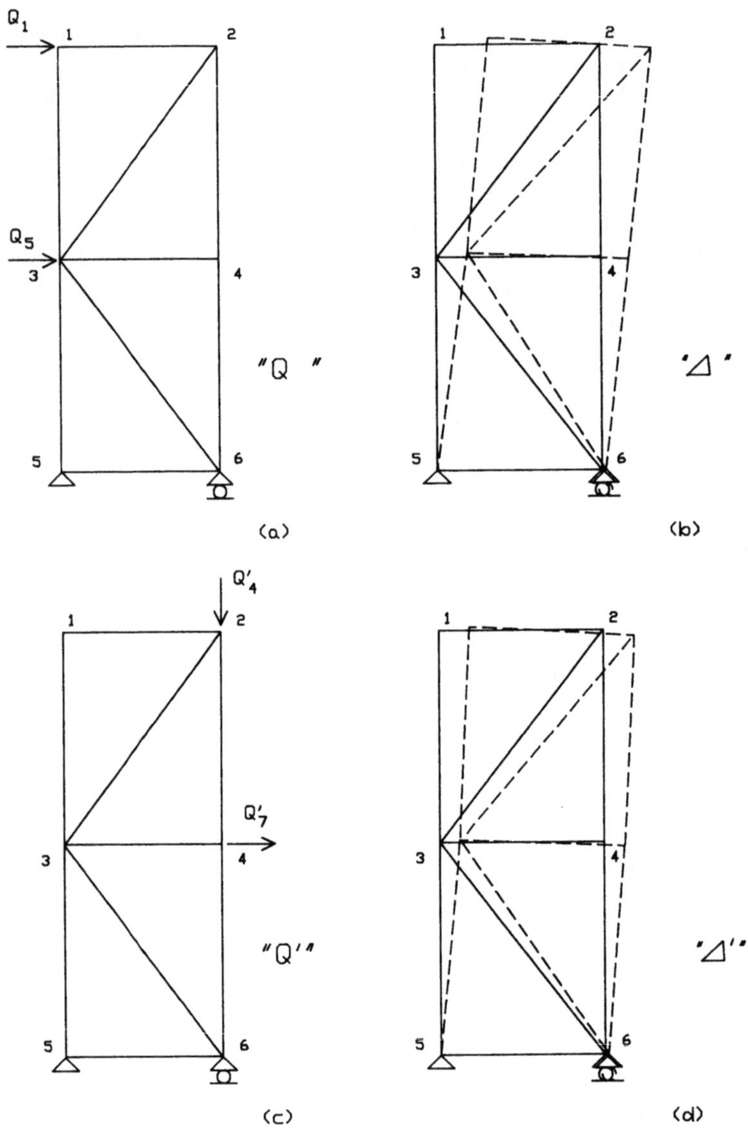

Figure 7.8 *Reciprocal theorems applied to a truss.*

so that equation (7.17) becomes

$$\sum Q\Delta' = \sum q \frac{q'L}{EA} \tag{7.18}$$

Since equations (7.16) and (7.17) both have the same right-hand sides, it follows that the left-hand sides, both expressing external work quantities,

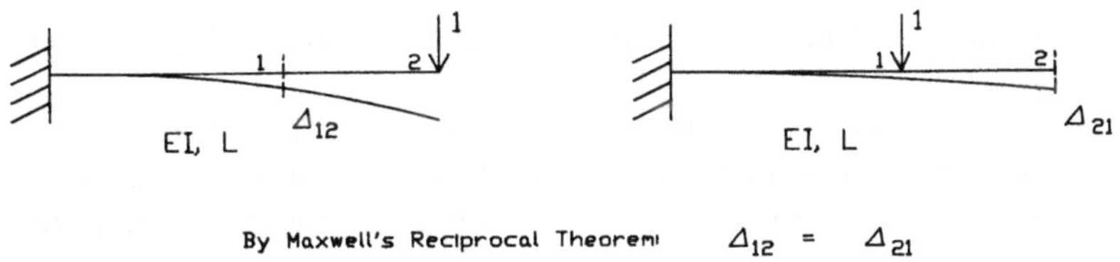

Figure 7.9 Maxwell's reciprocal theorem.

are equal, thus proving the theorem. Specifically, with the suggested values of the Q and Q' force systems, then the reciprocal theorem states that

$$Q_1\Delta_1' + Q_5\Delta_5' = Q_4'\Delta_4 + Q_7'\Delta_7$$

More specifically, Maxwell presented the reciprocal theorem in a form where the force systems are represented by a unit force acting in the direction of an unrestrained degree of freedom. This condition leads directly to a reciprocal relationship between specified displacements acting on a structure. Maxwell's reciprocal theorem may be stated as follows:

> *For a linear elastic structure, the displacement Δ_{ij} at a point i, in the direction of a unit load at i but caused by a unit force at j, is equal to the displacement Δ_{ji} at the point j, in the direction of the unit force at j but caused by the unit force at i*

The theorem follows directly from Betti's Law and it can be readily verified by considering the simple example of figure 7.9 and using the principle of virtual displacements.

7.5 PROOF OF THE RELATIONSHIP BETWEEN THE STATICS MATRIX AND THE KINEMATICS MATRIX

In section 3.2 of chapter 3, where the matrix analysis of trusses was introduced in a particular application of the stiffness method, it was stated that the kinematics matrix, B, was necessarily the transpose of the statics matrix, A. This relationship can now be proved using the principle of virtual forces. Once again, it is convenient to use a plane truss as the structure for consideration.

Any plane truss in equilibrium under the action of a set of loads, P, develops the set of internal forces, f. The action also produces the set of element extensions, e, and results in the displacements of the structure, d. Equations (3.3) and (3.5) give the general relationships between the load and force vectors and the element extension and displacement vectors for a truss as

$$P = A \cdot f \quad \text{and} \quad e = B \cdot d$$

where A is the statics matrix and B is the kinematics matrix or displacement transformation matrix.

It is required to prove that $B = A^{\mathrm{T}}$. Suppose that the general set of loads P has the particular form P' such that all the internal forces f, except in one element, are zero. This is shown in part in figure 7.10(a), and equation (3.3) is then

$$\begin{Bmatrix} P'_1 \\ P'_2 \\ P'_3 \\ \vdots \\ P'_k \\ \vdots \\ P'_n \end{Bmatrix} = \begin{bmatrix} a_{11} & a_{12} & a_{13} & \\ a_{21} & a_{22} & \cdots & \\ \vdots & & A_{n\times m} & \\ a_{n1} & & & a_{nm} \end{bmatrix} \begin{Bmatrix} 0 \\ 0 \\ \vdots \\ f_j \\ \vdots \\ 0 \end{Bmatrix} \tag{7.19}$$

A typical result from multiplying out equation (7.19) is

$$P'_k = a_{kj} f_j \tag{7.20}$$

Suppose now that a compatible deformation of the structure is introduced, caused by another load system P'', such that d_k is the only non-zero term in the displacement vector. This action, which is shown in part in figure 7.10(b), will result in the particular form of the element extension vector, e'', and equation (3.5) is then

$$\begin{Bmatrix} e''_1 \\ e''_2 \\ e''_3 \\ \vdots \\ e''_j \\ \vdots \\ e''_m \end{Bmatrix} = \begin{bmatrix} b_{11} & b_{12} & b_{13} & \\ b_{21} & b_{22} & \cdots & \\ \vdots & & B_{m\times n} & \\ b_{m1} & & & b_{mn} \end{bmatrix} \begin{Bmatrix} 0 \\ 0 \\ \vdots \\ d_k \\ \vdots \\ 0 \end{Bmatrix} \tag{7.21}$$

and a typical result from multiplying out equation (7.21) is

$$e''_j = b_{jk} d_k \tag{7.22}$$

The principle of virtual work may now be applied with figure 7.10(b) representing a virtual displacement system to be applied in conjunction with the force system of figure 7.10(a). Equation (7.8) then gives

$$P'_k d_k = f_j e''_j \tag{7.23}$$

as a relationship between external virtual work and internal virtual work. Substituting from equations (7.20) and (7.22), for P'_k and e''_j respectively,

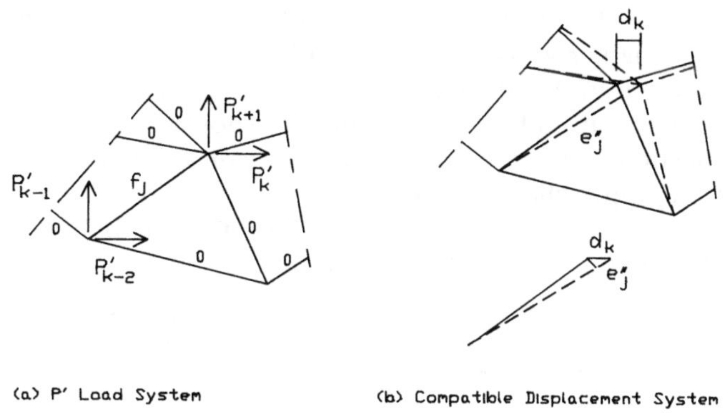

Figure 7.10 Force and displacement system acting on a truss.

into equation (7.23) results in

$$a_{kj}f_jd_k = f_jb_{jk}d_k$$

$$\therefore \quad a_{kj} = b_{jk}$$

which is the required proof.

7.6 PROBLEMS FOR SOLUTION

7.1 Using the principle of virtual displacements, verify the general expressions for shear force and bending moment in a simply supported beam under a single concentrated load as given in section 2.6 of chapter 2.

7.2 Calculate all the reactions of the beam of figure 7.3(a) using the principle of virtual displacements.

7.3 Calculate the horizontal deflection at node 1 and the rotation of the joint at node 2 for the rigid-jointed frame of figure P7.1.

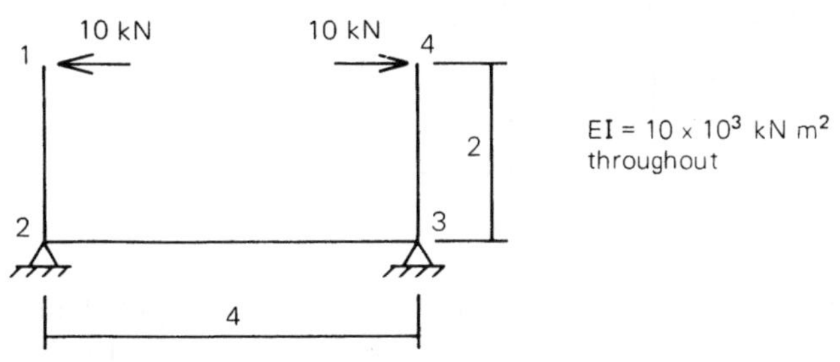

Figure P7.1.

7.4 Calculate the slope and deflection at the end of the cantilever beam of problem **4.1** of chapter 4, using the principle of virtual forces.

7.5 For the truss of figure P7.2, calculate the following quantities:
(a) vertical deflection at node 7 due to the load shown.
(b) vertical deflection at node 7 due to the effect of a temperature rise of 20°C in the upper chord elements only.
The area of each element is 500 mm and the coefficient of linear expansion $\alpha = 0.000\,012/°C$. $E = 200\ \mathrm{kN/mm^2}$.

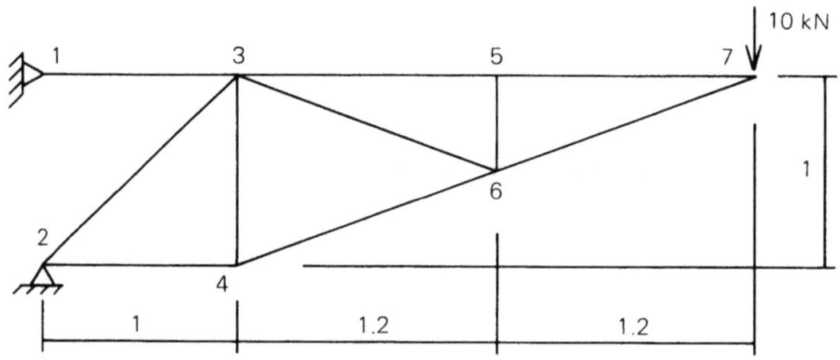

Figure P7.2.

7.6 Show that the deflection at the centre of a simply supported beam of length, L, and depth, h, caused by a rise of t in the temperature gradient over the depth is given by $(\alpha t L^2)/8h$, where α is the coefficient of linear expansion.

7.7 The bridge truss of figure P7.3 has steel elements with $E = 2 \times 10^5$ MPa, the area of each chord element $= 2500\ \mathrm{mm^2}$ and that of each web element $= 1500\ \mathrm{mm^2}$. Elements 1–2 and 6–8 are considered as web elements.
Calculate the vertical deflection of node 5, and the relative displacement between nodes 2 and 5 along the diagonal line 2–5.

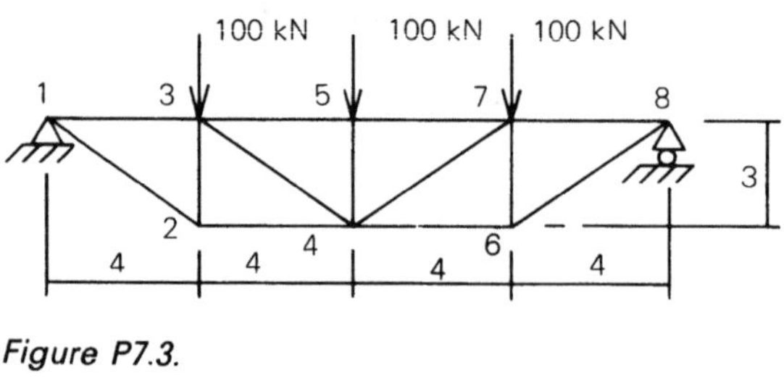

Figure P7.3.

7.8 A camber is to be introduced into the truss of problem **7.7** by modifying the lengths of the elements so that, under the loads shown, node 5 has the same elevation as nodes 1 and 8. Determine a suitable scheme for producing the required camber.

7.9 Evaluate the horizontal displacement of node 6 for the frame of figure P7.4.

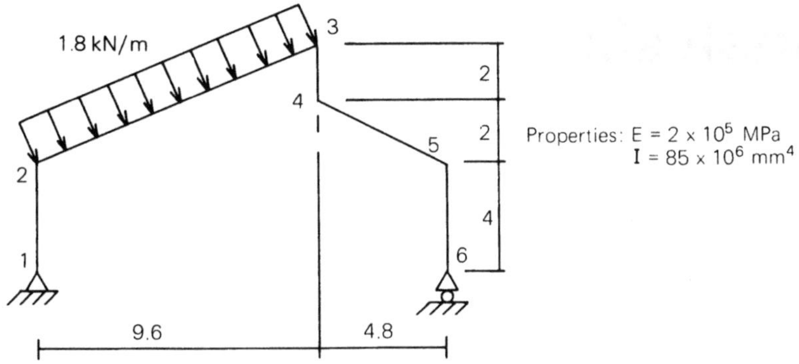

Figure P7.4.

7.10 Calculate the total deflection (in three components) of the free end of the cantilevered roadway sign structure shown in figure P7.5.

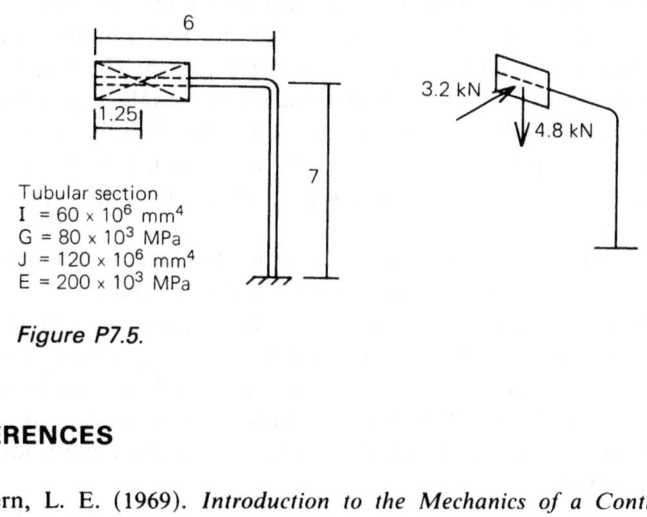

Figure P7.5.

REFERENCES

Malvern, L. E. (1969). *Introduction to the Mechanics of a Continuous Medium*, Prentice Hall, Englewood Cliffs, New Jersey.

Oden, J. T. and Ripperger, E. A. (1981). *Mechanics of Elastic Structures*, 2nd edn, McGraw-Hill, New York.

FURTHER READING

Oden, J. T. and Ripperger, E. A., *Mechanics of Elastic Structures*, 2nd edn, McGraw-Hill, New York, 1981.

Chapter 8
The Flexibility Method of Analysis

It is a feature of the analysis of structures that each technique can find expression in an alternative or dual form. This is sometimes referred to as the 'duality' of structural analysis and it is frequently useful for a student to study the dual problem. Obviously, certain problems are solved more directly by one method than by the other, but the fact remains that the alternative method is available. During the latter part of the last century and the early part of this century, methods of structural analysis proliferated. It is now recognized that all of the methods of structural analysis are based on either a stiffness approach or a flexibility approach, but this was not widely accepted until the advent of the digital computer which prompted further investigative work.

The stiffness method is also known as the displacement method or sometimes as the equilibrium method. The flexibility method, on the other hand, is also known as the force method or the compatibility method. Each of these titles is descriptive and its appropriateness becomes apparent on a reflection of what is involved in the application of the method. The flexibility method can be used effectively with computers, but it is less amenable to matrix formulation and therefore to computer methods than is the stiffness method.

8.1 BASIC CONCEPTS OF THE FLEXIBILITY METHOD

In presenting the basic concepts of the flexibility method, a deliberate comparison and contrast will be made with the stiffness method. This is both for convenience and to give emphasis to the dual nature of the problem.

Although either method can be used with statically determinate or indeterminate structures, this study will concentrate on the latter. Structures which are highly kinematically determinate represent convenient starting points for the analysis of structures by the stiffness method. In fact, the primary structure of the stiffness method is the restrained form where certain degrees of freedom have been suppressed. In the flexibility method, the primary structure is the released form, where unknown actions, either internal or external, have been released.

The primary unknowns in the stiffness method are the displacements at the nodes, while in the flexibility method, the primary unknowns are the released unknown actions. The initial equations affording the solution in the stiffness method can be regarded as equilibrium equations. The corresponding equations in the flexibility method are based on compatibility requirements or the need for consistent deformation. The following study of a propped cantilever beam under a uniformly distributed load further illustrates the concepts of the flexibility method and highlights a comparison with the stiffness method.

Figure 8.1 shows a series of operations that may be carried out on a propped cantilever in order to determine the moment at the built-in end. Ignoring the fact that this is a standard solution that is widely known, the direct stiffness method may be readily applied to determine the rotation θ_2 and hence the required moment. The steps of the analysis are shown in the series of operations of figure 8.1, commencing with the restrained primary structure where the end moments are taken from a standard solution. The action necessary to introduce a unit displacement at the imposed restraint at node 2 was covered in chapter 4. It can be recognised as a coefficient of the structure stiffness matrix, or simply as a stiffness coefficient. Moment equilibrium at node 2 is then satisfied by the equation

$$\theta_2 k_{11} - \frac{qL^2}{12} = 0 \tag{8.1}$$

Equation (8.1) is in fact of the form $P_F = K_F d_F$, as previously discussed as the basis of the stiffness method. In this case the stiffness matrix has one term only. The required moment is found by back-substitution and superposition. A careful study of the operations will show that the technique has all the steps of the direct stiffness method as explained in chapter 4.

Using the flexibility approach, the operations as shown in figure 8.1 commence with identifying a suitable unknown action. In this case, the redundant reaction at node 2 has been selected and designated X_2. In the released primary structure the displacement at node 2 can be readily found, either as a standard solution or by direct calculation. With a unit force applied at the release, the displacement designated f_{11} can also be found. Again this may or may not be a standard solution, but in any event it can

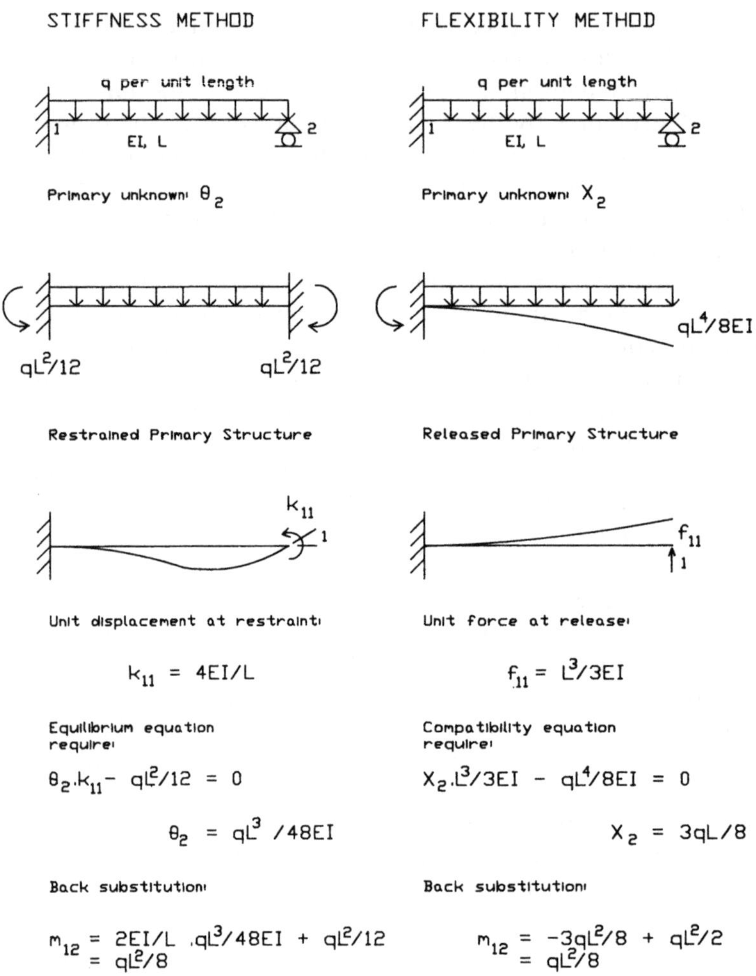

Figure 8.1 *Comparison of methods of structural analysis.*

be calculated. The displacement at a release due to unit action is known as a flexibility coefficient, the general nature of which will be presented in section 8.1.2. For unyielding supports, compatibility is then satisfied by the equation

$$X_2 \frac{L^3}{3EI} - \frac{qL^4}{8EI} = 0 \tag{8.2}$$

The unknown action, X_2, can be determined from equation (8.2) and the required moment, m_{12}, is again found by back-substitution and superposition.

8.1.1 Analysis of Structures with One Degree of Statical Indeterminacy

The flexibility method is particularly convenient for the analysis of structures with a degree of statical indeterminacy of one. In such cases, the selected redundant action can be found directly from a compatibility equation similar to that of equation (8.2). A significant feature of the method is that the analyst has a choice in the selection of the action to be released and that the action may be either internal or external. The example of figure 8.1 can be repeated, selecting the moment action m_{12} as the redundant. Actually such an approach was used in section 3.8 of chapter 3, as the basis for determining the standard solutions for fixed end actions on beams.

The tied portal frame of example 8.1 has a degree of statical indeterminacy of one. The tie may be necessary when poor soil conditions on one side of the site prevent the development of a lateral restraint at one footing. It is convenient to take the internal action in the tie as the primary unknown. Cutting the tie represents the one release necessary to give a statically determinate primary structure.

Example 8.1: Flexibility Analysis of a Tied Portal Frame

Given data:

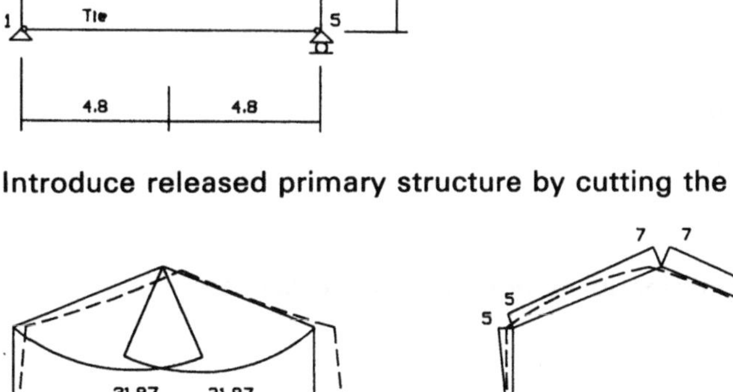

Introduce released primary structure by cutting the tie:

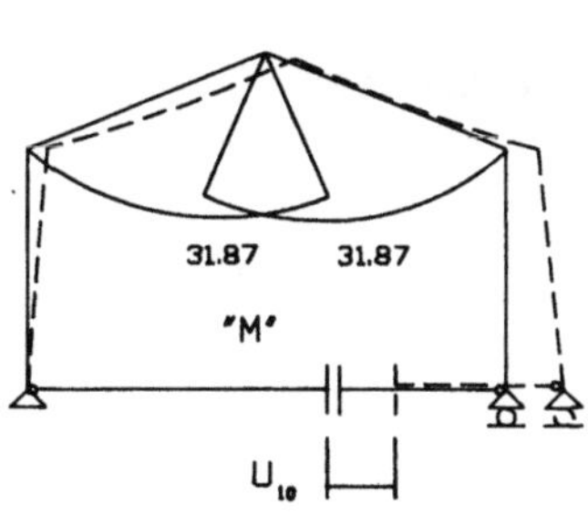

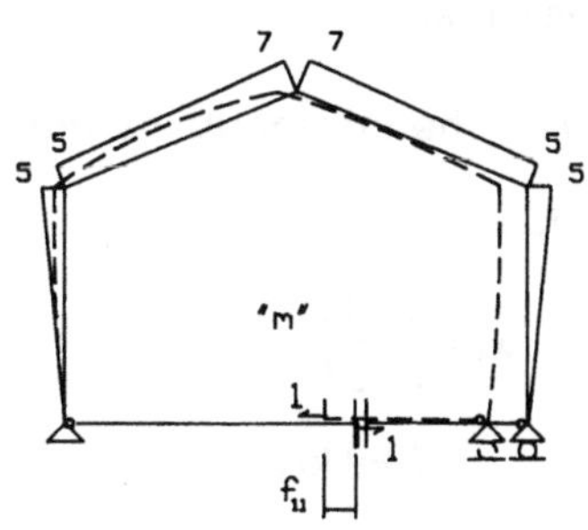

Calculate displacement at the release due to the loads (particular solution) as

$$U_{10}=\int m\frac{M}{EI}\,dx+\sum f\frac{FL}{EA}$$

For rafter 2–3:

$$M(x)=12.88\left(\frac{12}{13}\right)x-1.2x^2\left(\frac{12}{13}\right)$$

and

$$m(x)=-\left(5+\frac{2}{5.2}x\right)$$

Ignoring axial deformation in the columns and rafters, then

$$U_{10}=\frac{2}{EI}\int_0^{5.2}-(5+0.384x)(11.89x-1.107x^2)\,dx$$

$$=-1361.23/EI$$

$$=-68.06\text{ mm}$$

Calculate the displacement at the release due to unit action as

$$f_{11}=\int m\frac{m}{EI}\,dx+\sum f\frac{fL}{EA}$$

Using standard integrals for moment integration over the frame, then

$$f_{11}=\frac{461.25}{EI}+\frac{9.6}{EA}$$

$$=23.06+0.48$$

$$=23.54\text{ mm}$$

Applying the compatibility equation at the release:

$$U_{10}+f_{11}x_1=0$$

$$x_1=\frac{68.06}{23.54}$$

$$=2.89\text{ kN}$$

The final solution is given as a combination of the actions of the 'M' diagram and x_1 times the actions of the 'm' diagram.

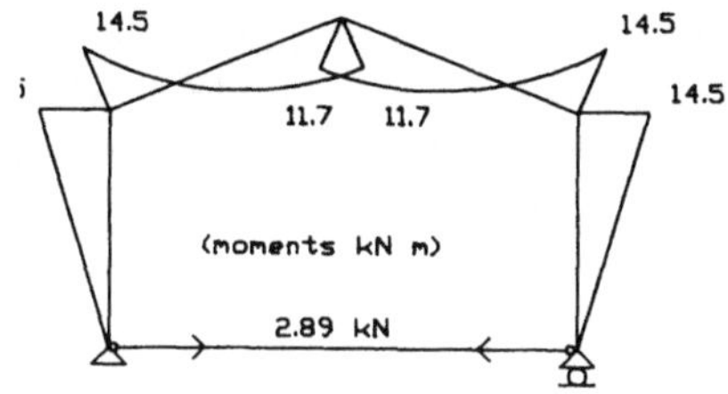

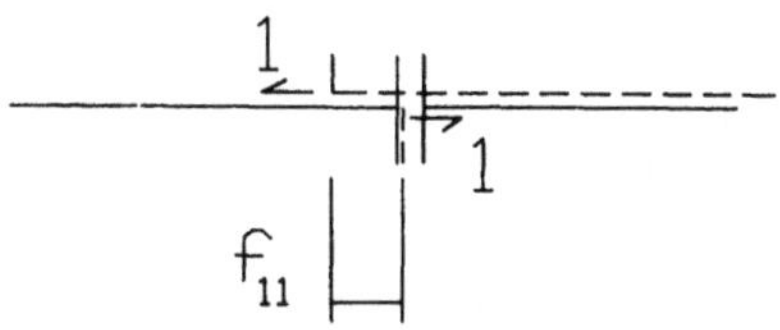

Under the action of the loads, the primary structure of example 8.1 deflects, giving a displacement at the release which may be designated as U_{10}. Since the frame is now statically determinate, all of the actions of the frame, including U_{10}, can be readily calculated. Such results may be regarded as the particular solution to the problem.

With unit force applied at the release, the frame develops the actions shown in the '*m*' diagram of example 8.1 resulting in the flexibility coefficient f_{11}. (The action at the release is shown enlarged at the end of the example.) The actual force in the tie may be denoted as X_1, so that the compatibility requirements of the structure are satisfied by the equation

$$U_{10}+f_{11} \cdot X_1=0 \tag{8.3}$$

The results show that the superposition of the actions of the structure of the 'M' diagram, and X_1 times the actions of the structure of the 'm' diagram, satisfy the requirements of the given structure. The latter set of results may be referred to as the complementary solution to the problem. The displacements U_{10} and f_{11} have been calculated using the principle of virtual forces, and they presented no difficulty since they represent displacements of a statically determinate structure. Actually, f_{11} should be seen as a displacement per unit force, so that equation (8.3) is dimensionally correct.

It may be noted that the axial effects in the tie have contributed little to the flexibility coefficient in example 8.1. Nevertheless it is important to note this effect and to appreciate how the elasticity of the tie is taken into consideration. If the tie were axially rigid, then the solution would be that for a portal frame with pinned bases. Of course the technique can be applied to such a frame, in which case the released action can simply be taken as the horizontal reaction at one support. The structures referred to in problem **8.1** of section 8.3 all fall within the same category as the structure of example 8.1. Although all the structures are also of a composite nature, comprising clearly defined flexural and axial force elements, this is not a necessary condition. The structures simply represent an interesting class of structure that can be analysed by the flexibility method.

The method as outlined is sometimes referred to as the method of consistent deformations. This arises from the nature of equation (8.3) which can clearly be seen to ensure consistent deformation in the structure. As will be seen in section 8.2, the flexibility method may be developed with more rigour to include the subsequent calculation of the displacements of the structure.

8.1.2 Application to Higher-order Statically Indeterminate Structures

The flexibility method can be effectively applied to structures with a degree of statical indeterminacy greater than one. The primary structure still remains statically determinate, so that it is necessary to introduce releases of action equal to the degree of statical indeterminacy. This leads to a set of values, U_{i0}, for the particular solution since displacements will occur at each release under the load on the structure. There is also a corresponding redundant action, X_i, associated with each release and the influence of unit force applied in turn at each release must be considered. The application of unit force at any one release may cause a displacement at any, or all, of the remaining releases. It is this interaction which characterises the difference between the analysis of first-order and higher-order statically indeterminate structures by the flexibility method. In the following discussion, the terms action and force are used in a generalised sense to include moments, shear and axial force. A displacement may of course be either a translation or a rotation.

In general, the flexibility coefficient may be defined as the displacement f_{ij}, which is the displacement at release i, caused by the action of unit force at release j. Equation (8.3), expressing compatibility or consistent deformation, now takes the more general form of

$$
\begin{aligned}
U_{10}+f_{11}X_1+f_{12}X_2+\cdots+f_{1n}X_n &= 0\\
U_{20}+f_{21}X_1+f_{22}X_2+\cdots+f_{2n}X_n &= 0\\
\vdots \qquad\qquad\qquad\qquad &\ \ \vdots\\
U_{n0}+f_{n1}X_1+f_{n2}X_2+\cdots+f_{nn}X_n &= 0
\end{aligned}
\tag{8.4}
$$

where the structure has a degree of statical indeterminacy of n.

Equation (8.4) is simply a set of simultaneous linear equations which can be solved for X_i once the flexibility coefficients and the particular solution have been found. The released actions may again be either internal or external and relate to any action of axial force, shear force or bending or torsional moment. The final solution to the problem is again given as a combination of the particular solution and each of the solutions for unit force in the redundant action, scaled by the now known redundant action X_i. The technique is illustrated by examples 8.2 and 8.3. In each case, the structures are a redundant form of the structures used in examples 7.1 and 7.2 of chapter 7.

Example 8.2: Flexibility Analysis of a Higher-order Indeterminate Frame

---

Given data:

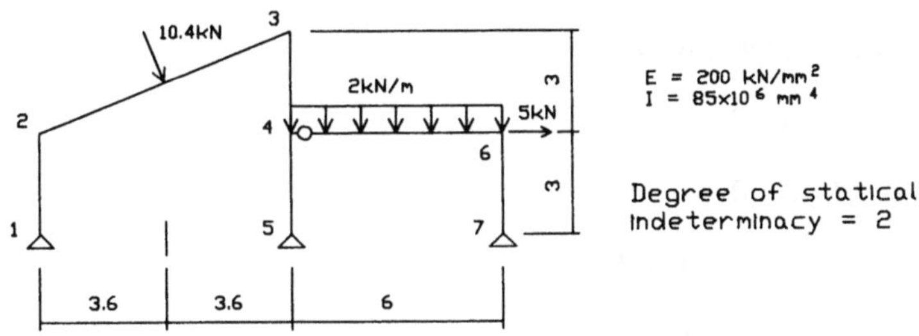

Introduce primary structure by releasing horizontal restraint at nodes 1 and 5:

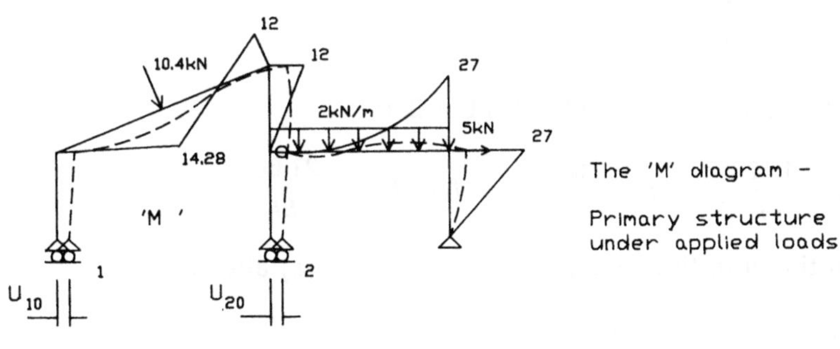

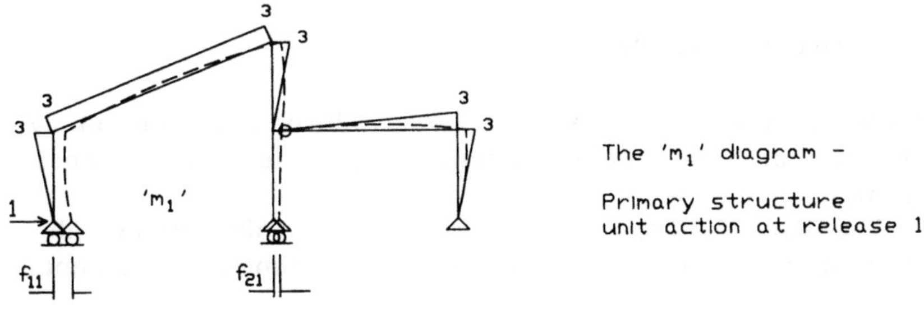

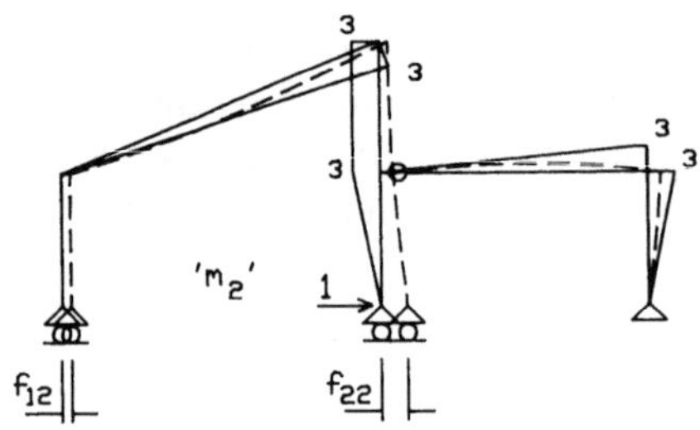

The 'm2' diagram –

Primary structure
unit action at release 2

Solution: Equation (8.4) gives

$$\begin{aligned} u_{10}+f_{11}x_1+f_{12}x_2&=0 \\ u_{20}+f_{21}x_1+f_{22}x_2&=0 \end{aligned} \qquad \text{(a)}$$

From example 7.1 of chapter 7:

$$u_{10}=\frac{128.12}{EI};\ u_{20}=\frac{178.04}{EI}$$

The flexibility coefficients may be calculated from

$$f_{ij}=\int m_i\frac{m_j}{EI}\,dx$$

Using the standard integrals of table B.1:

$$f_{11}=\frac{115.20}{EI};\qquad f_{22}=\frac{86.40}{EI};\qquad f_{12}=f_{21}=-\frac{21.60}{EI}$$

Substituting into the simultaneous equations (a) above, then

$$128.12+115.20x_1-21.60x_2=0$$

$$178.04-21.60x_1+86.40x_2=0$$

from which, $x_1=-1.57$ kN and $x_2=-2.45$ kN (the negative sign simply indicates that the sense of the reaction is opposite to that assumed with the unit action).

The frame bending moment diagram is then given as a combination of the moments of the 'M' diagram, x_1 times the moments of the 'm$_1$' diagram and x_2 times the moments of the 'm$_2$' diagram.

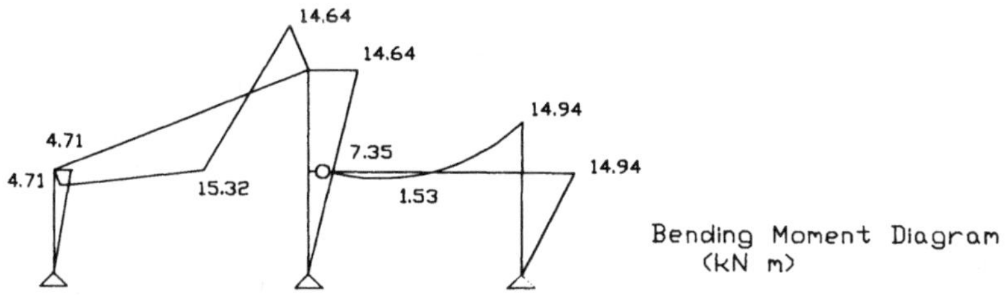

In the structure of example 8.3, additional bracing has been introduced in both panels to give a redundant form. It is assumed that the bracing elements can cope adequately with both tension and compression forces so that the structure is truly a redundant one. (In many cases, wind bracing is designed to take the tension load with a slenderness ratio such that it buckles elastically under load reversal to become ineffective in compression. This is compensated for by having such bracing elements in two directions.) The choice of redundants in this example is very obvious; the released structure is found by cutting the redundant diagonal elements.

Example 8.3: Flexibility Analysis of a Higher-order Indeterminate Truss

Given data:

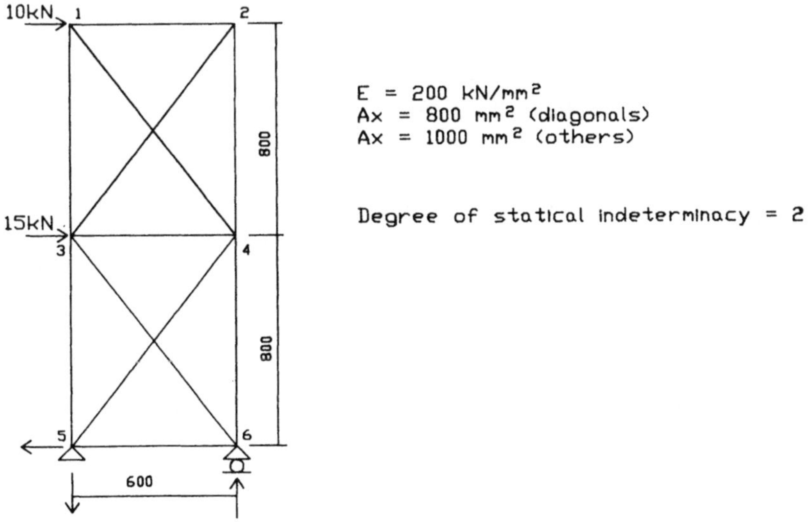

Introduce primary structure by cutting a diagonal in each panel.

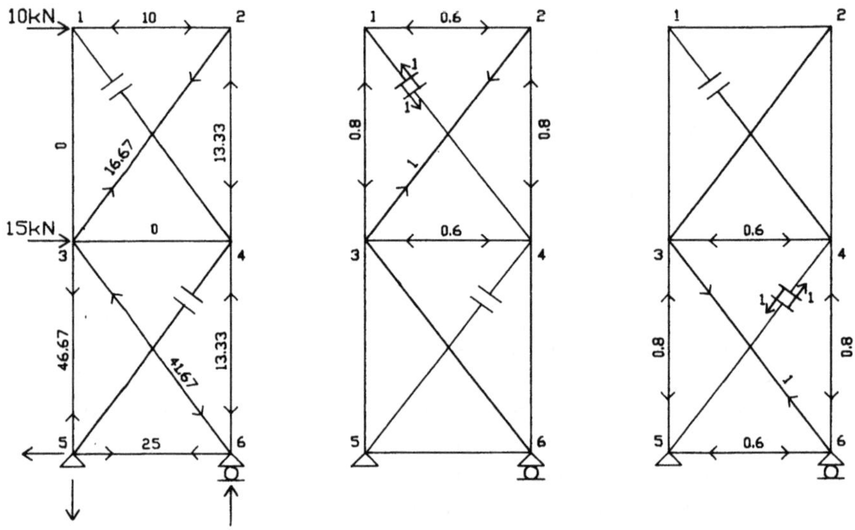

"F" "f_1" "f_2"

The 'F' diagram—primary structure under applied loads. The 'f_1' diagram—primary structure, unit action at release 1. The 'f_2' diagram—primary structure, unit action at release 2.

The particular solution, u_{10} and u_{20} and the flexibility coefficients can all be calculated as deflections in a determinate truss, hence

$$u_{i0}=\sum f_i\frac{FL}{EA} \quad \text{and} \quad f_{ij}=\sum f_i\frac{f_jL}{EA}$$

In table form, the calculations are:

Element	$\frac{L}{EA}\times10^3$	F	f_1	f_2	$f_1\frac{FL}{EA}\times10^3$	$f_2\frac{FL}{EA}\times10^3$	$f_1\frac{f_1L}{EA}\times10^3$	$f_2\frac{f_2L}{EA}\times10^3$	$f_1\frac{f_2L}{EA}\times10^3$
1–2	3	−10	−0.6		18		1.08		
1–3	4	0	−0.8		0		2.56		
2–3	6.25	16.67	1.0		104.19		6.25		
2–4	4	−13.33	−0.8		42.66		2.56		
3–4	3	0	−0.6	−0.6	0		1.08	1.08	1.08
3–5	4	46.67		−0.8		−149.34		2.56	
3–6	6.25	−41.67		1.0		−260.44		6.25	
4–6	4	−13.33		−0.8		42.66		2.56	
5–6	3	25		−0.6		−45.00		1.08	
1–4	6.25	0	1.0		0		6.25		
4–5	6.25	0		1.0		0		6.25	
				Σ	164.85	−412.12	19.78	19.78	1.08

Equation (8.4) gives

$$u_{10}+f_{11}x_1+f_{12}x_2=0$$
$$u_{20}+f_{21}x_1+f_{22}x_2=0 \qquad \text{(a)}$$

Substituting into equations (a) results in

$$164.85+19.78x_1+1.08x_2=0$$
$$-412.12+1.08x_1+19.78x_2=0$$

from which $x_1=-9.5$ kN and $x_2=21.4$ kN

The final forces in the truss are given by combining the set of forces, 'F', with x_1 times the set of forces, 'f_1,' and x_2 times the set of forces, 'f_2'.

Element	F	x_1f_1	x_2f_2	Final force (kN)
1–2	−10	5.7		−4.3
1–3	0	7.6		7.6
2–3	16.67	−9.5		7.17
2–4	−13.33	7.6		−5.73
3–4	0	5.7	−12.84	−7.14
3–5	46.67		−17.12	29.55
3–6	−41.67		21.4	−20.27
4–6	−13.33		−17.12	−30.45
5–6	25		−12.84	12.16
1–4	0	−9.5		−9.5
4–5	0		21.4	21.4

From the preceding examples it becomes apparent that there is a matter of choice in the selection of the redundant actions of a structure, for the purposes of analysis by the flexibility method. While the technique is not dependent on the selection of any particular redundant action, the selection does influence the ease with which the required displacements can be calculated. Often the idea of introducing a moment release, through a pin, in a flexural element in a frame is overlooked. In many cases the moment release will lead to a well conditioned set of simultaneous equations with simple calculations for the rotations. Of course a mixture of both force and moment releases is acceptable and sometimes desirable. The illustrative example in section 8.2 uses moment releases to give a statically determinate primary structure.

8.1.3 Deflection Calculations for Statically Indeterminate Frames

Since the flexibility method treats the redundant actions as the primary unknowns, deflections throughout the structure are not available without further analysis. Such a deflection analysis can be carried out using the principle of virtual forces as in chapter 7, since the technique is equally applicable to statically indeterminate structures. An apparent difficulty arises though since the moment diagram for the frame under a unit virtual force may require another indeterminate analysis to establish it. Fortunately this step can be avoided using the following observation.

The behaviour of any indeterminate frame, including the deflections, is equivalent to the behaviour of the primary structure under the action of the loads and the redundant actions at the releases. Any displacement can therefore be found by applying a unit virtual force at the required point on the primary structure, and integrating the resulting moment diagram with the moment diagram of the indeterminate frame. The primary structure is of course statically determinate so that the virtual force moment diagram is easily established. Further, the primary structure need not be the same primary structure as that used in the original indeterminate analysis, if one was used at all. The analysis may have been based on a stiffness method or obtained from a computer program, and deflections other than those given by such methods may be required. It is only necessary to use a statically determinate primary structure as a basis for the virtual force system when using the principle of virtual forces to calculate deflections.

The procedure is shown in example 8.4 using the results of the analysis of example 8.2. It is worth noting that in a routine structural analysis by computer, the deflections throughout the structure are reported at selected nodes depending on how the structure is modelled. Subsequently the design engineer may require a deflection at some other point in the structure, and the technique outlined here could be used to advantage.

Example 8.4: Deflection of a Statically Indeterminate Frame

Given the following results:

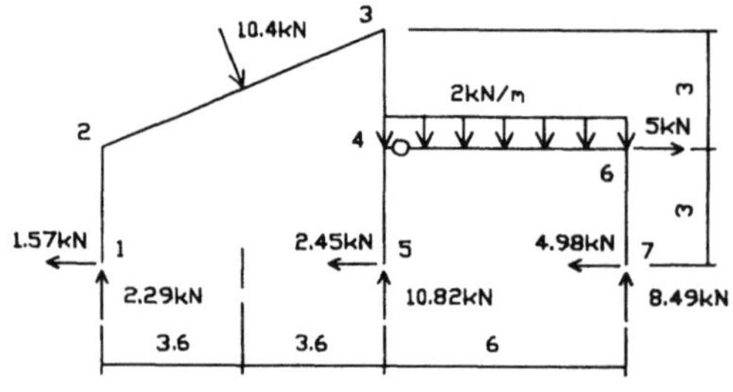

E = 200 kN/mm^2
I = 85×10^6 mm^4

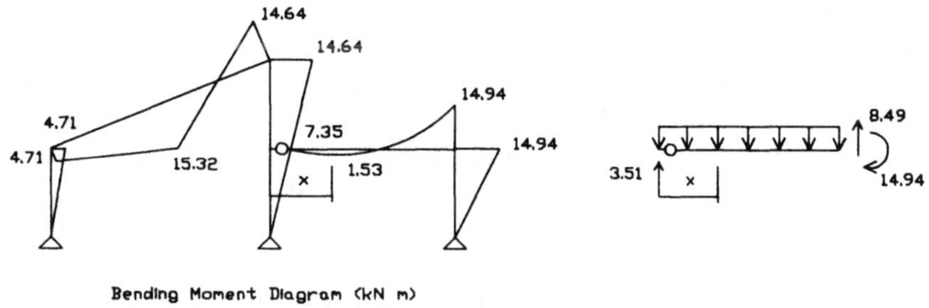

To calculate the horizontal deflection at node 6, apply a unit horizontal load there on a released primary structure:

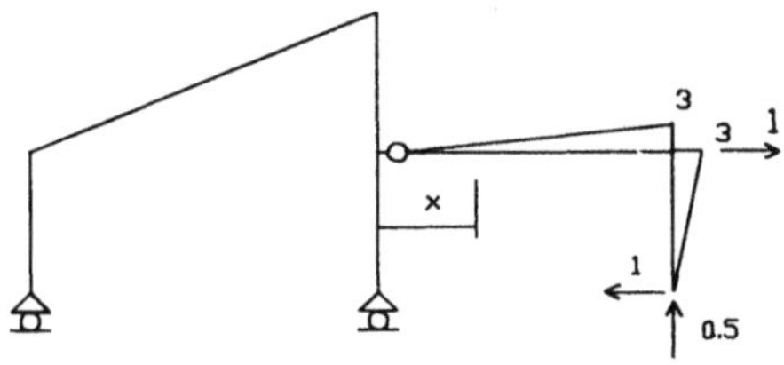

From the principle of virtual forces, the deflection is given by

$$\Delta_6 = \int m \frac{M}{EI} dx$$

$$= \frac{1}{EI} \int_0^6 -\frac{1}{2} x \left(3.51x - 2\frac{x^2}{2} \right) dx + \frac{3}{3EI} (3)(14.94)$$

$$= \frac{36}{EI} + \frac{44.82}{EI}$$

$$= \frac{80.82}{EI}$$

$$= 0.0047 \text{ m}$$

8.2 MATRIX FORMULATION OF THE FLEXIBILITY METHOD

The flexibility method of analysis has been presented in the previous sections, as a general philosophy for the analysis of statically indeterminate structures. In fact the technique has been widely used for many years under a variety of titles, without particular reference to its relationship to the matrix stiffness method. It is relevant to consider the technique in the context of this relationship, in order that both methods may be more clearly understood. Although displacements have been central to this chapter so far, the displacements have been those at the releases and not the general displacements at nominated nodes in the structure. The general displacements of a structure can be calculated by the flexibility method of analysis by an extension of the technique already presented. The method can then be expressed in matrix notation and its relationship to the matrix stiffness method will become more apparent.

Consider the frame of figure 8.2(a) which was previously analysed in chapter 4. The analysis can again be expressed in two parts so that any general loading can be considered as a set of nodal loads plus the fixed end actions. The nodal loads and displacements, as shown in figure 8.2(b), have a relationship given by the equation

$$\begin{Bmatrix} M_3 \\ M_4 \\ Q \end{Bmatrix} = \begin{bmatrix} k_{11} & k_{12} & k_{13} \\ k_{21} & k_{22} & k_{23} \\ k_{31} & k_{32} & k_{33} \end{bmatrix} \begin{Bmatrix} \theta_3 \\ \theta_4 \\ \Delta \end{Bmatrix} \tag{8.5}$$

Equation (8.5) may be written as

$$P_F = K_F \cdot d_F \tag{8.6}$$

where K_F is the structure stiffness matrix based on the three degrees of freedom nominated, ignoring axial deformation.

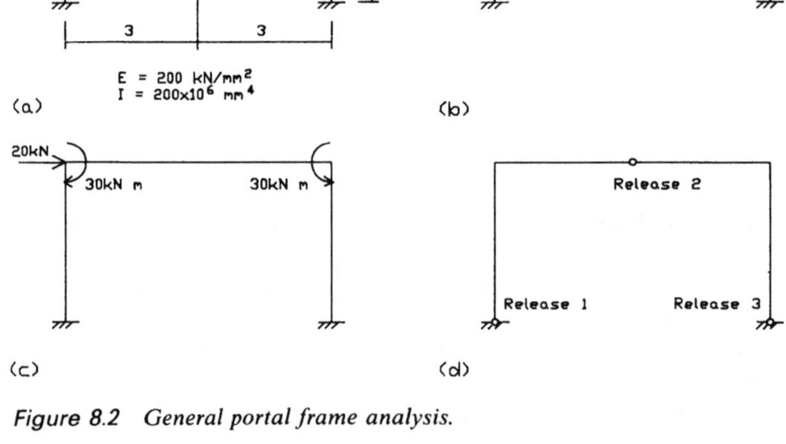

Figure 8.2 General portal frame analysis.

In parallel with the stiffness method, the flexibility method ought to express the inverse relationship to that given by equation (8.5), that is

$$\begin{Bmatrix} \theta_3 \\ \theta_4 \\ \Delta \end{Bmatrix} = \begin{bmatrix} g_{11} & g_{12} & g_{13} \\ g_{21} & g_{22} & g_{23} \\ g_{31} & g_{32} & g_{33} \end{bmatrix} \begin{Bmatrix} M_3 \\ M_4 \\ Q \end{Bmatrix} \tag{8.7}$$

Equation (8.7) may be written as

$$d_F = G \cdot P_F \tag{8.8}$$

where G is the structure flexibility matrix, necessarily the inverse of the matrix K_F. The basic concepts of the flexibility method can now be extended to derive the flexibility matrix of equation (8.8) and to determine the displacements of the frame.

Figure 8.2(c) shows the particular values of the load vector, P_F, which are relevant to the given problem, while figure 8.2(d) shows a released primary structure suitable for flexibility analysis. The initial task is to find the redundant actions which, in this case, are the moments at nodes 1 and 2 and at the mid point of element 3–4. (This latter point can be conveniently described as 'a'.) Proceeding on the basis of the previous examples, and using the moment diagrams of figure 8.3, then equation (8.4) can be written as

$$\begin{Bmatrix} U_{10} \\ U_{20} \\ U_{30} \end{Bmatrix} + \begin{bmatrix} f_{11} & f_{12} & f_{13} \\ f_{21} & f_{22} & f_{23} \\ f_{31} & f_{32} & f_{33} \end{bmatrix} \begin{Bmatrix} x_1 \\ x_2 \\ x_3 \end{Bmatrix} = \begin{Bmatrix} 0 \\ 0 \\ 0 \end{Bmatrix} \tag{8.9}$$

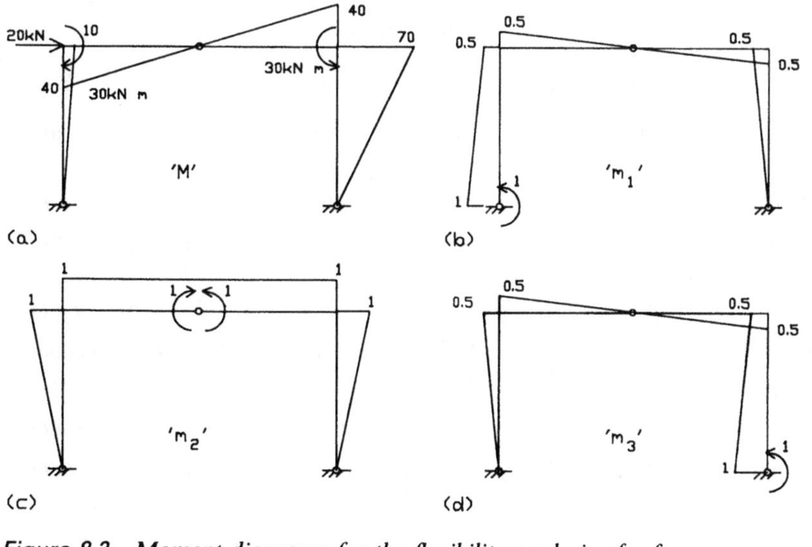

Figure 8.3 Moment diagrams for the flexibility analysis of a frame.

and conveniently expressed in the notation

$$U_R + G_{11} \cdot x = 0 \tag{8.10}$$

where U_R is a vector of displacements at the releases, x is the vector of redundant actions, and G_{11} is a flexibility matrix.

The displacement computations necessary to evaluate the terms of U_R and the flexibility coefficients are summarised in table 8.1, and substituting into equation (8.9) results in

$$\frac{1}{EI}\begin{Bmatrix} -80 \\ 80 \\ -120 \end{Bmatrix} + \frac{1}{EI}\begin{bmatrix} 2.9166 & 0.6666 & 1.5833 \\ 0.6666 & 5.6666 & -0.6666 \\ 1.5833 & -0.6666 & 2.9166 \end{bmatrix}\begin{Bmatrix} x_1 \\ x_2 \\ x_3 \end{Bmatrix} = \begin{Bmatrix} 0 \\ 0 \\ 0 \end{Bmatrix} \tag{8.11}$$

The simultaneous equations of equation (8.11) may be solved to give

$$x_1 = 13.22 \text{ kN m}; \qquad x_2 = -12.00 \text{ kN m}; \quad \text{and} \quad x_3 = 31.22 \text{ kN m}$$

from which the bending moment diagram for the frame of figure 8.2(a) can be readily established, recalling that the fixed end moments must be added to element 3–4.

The displacements of the frame at any point can also be found by superposition of the influences of the actions on the primary structure. The equations which permit this can be seen as an extension to the set of equations already used, in that they again express consistent deformation of the structure. While any displacements can be found, it is nodal displace-

Table 8.1 Displacement computations for flexibility analysis (a) [based on principle of virtual work]

$\int g_1(x)g_2(x)\,dx$	Element length: 1–3	3–a	a–4	4–2	Σ
Mm_1	$-\frac{4}{6}(10(1+1))$	$-\frac{3}{3\cdot 2}\cdot 40\cdot\frac{1}{2}$	$-\frac{3}{3\cdot 2}\cdot 40\cdot\frac{1}{2}$	$-\frac{4}{3}\cdot 70\cdot\frac{1}{2}$	−80
Mm_2	$-\frac{4}{3}\cdot 10\cdot 1$	$-\frac{3}{3\cdot 2}\cdot 40\cdot 1$	$\frac{3}{3\cdot 2}\cdot 40\cdot 1$	$\frac{4}{3}\cdot 70\cdot 1$	80
Mm_3	$-\frac{4}{3}\cdot 10\cdot\frac{1}{2}$	$-\frac{3}{3\cdot 2}\cdot 40\cdot\frac{1}{2}$	$-\frac{3}{3\cdot 2}\cdot 40\cdot\frac{1}{2}$	$-\frac{4}{6}(70(1+1))$	−120
m_1m_1	$\frac{4}{6}(2\frac{1}{2}+1)$	$\frac{3}{3\cdot 2}\cdot\frac{1}{2}\cdot\frac{1}{2}$	$\frac{3}{3\cdot 2}\cdot\frac{1}{2}\cdot\frac{1}{2}$	$\frac{4}{3}\cdot\frac{1}{2}\cdot\frac{1}{2}$	2.91666
m_1m_2	$\frac{4}{6}\cdot 1\cdot 2$	$\frac{3}{3\cdot 2}\cdot 1\cdot\frac{1}{2}$	$-\frac{3}{3\cdot 2}\cdot 1\cdot\frac{1}{2}$	$-\frac{4}{3}\cdot 1\cdot\frac{1}{2}$	0.66666
m_1m_3	$\frac{4}{6}\cdot\frac{1}{2}\cdot 2$	$\frac{3}{3\cdot 2}\cdot\frac{1}{2}\cdot\frac{1}{2}$	$\frac{3}{3\cdot 2}\cdot\frac{1}{2}\cdot\frac{1}{2}$	$\frac{4}{6}\cdot\frac{1}{2}\cdot 2$	1.58333
m_2m_2	$\frac{4}{3}\cdot 1\cdot 1$	$\frac{3}{2}\cdot 1\cdot 1$	$\frac{3}{2}\cdot 1\cdot 1$	$\frac{4}{3}\cdot 1\cdot 1$	5.66666
m_2m_3	$\frac{4}{3}\cdot 1\cdot\frac{1}{2}$	$\frac{3}{2\cdot 2}\cdot 1\cdot\frac{1}{2}$	$-\frac{3}{2\cdot 2}\cdot 1\cdot\frac{1}{2}$	$-\frac{4}{6}\cdot 1\cdot 2$	−0.66666
m_3m_3	$\frac{4}{3}\cdot\frac{1}{2}\cdot\frac{1}{2}$	$\frac{3}{3\cdot 2}\cdot\frac{1}{2}\cdot\frac{1}{2}$	$\frac{3}{2\cdot 2}\cdot\frac{1}{2}\cdot\frac{1}{2}$	$\frac{4}{6}(2\frac{1}{2}+1)$	2.91666

ments of the vector d_F that are of prime interest. The nodal displacements are a combination of the nodal displacements of the primary structure under load (a particular solution) and the nodal displacements of the primary structure due to each of the redundant actions (a complementary solution). Evaluating the nodal displacements in this way leads to another set of equations involving particular values, flexibility coefficients and the redundants:

$$\begin{aligned} U_{\theta_3 0}+f_{41}x_1+f_{42}x_2+f_{43}x_3 &= \theta_3 \\ U_{\theta_4 0}+f_{51}x_1+f_{52}x_2+f_{53}x_3 &= \theta_4 \\ U_{\Delta 0}+f_{61}x_1+f_{62}x_2+f_{63}x_3 &= \Delta \end{aligned} \tag{8.12}$$

The flexibility coefficients of equation (8.12) are again found by the principle of virtual forces using the moment diagrams of figures 8.3 and 8.4, which show the effect of unit action at the degrees of freedom. Equations (8.12) may be written in matrix notation as

$$U_D + G_{01} \cdot x = d_F \tag{8.13}$$

where U_D is a vector of the displacements at the nodes of the primary structure under the particular applied loads, and G_{01} is an extension of the matrix of flexibility coefficients previously defined.

The terms of U_D and G_{01} are given by displacement computations using the moment diagrams of figures 8.3 and 8.4 and the results are

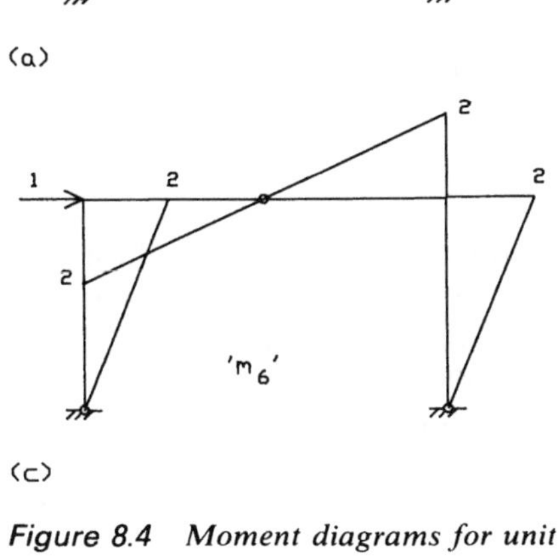

Figure 8.4 Moment diagrams for unit actions at degrees of freedom.

Table 8.2 Displacement computations for flexibility analysis (b)

	Element length				
$\int g_1(x)g_2(x)\,dx$	1–3	3–a	a–4	4–2	Σ
Mm_4	$\frac{4}{3}\cdot 10\cdot\frac{1}{2}$	$-\frac{3}{3\cdot 2}\cdot 40\cdot\frac{1}{2}$	$-\frac{3}{3\cdot 2}\cdot 40\cdot\frac{1}{2}$	$-\frac{4}{3}\cdot 70\cdot\frac{1}{2}$	−60
Mm_5	$-\frac{4}{3}\cdot 10\cdot\frac{1}{2}$	$-\frac{3}{3\cdot 2}\cdot 40\cdot\frac{1}{2}$	$-\frac{3}{3\cdot 2}\cdot 40\cdot\frac{1}{2}$	$\frac{4}{3}\cdot 70\cdot\frac{1}{2}$	20
Mm_6	$\frac{4}{3}\cdot 10\cdot 2$	$\frac{3}{3\cdot 2}\cdot 40\cdot 2$	$\frac{3}{3\cdot 2}\cdot 40\cdot 2$	$\frac{4}{3}\cdot 70\cdot 2$	293.333
m_4m_1	$-\frac{4}{6}(\frac{1}{2}(1+1))$	$\frac{3}{3\cdot 2}\cdot\frac{1}{2}\cdot\frac{1}{2}$	$\frac{3}{3\cdot 2}\cdot\frac{1}{2}\cdot\frac{1}{2}$	$\frac{4}{3}\cdot\frac{1}{2}\cdot\frac{1}{2}$	−0.08333
m_4m_2	$-\frac{4}{3}\cdot 1\cdot\frac{1}{2}$	$\frac{3}{2\cdot 2}\cdot 1\cdot\frac{1}{2}$	$-\frac{3}{2\cdot 2}\cdot 1\cdot\frac{1}{2}$	$-\frac{4}{3}\cdot 1\cdot\frac{1}{2}$	−1.33333
m_4m_3	$-\frac{4}{3}\cdot\frac{1}{2}\cdot\frac{1}{2}$	$\frac{3}{3\cdot 2}\cdot\frac{1}{2}\cdot\frac{1}{2}$	$\frac{3}{3\cdot 2}\cdot\frac{1}{2}\cdot\frac{1}{2}$	$\frac{4}{6}(0.5(1+1))$	0.58333
m_5m_1	$\frac{4}{6}(0.5(1+1))$	$\frac{3}{3\cdot 2}\cdot\frac{1}{2}\cdot\frac{1}{2}$	$\frac{3}{3\cdot 2}\cdot\frac{1}{2}\cdot\frac{1}{2}$	$-\frac{4}{3}\cdot\frac{1}{2}\cdot\frac{1}{2}$	0.58333
m_5m_2	$\frac{4}{3}\cdot 1\cdot\frac{1}{2}$	$\frac{3}{3\cdot 2}\cdot 1\cdot\frac{1}{2}$	$-\frac{3}{3\cdot 2}\cdot 1\cdot\frac{1}{2}$	$\frac{4}{3}\cdot 1\cdot\frac{1}{2}$	1.33333
m_5m_3	$\frac{4}{3}\cdot\frac{1}{2}\cdot\frac{1}{2}$	$\frac{3}{3\cdot 2}\cdot\frac{1}{2}\cdot\frac{1}{2}$	$\frac{3}{3\cdot 2}\cdot\frac{1}{2}\cdot\frac{1}{2}$	$-\frac{4}{6}(\frac{1}{2}(1+1))$	−0.08333
m_6m_1	$-\frac{4}{6}\cdot 2(1+1)$	$-\frac{3}{6}\cdot 2\cdot\frac{1}{2}$	$-\frac{3}{6}\cdot 2\cdot\frac{1}{2}$	$-\frac{4}{3}\cdot 2\cdot\frac{1}{2}$	−5
m_6m_2	$-\frac{4}{3}\cdot 1\cdot 2$	$-\frac{3}{2\cdot 2}\cdot 2\cdot 1$	$\frac{3}{2\cdot 2}\cdot 2\cdot 1$	$\frac{4}{3}\cdot 1\cdot 2$	0
m_6m_3	$-\frac{4}{3}\cdot 2\cdot\frac{1}{2}$	$-\frac{3}{6}\cdot 2\cdot\frac{1}{2}$	$-\frac{3}{6}\cdot 2\cdot\frac{1}{2}$	$-\frac{4}{6}(2(1+1))$	−5

summarised in table 8.2. Equation (8.13) then becomes

$$\frac{1}{EI}\begin{Bmatrix}-60\\ 20\\ 293.3\end{Bmatrix}+\frac{1}{EI}\begin{bmatrix}-0.0833 & -1.3333 & 0.5833\\ 0.5833 & 1.3333 & -0.0833\\ -5.0 & 0.0 & -5.0\end{bmatrix}\begin{Bmatrix}13.22\\ -12.00\\ 31.22\end{Bmatrix}=\begin{Bmatrix}\theta_3\\ \theta_4\\ \Delta\end{Bmatrix} \tag{8.14}$$

that is

$$\frac{1}{EI}\begin{Bmatrix}-26.9\\ 9.11\\ 71.13\end{Bmatrix}=\begin{Bmatrix}\theta_3\\ \theta_4\\ \Delta\end{Bmatrix}$$

and the results compare favourably with those obtained in chapter 4.

8.2.1 Forming the Flexibility Matrix

Although the problem of the previous section has now been solved for both the actions and the displacements of the frame, the flexibility matrix of equation (8.8) has not been explicitly formed. This can be achieved by further examination of the matrices used to this stage.

Equations (8.10) and (8.13) may be combined and written as

$$\begin{Bmatrix}U_R\\ \hline U_D\end{Bmatrix}+\begin{bmatrix}G_{11}\\ \hline G_{01}\end{bmatrix}\cdot\{x\}=\begin{Bmatrix}0\\ \hline d_F\end{Bmatrix} \tag{8.15}$$

However equation (8.15) is still a function of the particular values of U_R and U_D. For other load cases, these particular values may be recalculated but, more conveniently, they can be expressed in terms of unit actions and related to the load vector. Recalling the definition of a flexibility coefficient and referring to the moment diagrams of figures 8.3 and 8.4, it may be seen that

$$\begin{aligned} U_{10} &= M_3 f_{14} + M_4 f_{15} + Q f_{16} \\ U_{20} &= M_3 f_{24} + M_4 f_{25} + Q f_{26} \\ U_{30} &= M_3 f_{34} + M_4 f_{35} + Q f_{36} \end{aligned} \tag{8.16}$$

Equation (8.16) may be written in the notation of

$$U_R = G_{10} \cdot P_F \tag{8.17}$$

where it may be noted from the definition of the flexibility coefficients that G_{10} is the transpose of G_{01}. Similarly it is seen that

$$\begin{aligned} U_{\theta_3 0} &= M_3 f_{44} + M_4 f_{45} + Q f_{46} \\ U_{\theta_4 0} &= M_3 f_{54} + M_4 f_{55} + Q f_{56} \\ U_{\Delta 0} &= M_3 f_{64} + M_4 f_{65} + Q f_{66} \end{aligned} \tag{8.18}$$

which can be expressed as

$$U_D = G_{00} \cdot P_F \tag{8.19}$$

The flexibility coefficients of the matrix G must be calculated in a manner similar to those previously determined, and the results for these coefficients are summarised in Table 8.3.

Substituting equations (8.17) and (8.19) into equation (8.15) gives

$$\left[\frac{G_{10}}{G_{00}}\right] \cdot \{P_F\} + \left[\frac{G_{11}}{G_{01}}\right] \cdot \{x\} = \left\{\frac{0}{d_F}\right\} \tag{8.20}$$

Table 8.3 Displacement computation for flexibility analysis (c)

$\int g_1(x) g_2(x)\, dx$	Element length				Σ
	1–3	3–a	a–4	4–2	
$m_4 m_4$	$\frac{4}{3}\cdot\frac{1}{2}\cdot\frac{1}{2}$	$\frac{3}{3\cdot 2}\cdot\frac{1}{2}\cdot\frac{1}{2}$	$\frac{3}{3\cdot 2}\cdot\frac{1}{2}\cdot\frac{1}{2}$	$\frac{4}{3}\cdot\frac{1}{2}\cdot\frac{1}{2}$	0.91666
$m_4 m_5$	$-\frac{4}{3}\cdot\frac{1}{2}\cdot\frac{1}{2}$	$\frac{3}{3\cdot 2}\cdot\frac{1}{2}\cdot\frac{1}{2}$	$\frac{3}{3\cdot 2}\cdot\frac{1}{2}\cdot\frac{1}{2}$	$-\frac{4}{3}\cdot\frac{1}{2}\cdot\frac{1}{2}$	−0.41666
$m_4 m_6$	$\frac{4}{3}\cdot 2\cdot\frac{1}{2}$	$-\frac{3}{3\cdot 2}\cdot 2\cdot\frac{1}{2}$	$-\frac{3}{3\cdot 2}\cdot 2\cdot\frac{1}{2}$	$-\frac{4}{3}\cdot 2\cdot\frac{1}{2}$	−1.000
$m_5 m_5$	$\frac{4}{3}\cdot\frac{1}{2}\cdot\frac{1}{2}$	$\frac{3}{3\cdot 2}\cdot\frac{1}{2}\cdot\frac{1}{2}$	$\frac{3}{3\cdot 2}\cdot\frac{1}{2}\cdot\frac{1}{2}$	$\frac{4}{3}\cdot\frac{1}{2}\cdot\frac{1}{2}$	0.91666
$m_5 m_6$	$-\frac{4}{3}\cdot\frac{1}{2}\cdot 2$	$-\frac{3}{3\cdot 2}\cdot\frac{1}{2}\cdot 2$	$-\frac{3}{3\cdot 2}\cdot 2\cdot\frac{1}{2}$	$\frac{4}{3}\cdot\frac{1}{2}\cdot 2$	−1.000
$m_6 m_6$	$\frac{4}{3}\cdot 2\cdot 2$	$\frac{3}{3\cdot 2}\cdot 2\cdot 2$	$\frac{3}{3\cdot 2}\cdot 2\cdot 2$	$\frac{4}{3}\cdot 2\cdot 2$	14.66666

Re-arranging, equation (8.20) may be conveniently written as

$$\left[\begin{array}{c|c} G_{00} & G_{01} \\ \hline G_{10} & G_{11} \end{array}\right] \cdot \left\{\begin{array}{c} P_F \\ \hline x \end{array}\right\} = \left\{\begin{array}{c} d_F \\ \hline 0 \end{array}\right\} \tag{8.21}$$

representing equations (8.10) and (8.13) in a compact and general form.

Multiplying out equation (8.21) gives the equations

$$G_{00} \cdot P_F + G_{01} \cdot x = d_F \tag{8.22a}$$

$$G_{10} \cdot P_F + G_{11} \cdot x = 0 \tag{8.22b}$$

and from equation (8.22b)

$$x = -G_{11}^{-1} \cdot G_{10} \cdot P_F$$

Substituting for x into equation (8.22a) gives

$$G_{00} \cdot P_F - G_{01} \cdot G_{11}^{-1} \cdot G_{10} \cdot P_F = d_F$$

that is

$$[G_{00} - G_{01} \cdot G_{11}^{-1} \cdot G_{10}] \cdot P_F = d_F \tag{8.23}$$

Comparing equation (8.23) with equation (8.8), it is seen that the structure flexibility matrix, G, is given by

$$G = [G_{00} - G_{01} \cdot G_{11}^{-1} \cdot G_{10}]$$

In particular, for the frame of figure 8.2(a):

$$G_{11}^{-1} = EI \begin{bmatrix} 0.53611 & -0.10000 & -0.31389 \\ -0.10000 & 0.20000 & 0.10000 \\ -0.31389 & 0.10000 & 0.53611 \end{bmatrix}$$

$$G_{01} \cdot G_{11}^{-1} \cdot G_{10} = \frac{1}{EI} \begin{bmatrix} 0.39444 & -0.33889 & -0.55555 \\ -0.33889 & 0.39444 & -0.55555 \\ -0.55555 & -0.55555 & 11.11050 \end{bmatrix}$$

$$G_{00} = \frac{1}{EI} \begin{bmatrix} 0.91666 & -0.41666 & -1.00000 \\ -0.41666 & 0.91666 & -1.00000 \\ -1.00000 & -1.00000 & 14.66666 \end{bmatrix}$$

So that

$$G = \frac{1}{EI} \begin{bmatrix} 0.52222 & -0.07777 & -0.44444 \\ -0.07777 & 0.52222 & -0.44444 \\ -0.44444 & -0.44444 & 3.55555 \end{bmatrix}$$

and it may readily be confirmed that the matrix G is the inverse of the matrix K_F given in example 4.3 in chapter 4.

While the preceding material has been developed from a specific example, the approach is quite general and equation (8.21) summarises the features of the matrix formulation of the flexibility method. The various matrices as defined in equation (8.21) simply take on a size directly related to the degrees of freedom introduced, and to the degree of statical indeterminacy of the structure. For a statically determinate structure, equation (8.22a) is simply

$$G_{00} \cdot P_F = d_F \tag{8.24}$$

For a statically indeterminate structure, equation (8.21) may be seen as an extension of equation (8.24), where the load vector has been augmented by the redundant actions and both the flexibility matrix and the displacement vector have been suitably expanded.

Where displacements of the structure are not sought, equation (8.21) reduces to

$$[G_{10} \mid G_{11}] \cdot \begin{Bmatrix} P_F \\ \hline X \end{Bmatrix} = \{0\} \tag{8.25}$$

Expanding equation (8.25) gives

$$G_{10} \cdot P_F + G_{11} \cdot X = 0$$

which, using equation (8.17), is

$$U_R + G_{11} \cdot X = 0 \tag{8.26}$$

Equation (8.26) is the matrix formulation of the set of equations expressed in equation (8.4).

As has been previously mentioned, the selection of the redundant actions in any structure is a matter of choice. It is difficult to standardise this selection unless the structure takes on a regular, repetitive form. This is one of the reasons why the flexibility method is not widely used in the computer analysis of structures.

8.2.2 Analysis for Temperature and Support Movement

The general nature of equation (8.26) may be demonstrated by two further conditions imposed on the frame of figure 8.2(a). The flexibility method of analysis can readily handle problems with either temperature load, initial lack of fit of elements, or support movements on an indeterminate structure. The general philosophy is simply that the vector U_R of equation (8.26) represents the displacements at the releases, whatever the cause. Since the released structure is necessarily statically determinate, the displacements U_R can be calculated for various effects as discussed in section 7.3.3.

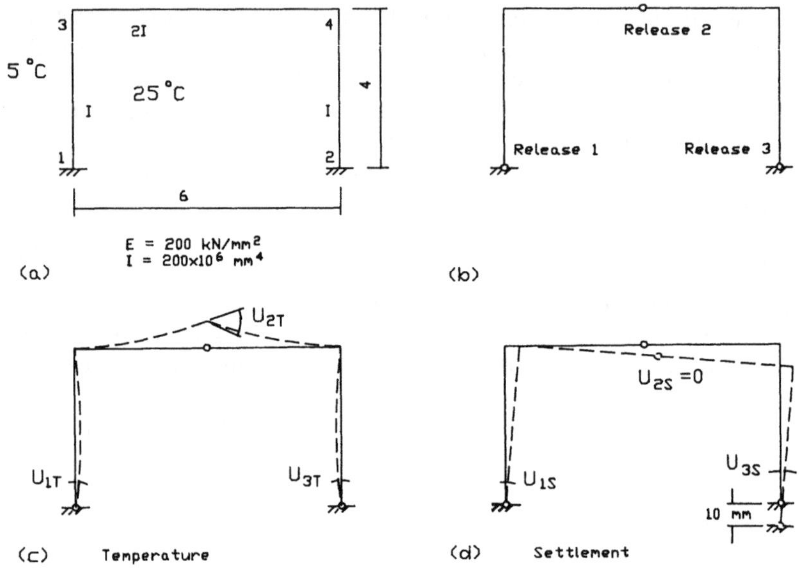

Figure 8.5 *Displacements at releases.*

Figure 8.5(a) shows the portal frame that was the subject of the study of the previous section. In this instance, the frame is subjected to a temperature variation across the elements due to the internal and external temperatures shown, and to a vertical settlement of the support at node 2 of 10 mm. An analysis is to be carried out to find the moments in the frame due to these conditions considered separately. In addition to the data given, it is assumed that the columns have a depth of section of 400 mm; the beam has a depth of 480 mm; and that all elements are of steel with a coefficient of linear expansion of 14×10^{-6} per °C. Further, it is assumed that the construction temperature of the frame was the mean temperature of 15°C, so that no longitudinal expansion of the elements will occur, although this effect could be readily incorporated into the problem.

The temperature gradient across each element of the released structure of figure 8.5(b) causes the rotations at the releases as shown in figure 8.5(c). These rotations may be calculated by virtual work using the expression given in section 7.3.3 as

$$\theta_i = \int m_i \, d\phi$$

where $d\phi = (2\alpha \Delta T / d)\, dx$, as specified in figure 7.7.

The unit moment diagrams of figures 8.3(b), (c) and (d) form the basis for the evaluation of the integral expression in conjunction with the constant curvature caused by the temperature gradient, so that

$$U_{1T}=\int m_1\,d\phi;\qquad U_{2T}=\int m_2\,d\phi;\quad\text{and}\quad U_{3T}=\int m_3\,d\phi$$

For the columns

$$d\phi=\frac{0.000014(10)}{200}\,dx$$

and for the beams

$$d\phi=\frac{0.000014(10)}{240}\,dx$$

from which

$$U_{1T}=-0.0014\text{ radians}$$

$$U_{2T}=-0.0063\text{ radians}$$

and

$$U_{3T}=0.0014\text{ radians}$$

The specific form of equation (8.26) relevant to the problem is then

$$\begin{Bmatrix}-0.0014\\-0.0063\\0.0014\end{Bmatrix}+\frac{1}{EI}\begin{bmatrix}2.9166 & 0.6666 & 1.5833\\0.6666 & 5.6666 & -0.6666\\1.5833 & -0.6666 & 2.9166\end{bmatrix}\begin{Bmatrix}x_1\\x_2\\x_3\end{Bmatrix}=\begin{Bmatrix}0\\0\\0\end{Bmatrix}$$

and the equations may be solved to give $x_1=22.4$ kN m; $x_2=39.2$ kN m; and $x_3=-22.4$ kN m—resulting in the bending moment diagram of figure 8.6(a).

The rotations at the releases due to the support settlement are shown in figure 8.5(d) and the values may be deduced either directly from the geometry of the frame or by virtual work. In this case, $U_{1s}=$

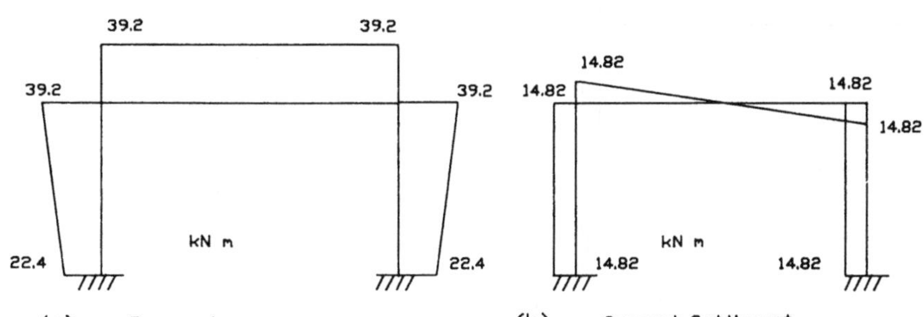

Figure 8.6 Moment diagrams for temperature and settlement.

−0.00166 radians; $U_{2s}=0$; and $U_{3s}=-0.00166$ radians. Substituting for U_R into equation (8.11) and solving for the redundant moments gives $x_1=$ 14.82 kN m; $x_2=0$; and $x_3=14.82$ kN m—from which the bending moment diagram of figure 8.6(b) has been drawn.

In both cases, the results are dependent on the specific values of the flexural rigidity, *EI*, of the elements of the frame. This is in contrast to the moments due to applied loads which, in a statically indeterminate frame, may be calculated when only relative values of flexural rigidity are specified.

8.2.3 Element Flexibility Matrices

Although this chapter started with a comparison with the stiffness method of analysis, the flexibility method has been presented without introducing element behaviour, or the element flexibility matrix. However, element flexibility matrices are recognized and one has already been used in chapter 3, although it was not noted as such.

While a series of element flexibility matrices can be introduced to cover different element types, they do not exist as the inverse of all the corresponding element stiffness matrices. For example, it is not possible to invert the beam element stiffness matrix of equation (3.25), since all of the displacements cannot be unspecified; boundary condtions must be introduced, Nevertheless, if one end of the beam element were fully restrained to give a cantilever beam as shown in figure 8.7, then a beam element flexibility matrix could be found.

With reference to figure 8.7, the relationship sought has the form:

$$\begin{Bmatrix} d_2 \\ \theta_2 \end{Bmatrix} = \begin{bmatrix} g_{11} & g_{12} \\ g_{21} & g_{22} \end{bmatrix} \begin{Bmatrix} v_{21} \\ m_{21} \end{Bmatrix} \tag{8.27}$$

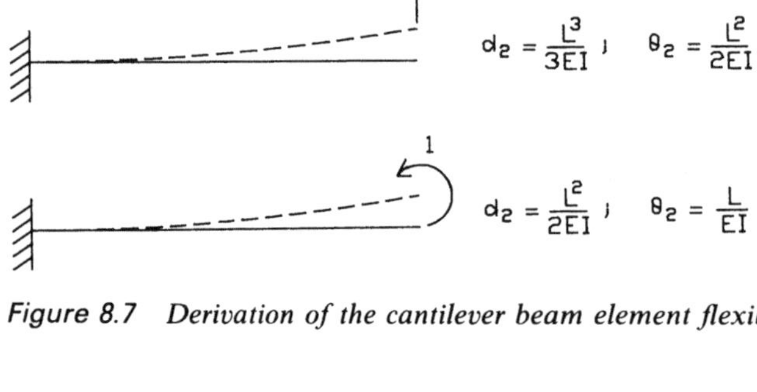

Figure 8.7 Derivation of the cantilever beam element flexibility matrix.

The elements of the flexibility matrix of equation (8.27) may be found by a process similar to that used in the direct stiffness method. If v_{21} is given a unit value while m_{21} is zero, then the resulting displacements at node 2 of the beam are equivalent to the elements g_{11} and g_{21} of the flexibility matrix. Similarly, if m_{21} is given a unit value while v_{21} is zero, then the elements g_{12} and g_{22} can be found. It is this influence of unit actions on a structure which characterises the flexibility method. The results of such actions are shown in figure 8.7 and they may be verified by using the principle of virtual forces, or any other technique which gives the displacements of a statically determinate structure.

Equation (8.27), expressing a beam element displacement-force relationship, then gives the cantilever beam element flexibility matrix as

$$G=\begin{bmatrix} \dfrac{L^3}{3EI} & \dfrac{L^2}{2EI} \\ \dfrac{L^2}{2EI} & \dfrac{L}{EI} \end{bmatrix} \tag{8.28}$$

It can be readily shown that this matrix is the inverse of the corresponding 2 by 2 sub-matrix of the beam element stiffness matrix of equation (3.25). The matrix of equation (8.28) can also be augmented to accommodate axial load effects. Using the notation of equation (6.11), the required matrix is

$$G=\begin{bmatrix} \dfrac{L}{EA} & 0 & 0 \\ 0 & \dfrac{L^3}{3EI} & \dfrac{L^2}{2EI} \\ 0 & \dfrac{L^2}{2EI} & \dfrac{L}{EI} \end{bmatrix} \tag{8.29}$$

since the inverse of the axial force–extension relationship,

$$p_{21}=\frac{EA}{L}s_2$$

is clearly

$$s_2=\frac{L}{EA}p_{21}$$

An alternative and useful beam element flexibility matrix is given by applying other boundary conditions to the unrestrained beam element. In figure 8.8, the end translations of the beam element have been restrained

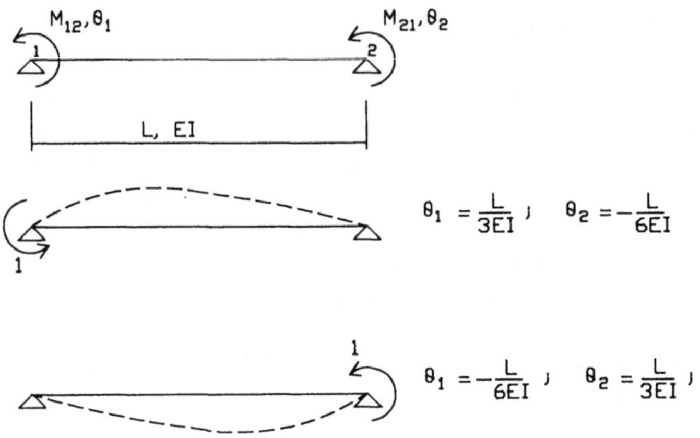

Figure 8.8 *Derivation of a simply supported beam element flexibility matrix.*

and unit actions applied to give the rotation responses. This results in the beam element flexibility matrix of the form:

$$G = \begin{bmatrix} \dfrac{L}{3EI} & -\dfrac{L}{6EI} \\ -\dfrac{L}{6EI} & \dfrac{L}{3EI} \end{bmatrix} \tag{8.30}$$

which is the inverse of the beam element stiffness matrix when only rotations are admitted as degrees of freedom. The matrix given by equation (8.30) was in fact introduced in chapter 3 in equation (3.17).

8.2.4 Fixed End Actions by Flexibility Analysis

Fixed end actions play an important role in the stiffness method in accommodating general transverse loads on the elements. The topic was introduced in chapter 3, particularly in section 3.8 where it can now be seen that the principles of the flexibility method were used. Consider again the case of a beam built in at both ends and under a uniformly distributed load q, as shown in figure 8.9(a). Ignoring the advantages that the symmetry of the beam offers, a suitable released primary structure is the cantilever beam of figure 8.9(b). The particular solution then consists of the translation and the rotation of the free end of the beam. The fixed end actions sought are the redundants x_1 and x_2; the vertical reaction and the reactive end moment respectively, at node 2.

The problem is of the form given by equation (8.26), that is

$$U_R + G_{11} \cdot X = 0$$

where the vector U_R represents the displacements at the releases due to the

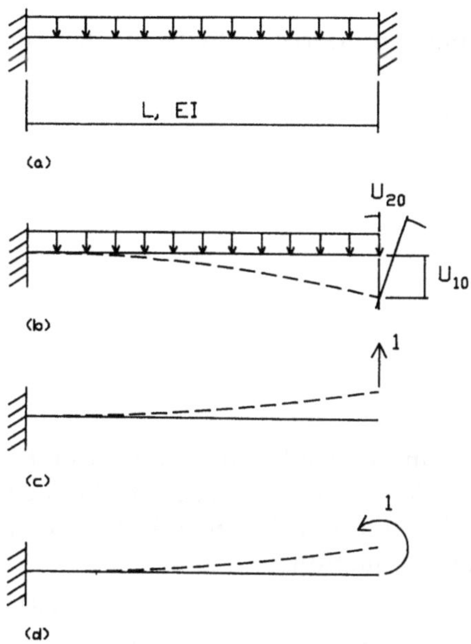

Figure 8.9 *Fixed end actions by flexibility analysis.*

applied loads, namely, the particular solution. Using the principle of virtual forces, the components of U_R are easily obtained as

$$U_{\mathrm{R}} = \begin{Bmatrix} -\dfrac{qL^4}{8EI} \\ -\dfrac{qL^3}{6EI} \end{Bmatrix}$$

The matrix G_{11} is already defined as the cantilever beam element flexibility matrix of equation (8.28).

Equation (8.26) is then

$$\begin{Bmatrix} -\dfrac{qL^4}{8EI} \\ -\dfrac{qL^3}{6EI} \end{Bmatrix} + \begin{bmatrix} \dfrac{L^3}{3EI} & \dfrac{L^2}{2EI} \\ \dfrac{L^2}{2EI} & \dfrac{L}{EI} \end{bmatrix} \begin{Bmatrix} x_1 \\ x_2 \end{Bmatrix} = \begin{Bmatrix} 0 \\ 0 \end{Bmatrix}$$

Solving for X then gives

$$\begin{Bmatrix} X_1 \\ X_2 \end{Bmatrix} = -\begin{bmatrix} \dfrac{L^3}{3EI} & \dfrac{L^2}{2EI} \\ \dfrac{L^2}{2EI} & \dfrac{L}{EI} \end{bmatrix}^{-1} \begin{Bmatrix} -\dfrac{qL^4}{8EI} \\ -\dfrac{qL^3}{6EI} \end{Bmatrix}$$

The required inverse is the corresponding sub-matrix of the beam element stiffness matrix of equation (3.25); thus

$$\begin{Bmatrix} x_1 \\ x_2 \end{Bmatrix} = -\begin{bmatrix} \dfrac{12EI}{L^3} & -\dfrac{6EI}{L^2} \\ -\dfrac{6EI}{L^2} & \dfrac{4EI}{L} \end{bmatrix} \begin{Bmatrix} -\dfrac{qL^4}{8EI} \\ -\dfrac{qL^3}{6EI} \end{Bmatrix}$$

$$= \begin{Bmatrix} \dfrac{qL}{2} \\ -\dfrac{qL^2}{12} \end{Bmatrix}$$

which is the required result. The remaining fixed end actions can be recovered by equilibrium. Obviously other fixed end actions can be obtained with appropriate changes to the particular solution. In general, the vector of fixed end actions is given by the matrix equation

$$X = -G_{11}^{-1} \cdot U_{\mathrm{R}} \tag{8.31}$$

It should be noted that the displacements and element actions have been defined with regard to the same sign convention as that used with the standard beam element throughout this text.

8.3 PROBLEMS FOR SOLUTION

8.1 For each of the structures shown in figure P8.1, calculate the forces in the struts and ties, and draw the final bending moment diagram for all the flexural elements. Axial effects in the flexural elements may be ignored.

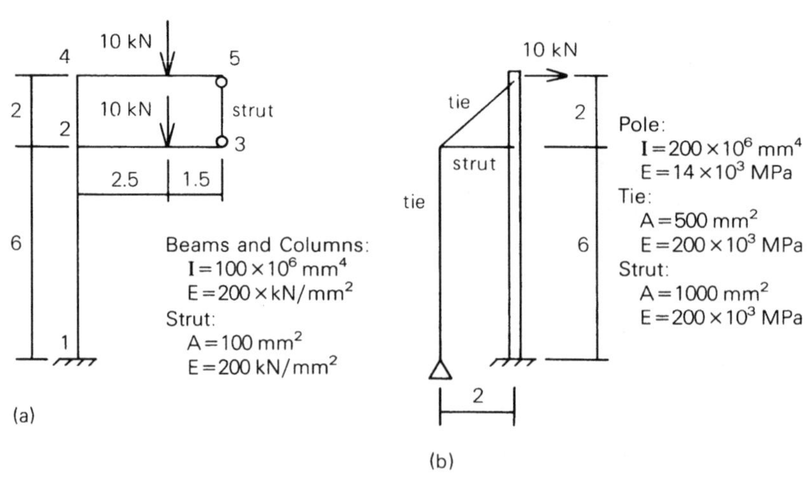

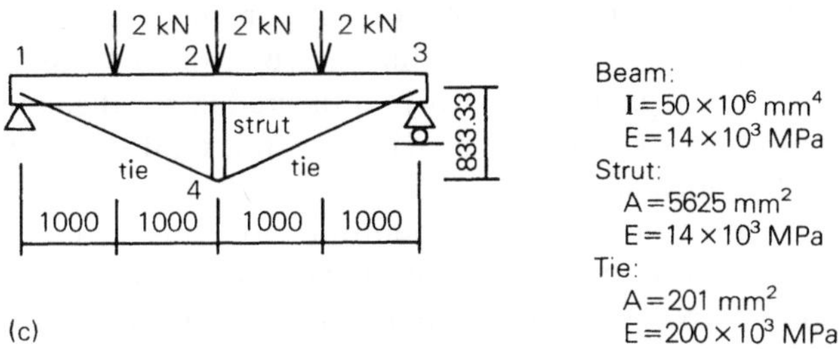

(c)

Figure P8.1.

8.2 Calculate the deflection at the centre of the beam of figure P8.1(c).

8.3 Using the flexibility method, analyse the structures shown in figure P8.2. Draw the bending moment and torsion diagrams as appropriate.

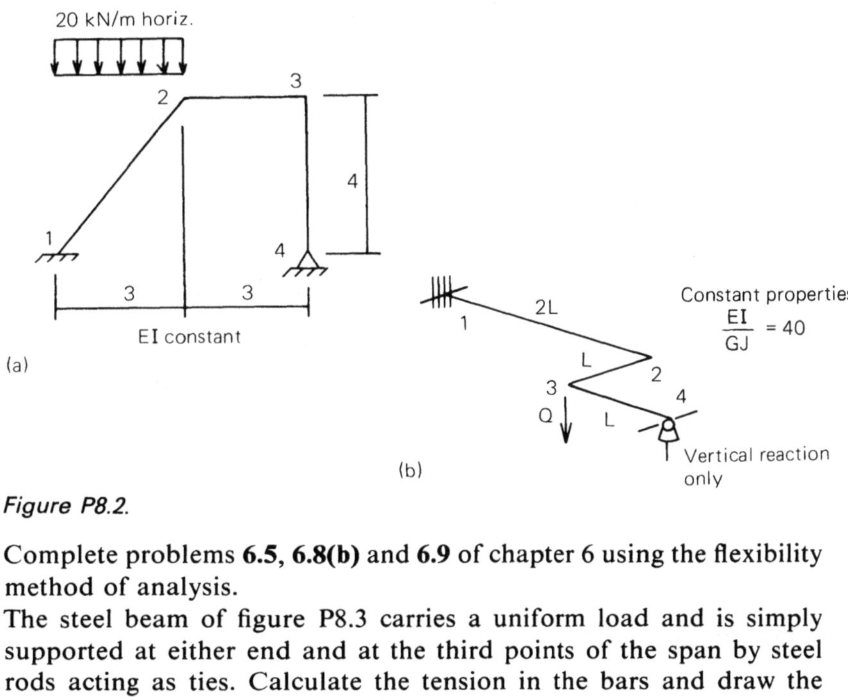

Figure P8.2.

8.4 Complete problems **6.5, 6.8(b)** and **6.9** of chapter 6 using the flexibility method of analysis.

8.5 The steel beam of figure P8.3 carries a uniform load and is simply supported at either end and at the third points of the span by steel rods acting as ties. Calculate the tension in the bars and draw the bending moment diagram for the beam.

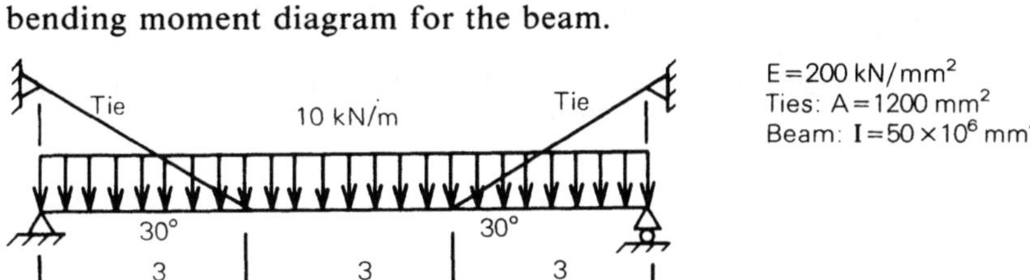

Figure P8.3.

8.6 Figure P8.4 shows a structural system for supporting a cantilevered roof of a grandstand. The principal column, ABC, raker beam, BD, and cantilevered rafter, CE, are all rigidly connected flexural elements, while the remaining struts and ties providing additional support are all pin-connected elements.

Analyse the system to determine the bending moments in the frame and the forces in the ties and struts due to the uniformly distributed load of intensity of 5 kN/m, acting normally along the rafter as shown.

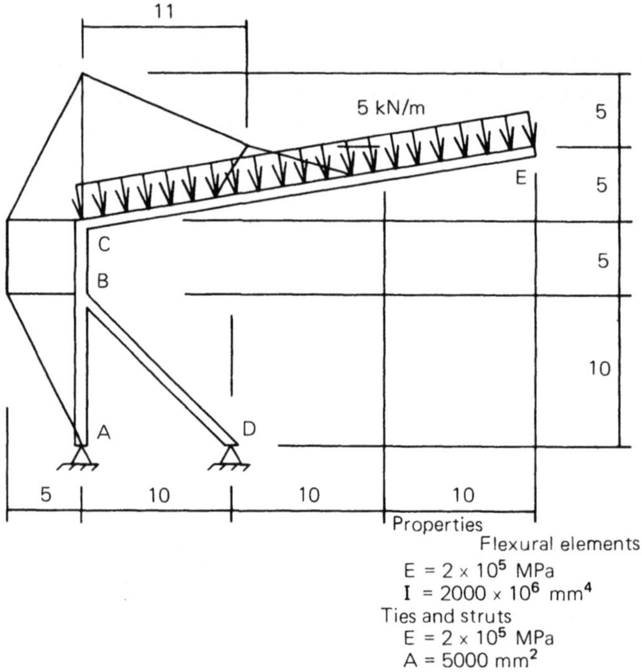

Figure P8.4.

FURTHER READING

Weaver, Jr, W. and Gere, J. M., *Matrix Analysis of Framed Structures*, 2nd edn, Van Nostrand, New York, 1960.

White, R. N., Gergely, P. and Sexsmith, R. G., *Structural Engineering*, Combined edn, Vols 1 and 2, Wiley, New York, 1976.

Chapter 9
The Approximate Analysis of Structures

Experienced structural engineers have expressed a concern that the adoption of computer-based methods of analysis and design will lead to a generation of structural engineers who have little feel for and understanding of structural behaviour. Engineering educators are aware of this and many have noted the blind fascination that students seem to have for numbers forming the output of a computer. There are several ways of tackling the problem. One solution lies in the way in which the interface between the user and the computer program is developed. The user should have a feeling of being in control and not being divorced from an automatic process. Alternatively, and more importantly, structural engineers must be trained continually to exercise engineering judgement on the results from computer programs or other methods of analysis. One way of achieving this last objective is to introduce a study of approximate methods of structural analysis. In one sense, all structural analysis is approximate since some idealisation of structural behaviour is involved in proposing the model for analysis. However, there are clear cases where a known principle is relaxed, or a condition is assumed, that makes the analysis approximate in the accepted sense of the word.

Engineering is not an exact science. There is still plenty of art to engineering design, and scope for the imaginative and innovative mind. Often the professed mathematical rigour tends to overshadow the true nature of structural engineering, and creative minds feel restricted. Skills with approximate analysis ought to build up confidence and encourage creative design. Apart from the broad philosophical reason given, there are two clear situations where approximate analysis is appropriate. The first of these is in the area of preliminary design. Details of the design process are discussed in chapter 10, but it is widely appreciated that at the early stages of design, a number of alternative solutions are usually proposed. The experienced engineer can quickly appraise the alternatives and size the elements in order that costs can be estimated. Although experience may tend to be mixed with approximate analysis, there is a clear case for the latter. Approximate analysis is also relevant later in the design process of statically indeterminate structures. Since the analysis is based on the behaviour of the element, element properties must be known. The analysis of a continuous beam, for example, is dependent on the relative stiffness of each span, expressed through the flexural rigidity, EI. An estimate of the beam properties is therefore necessary to start the analysis cycle. The second situation, which has already been highlighted, is the use of approximate methods to check the results of a more formal analysis.

9.1 APPROXIMATE ANALYSIS OF BEAMS AND RECTANGULAR FRAMES

Statically indeterminate beams and frames can be analysed by a consideration of the behaviour of a statically determinate version, known as the primary structure. This is the basis of the flexibility method of analysis discussed in chapter 8. For structures dominated by flexural actions, it is frequently convenient to introduce moment releases, in the form of pins, to reduce the structure to a determinate form. As a statically determinate system, equilibrium alone will then be sufficient for the analysis to proceed to give the actions in the primary structure. Compatibility is generally violated at the releases, and the actions there must subsequently be considered to give the correct analysis for the indeterminate structure. If, by chance, moment releases were selected at points in the structure where no moments were acting, and there were no other releases in the structure, then the solution of the primary structure would be the solution to the given structure.

There are indeed points of zero moment in beams and frames corresponding to points of inflection in the deflection curve; that is, points where the curvature and moment change sign. With experience it is possible to estimate the position of points of inflection based on sketching deflected shapes. Provided sufficient such points are nominated, the structure is effectively released to the statically determinate form where analysis by equilibrium can proceed.

The procedure outlined forms the basis of an approximate method of analysis, the accuracy of which is a function of how accurately the points of inflection are selected. It is important to stress that the analysis is still based on equilibrium principles and at no stage can either the whole or part of the structure violate equilibrium. Although the principle can be applied to any structural system, guidance on the selection of points is readily available for continuous beams and low-rise rectangular frames.

9.1.1 Flexural Elements and Points of Inflection

A study of the exact analysis of a number of standard beam problems is a useful starting point for locating points of inflection. Solutions to such problems have been given in chapter 3 and a useful summary may be found in appendix B. Working from a known solution, it is a simple matter to locate a point of inflection. An expression for bending moment as a function of position, say x, can be written down and equated to zero. The resulting equation can then be solved for x. Figure 9.1 shows the results for such an

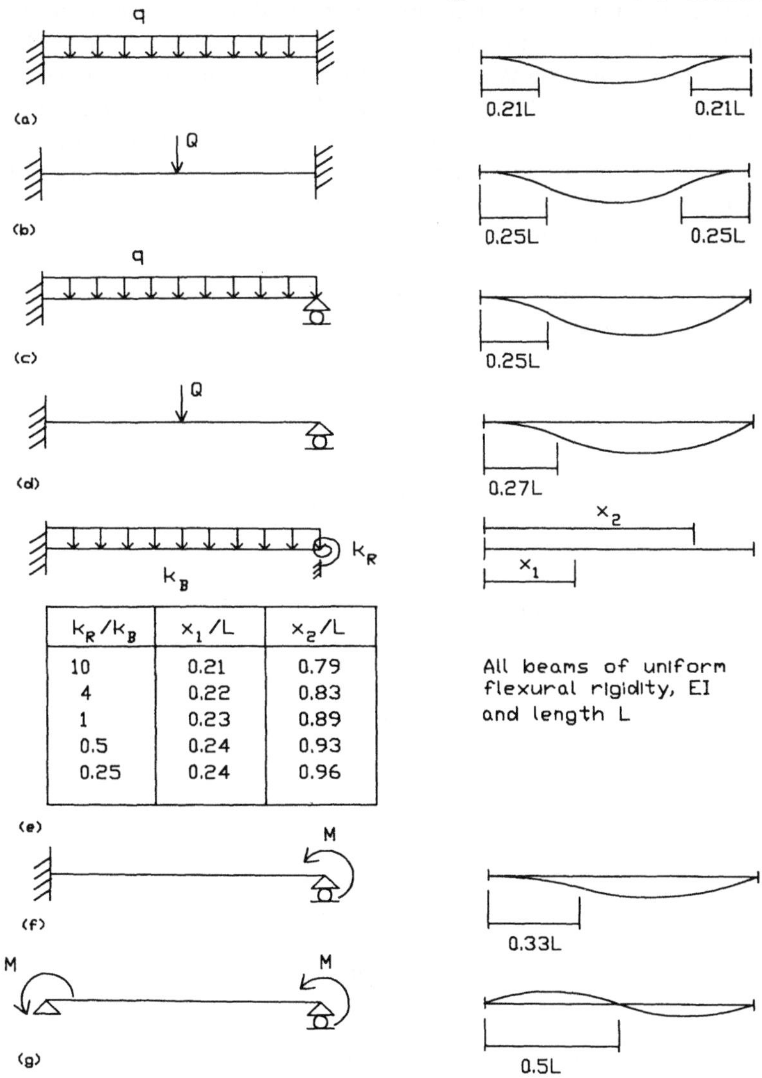

k_R/k_B	x_1/L	x_2/L
10	0.21	0.79
4	0.22	0.83
1	0.23	0.89
0.5	0.24	0.93
0.25	0.24	0.96

Figure 9.1 *Points of inflection in beams.*

analysis for the beams illustrated. Particular attention is drawn to case (e) in figure 9.1, where one end of the beam is restrained by a rotational spring of stiffness, k_R. Such a restraint may be provided by other elements framing into the joint such as a column, as in the case of a simple bent. One of the solutions for case (e) of figure 9.1 is given in example 9.1 to illustrate the point.

It is possible to use the data of figure 9.1 to provide guidance in selecting points of inflection in the deflected shape of continuous beams and frames. Coupled with reasonable sketches of the elastic curve, points of inflection can be chosen to give a basis for an approximate analysis. It may be noted that the range over which the inflection points move is small: from $0.21L$ to $0.27L$ for the beams under transverse load. Further, the points of inflection move away from the restrained beam end towards the end of least restraint, with one point of inflection becoming coincident with the pin end in cases (c) and (d). This is shown again in the table accompanying case (e) of figure 9.1, which also indicates that with the ratio of k_R/k_B equal to 10, both ends of the beam are effectively restrained.

Example 9.1: Beam Analysis

Given data:

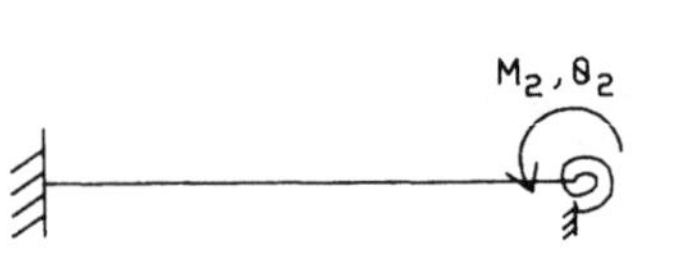

given $k_R/k_B = 0.25$

Analysis by Direct Stiffness Method:

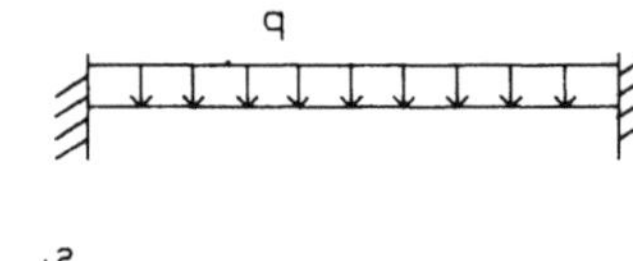

$$M_2 = k_{11} \cdot \theta_2$$

Fixed End Moment:

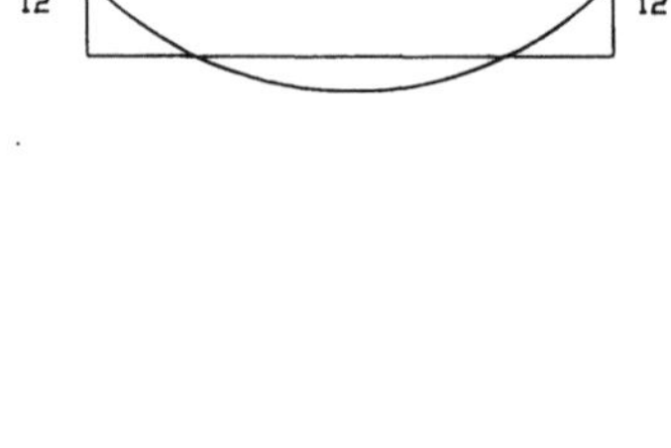

$$k_{11} = \frac{4EI}{L} + k_R$$

$$= \frac{5EI}{L}$$

$$\therefore \theta_2 = \frac{qL^3}{60EI}$$

Element actions:

$$\begin{Bmatrix} m_{12} \\ m_{21} \end{Bmatrix} = \frac{EI}{L}\begin{bmatrix} 4 & 2 \\ 2 & 4 \end{bmatrix}\begin{Bmatrix} 0 \\ \theta_2 \end{Bmatrix}$$

Final end moments:

$$m'_{12} = m_{12} + \frac{qL^2}{12} = 0.11\dot{6}qL^2$$

$$m'_{21} = m_{21} - \frac{qL^2}{12} = -0.01\dot{6}qL^2$$

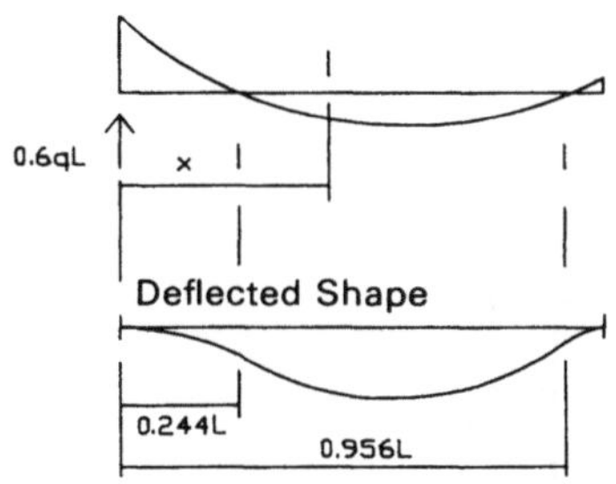

$$M(x) = 0.6qLx - qx^2/2 - 0.116qL^2$$

$M(x) = 0$ when

$$x^2 - 1.2Lx + 0.2333L^2 = 0$$

$$x = L(1.2 \pm 0.71)/2$$

$$= 0.244L \text{ or } 0.956L$$

Cases (e) and (f) indicate the behaviour of a beam under end moments only. In both cases, the end moments are anticlockwise and there is only one point of inflection in the span. Only one other case of end moment is possible, and that is where the end moments are acting in opposite directions. In this case the span will be in single curvature with no point of inflection. It is apparent then that no more than one point of inflection can occur in a span without transverse loads.

An approximate analysis can start with the sketching of the deflected shape of the structure. Boundary conditions must be preserved along with the angular relationship in a rigid joint of a frame. Axial shortening is ignored and the flexural element is regarded as maintaining its overall length in spite of the curvature. While the transverse scale of a deflected shape must be exaggerated, this should not be excessive. A good starting point is to nominate all known positions, such as those at points of support, followed by a sketch of the slope through those points. The loads and the relative stiffness of adjoining elements give an indication of the likely magnitude and direction of the slope. Two studies, indicating a procedure for sketching deflected shapes are shown in figure 9.2. The deflected shapes, coupled with the data of figure 9.1, suggest where points of inflection should be nominated. Since the bending moment at the point of inflection is zero, the appropriate number of equations of condition will have been introduced to lead to a statically determinate system. The analysis can then be carried out by

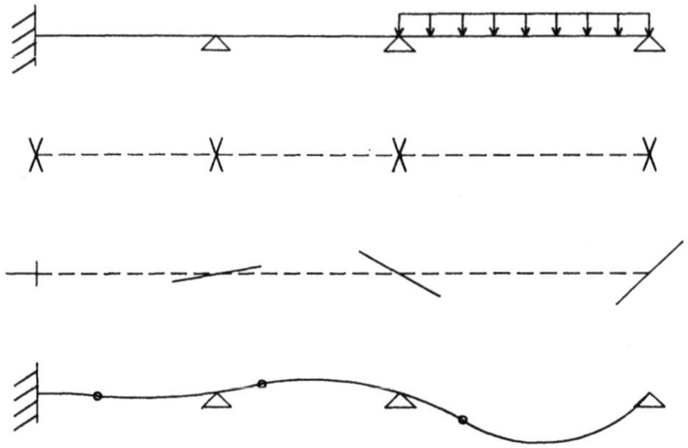

(a) A Continuous Beam Model

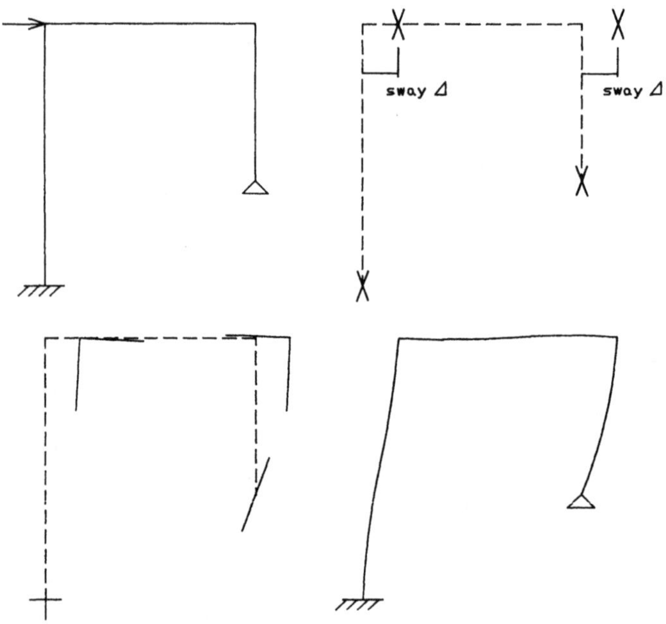

(b) Rigid Jointed Frame under Lateral Load

Figure 9.2 *Sketching deflected shapes.*

equilibrium alone. For the continuous beam of figure 9.2(a), under transverse load only the degree of statical indeterminacy is three, and three equations of condition have been introduced. In the first span, the point of inflection might be nominated at the third point of the span, while for the remaining spans the point of inflection could be taken at the quarter points.

9.1.2 Approximate Analysis of a Two-bay Rectangular Frame

A study of the behaviour of a two-bay rectangular frame under lateral load gives a further insight into the techniques of approximate analysis and the concepts can be extended into more complex frames. The frame of figure 9.3(a) has a degree of statical indeterminacy of three. If a flexibility analysis were to be carried out, then three releases could be introduced to give the statically determinate structure of figure 9.3(b). However, it is clear that the

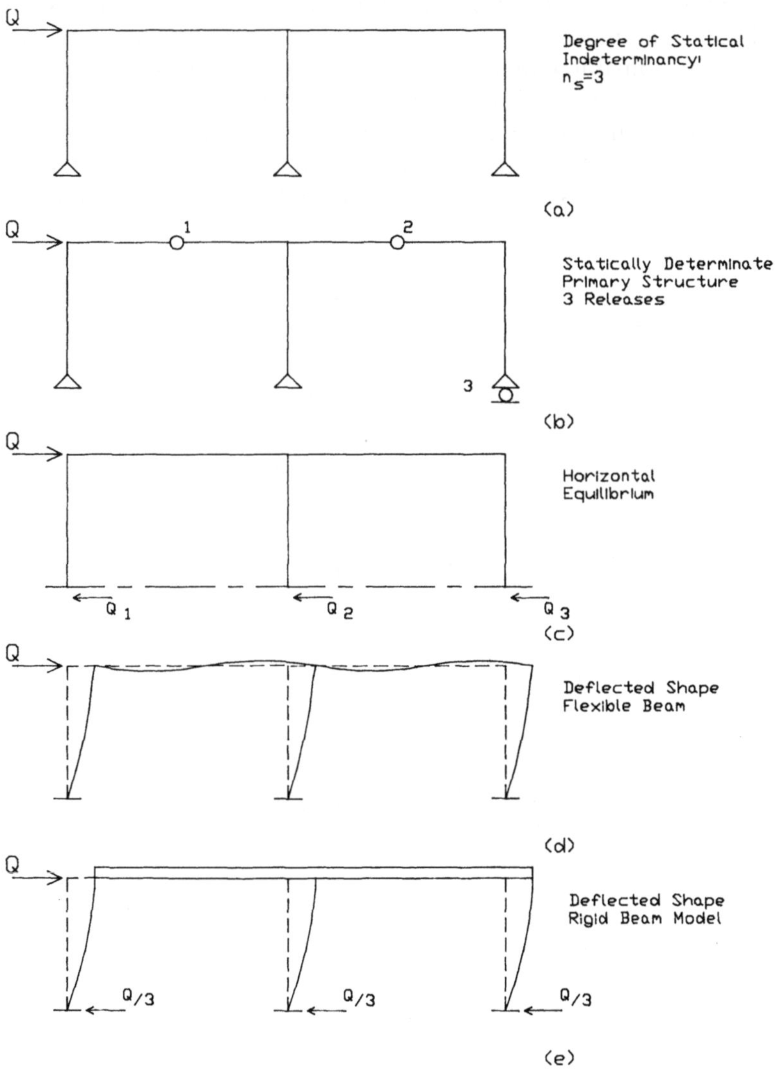

Figure 9.3 Behaviour of a two-bay frame.

resulting equilibrium analysis of the determinate form would not give an acceptably approximate result since the column on the extreme right would not be taking any load. Figure 9.3(c) demonstrates horizontal equilibrium with a horizontal shear plane taken through the structure. The equivalent forces also act as horizontal reactions at the supports. By introducing a distribution of the lateral force into the three columns, while retaining the points of inflection at the beam pins of figure 9.3(b), the structure is again determinate. Any combination of the three column shears that sum to the lateral force Q will satisfy equilibrium. For example, it might be assumed that each column takes one-third of the lateral load Q in shear, consistent with the rigid beam model of figure 9.3(e). The consequences of this assumption are demonstrated in figure 9.4(a), where the beam end moments follow from joint moment equilibrium requirements and the symmetry of the structure. In turn, these beam moments lead to the location of the points of inflection as shown at one-third of the beam spans from the central column. This assumption would be consistent with a situation where the

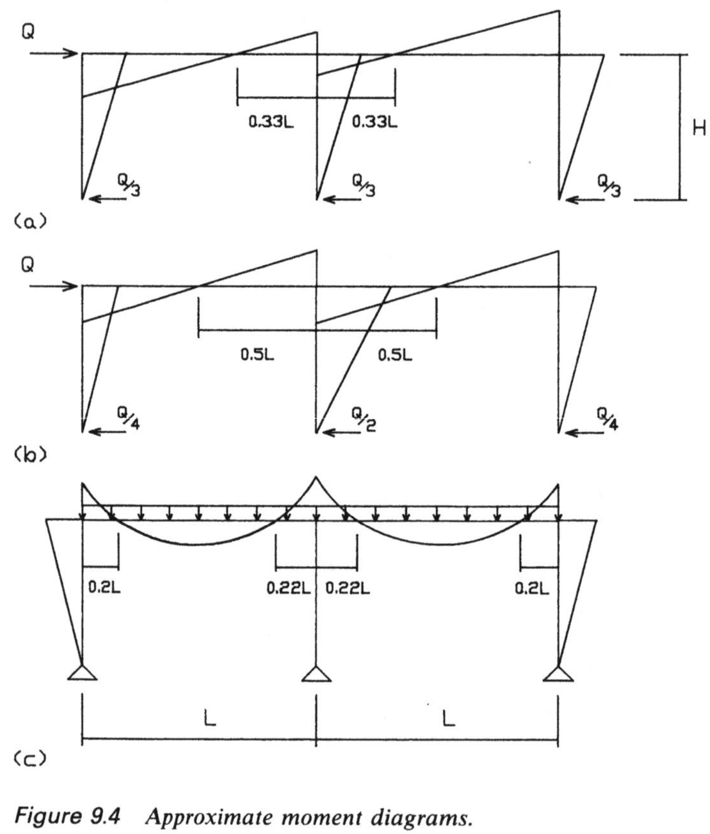

Figure 9.4 *Approximate moment diagrams.*

stiffness of the beams, expressed by the term EI/L, is significantly greater than the stiffness of the columns.

Alternatively, the symmetry of the problem suggests that the distribution of column shears should be as shown in figure 9.4(b) with the central column taking twice the shear of the exterior columns. Such an arrangement requires the points of inflection in the beams to be at mid span, in order that joint moment equilibrium can be satisfied. The rational basis for this distribution can be appreciated when the two-bay frame is considered as two similar single-bay frames, side by side sharing the lateral load. This is the basis for what is known as the portal method of approximate analysis, with the equivalent single bays taking an equal share of the shear at any level. For a multi-bay frame, all interior columns will take twice the shear load of the exterior columns. For a four-bay frame taking a total shear of 24 kN, the exterior columns will take 3 kN in shear, while the interior columns will all take 6 kN.

For a symmetrical single-bay rectangular frame similar to that of figure 9.3(a), the exact analysis coincides with the approximate analysis based on the three-pinned arch model, when the point of inflection at the mid span of the beam is taken as a pin. If one of the columns is stiffened, then it will attract more bending moment and a greater share of the lateral load. This is a general principle in the behaviour of statically indeterminate structures. For the two-bay frame, the approximate solution is exact when the interior column has twice the stiffness of the exterior columns and the frame is symmetrical.

The approximate analysis of frames under gravity loads follows on from the behaviour of continuous beams and the study of section 9.1.1. In figure 9.4(c), for example, symmetry suggests no joint rotation at the connection of the beams and the internal column, while the remaining joint rotations will be controlled by the relative stiffness of the beams and columns. Two limiting cases are of interest; as the structure approaches the rigid beam model and assuming the columns are axially rigid, then each beam span will behave like the propped cantilever of figure 9.1(c). As the flexural rigidity of the columns increases compared with that of the beams, then the beams tend to be fully restrained at both ends. A reasonable compromise is suggested by the nominated points of inflection in figure 9.4(c) with the form of the bending moment diagram indicated. It is interesting to note how the frame behaves differently under lateral and gravity loads and how different models, in the form of the pin releases at nominated points of inflection, are used in both cases. The model of figure 9.4(c), with four pins at the points of inflection, is strictly an unstable form since only three releases are necessary to make the structure determinate. With four releases the model is a mechanism, although equilibrium is possible under the gravity load. Example 9.2 illustrates the approach further and the results of both load cases can be combined to give the results for the frame as given.

Example 9.2: Approximate Analysis of a Two-bay Frame

Given data:

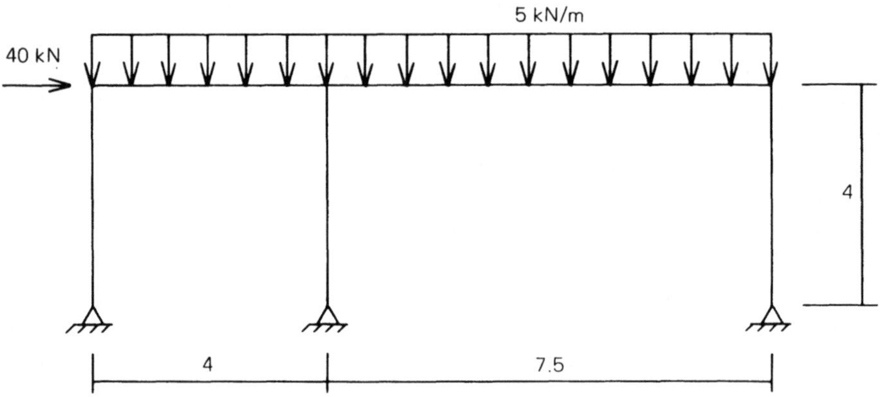

Case (a)—Gravity load only. Assume points of inflection at 0.2L from the ends of all beams, to give

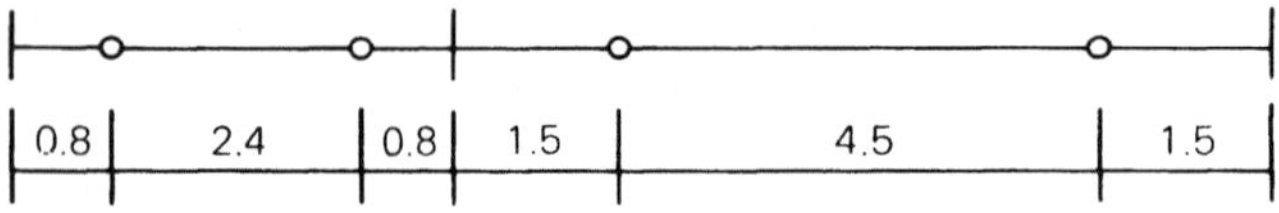

Hence the frame bending moment diagram and reactions are

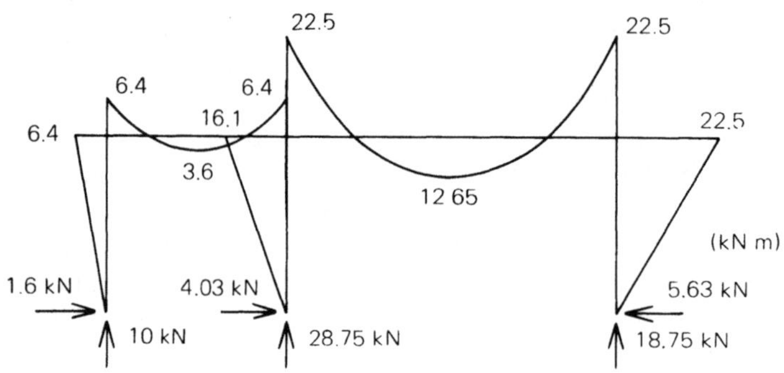

Case (b)—Lateral load only. Using the portal method to distribute the horizontal reaction gives the following result:

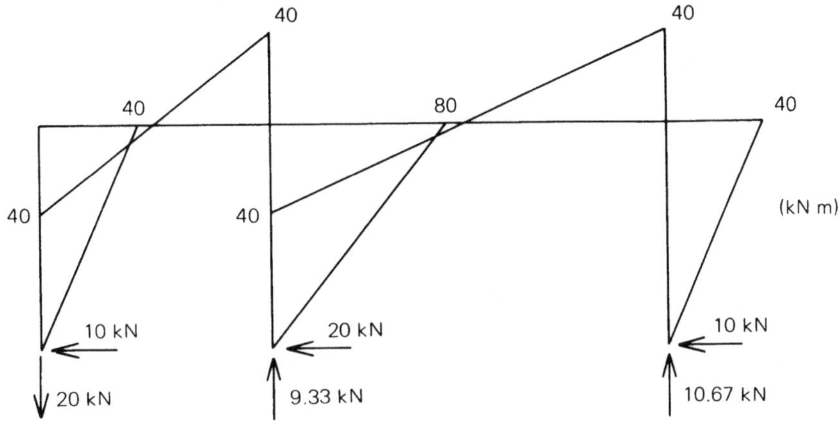

9.1.3 Approximate Analysis of Multi-storey Rectangular Frames

The principles suggested in the previous study can be extended to the approximate analysis of multi-storey frames by identifying frame behaviour in a similar manner. Under lateral load for instance, if points of inflection were assumed at the mid points of all the beams and columns in a multi-storey frame (except for the lowest storey columns if the bases were pinned), then the analysis can proceed from the top of the frame by a distribution of the shear at each level. However the behaviour of tall frames is influenced by axial forces in the columns and the portal method ignores this type of behaviour. A tall structure under lateral load, such as that shown in figure 9.5(a), is affected by overall bending of the structure as a beam cantilevered up from the base, in addition to the shear distortion or frame wracking. These two types of lateral deflection are shown in figures 9.5(b) and 9.5(c).

Considering the action of the cantilever bending only, the columns act like fibres in a beam and the section has a beam neutral axis through the centroid of the column group. Assuming a linear strain distribution, the resultant axial forces in the columns at any section can be calculated from equilibrium. This is illustrated in figure 9.6(b) which shows a free body diagram of the upper section of the frame of figure 9.6(a). The section x–x can be taken repeatedly at the mid point of the columns at all levels in the frame to determine the necessary axial forces. Coupled with the assumption of points of inflection at the mid point of all beams and columns as required, the approximate analysis can proceed from the upper level of the frame and a suitable distribution of the lateral forces will result. The technique

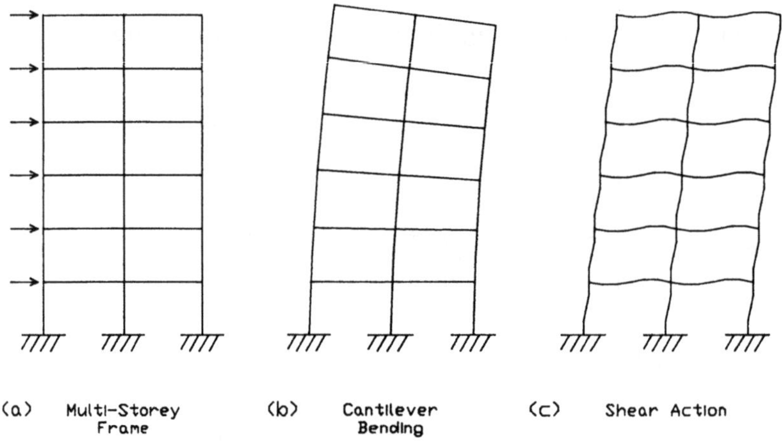

Figure 9.5 *Multi-storey frame behaviour under lateral load.*

is known as the cantilever method and it is discussed further, along with the portal method, in Norris and Wilbur (1960) and Benjamin (1959).

Before proceeding to an example using both techniques, it is relevant to return to the frame of figure 9.4 to study the influence of the assumptions of both approximate methods on the determinacy of the model. As observed earlier, the frame of figure 9.4(a) has a degree of statical indeterminacy of three. The two releases obtained by noting the effect of the points of inflection in the beams leave the structure with one degree of indeterminacy. There are in fact six unknown reactions and three equations of equilibrium plus

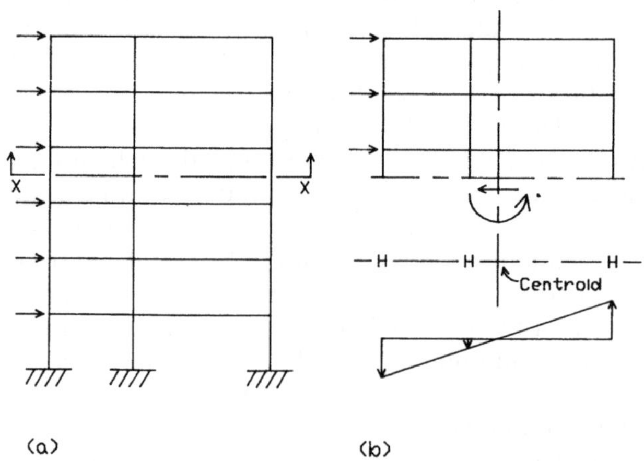

Figure 9.6 *Cantilever action in a multi-storey frame.*

two equations of condition available for the solution. The assumed distribution of the lateral forces of the portal method is equivalent to two more releases—effectively nominating two horizontal reactions—while the third is based on horizontal equilibrium. Similarly, the assumed distribution of the axial loads in the columns based on the cantilever method is equivalent to two more releases. In both cases more assumptions are made than are necessary, but the assumptions are consistent with the necessary ones. Further insight into the problem can be gained by analysing a determinate form of figure 9.4(b) with a roller release at one of the column bases carrying a nominated load.

The approximate analysis of a six-storey two-bay frame is demonstrated in example 9.3 using both the portal method and the cantilever method. The results are summarized with a comparison of the reactions in table 9.1 and the end moments of selected columns in table 9.2. The results from a general frame analysis program have been included as the exact results.

An initial comparison of the results of example 9.3 may not be encouraging, but it must be stressed that the results are derived from an approximate analysis. It is inappropriate to draw firm conclusions from such a limited study, although either method of approximate analysis does give the order of the moments and forces in the frame.

Example 9.3: Approximate Analysis of a Multi-storey Frame

Given data:

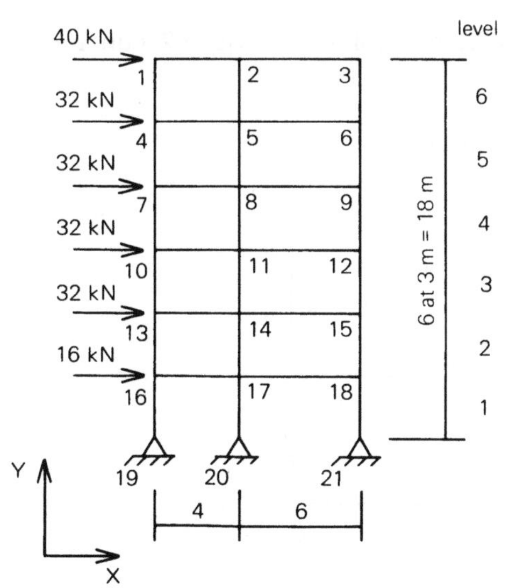

All beams: $I = 700 \times 10^6$ mm^4
All columns: $I = 300 \times 10^6$ mm^4
$A = 15\,000$ mm^2
E constant

Case (a)—Portal Method
Section at level 6 mid height gives

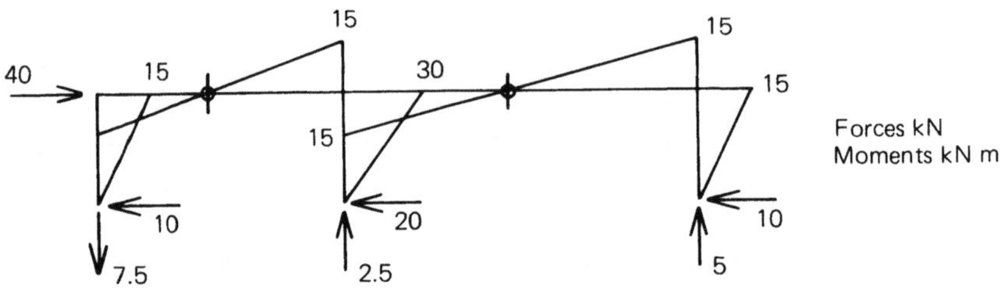

Section at level 5 mid height gives

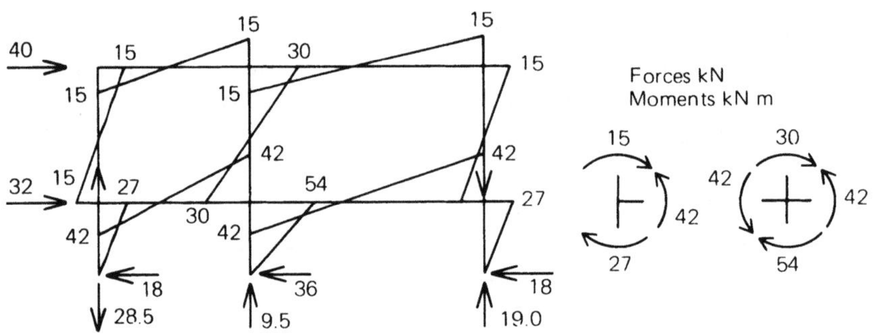

Proceeding in this way, the results of tables 9.1 and 9.2 can be obtained.
Case (b)—Cantilever Method
The column group centroid is given as

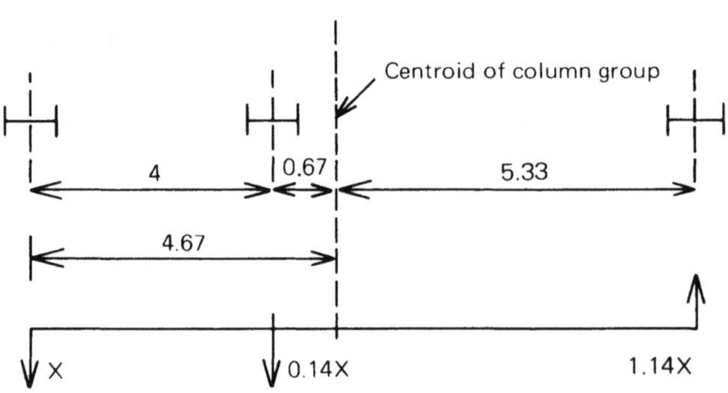

Axial Forces: Section at level 6 mid height

$$40(1.5) = 4.67x + (0.14x)0.67 + (1.14x)5.33$$

$$x = 5.54$$

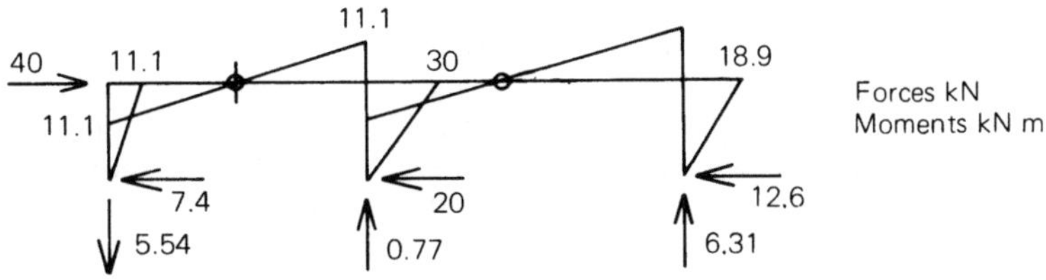

Axial Forces: Section at level 5 mid height

$$40(4.5) + 32(1.5) = 4.67x + (0.14x)0.67 + (1.14x)5.33$$

$$x = 21.03$$

and equilibrium then gives

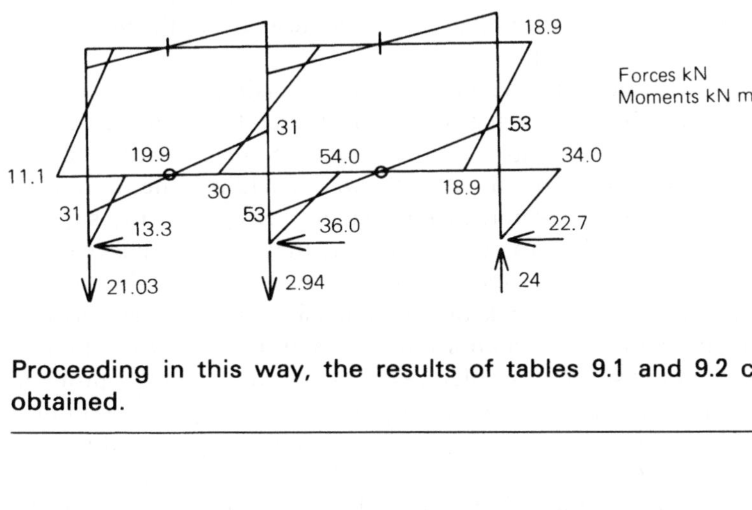

Proceeding in this way, the results of tables 9.1 and 9.2 can be obtained.

Table 9.1 Comparison of reactions (kN) from example 9.3

	X direction			Y direction		
Joint	Portal	Cant.	Exact	Portal	Cant.	Exact
19	−46.0	−33.9	−58.8	−264.0	−194.7	−274.8
20	−92.0	−92.0	−72.3	88.0	−27.3	106.0
21	−46.0	−58.0	−52.9	176.0	221.9	168.8

Table 9.2 Column end moments (kN m) from example 9.3

	Portal method		Cantilever method		Exact analysis	
Column	Top	Bottom	Top	Bottom	Top	Bottom
1–4	15.0	15.0	11.1	11.1	15.4	10.2
4–7	27.0	27.0	19.9	19.9	31.8	25.5
7–10	39.0	39.0	28.8	28.8	46.5	40.5
10–13	51.0	51.0	37.6	37.6	62.2	56.3
13–16	63.0	63.0	46.3	46.3	82.7	61.0
16–19	138.0	0.0	101.8	0.0	176.5	0.0

9.2 BOUNDS ON SOLUTIONS

In any problem-solving situation, the analyst can be guided by past experience with similar or related problems. Often a restatement of the problem in an alternative and more familiar form is helpful, as is some idea of the nature of the outcome. These ideas can certainly be applied to the analysis of structures and, as part of an approximate analysis, the mathematical model could be simplified.

For example, faced with the problem of analysing a two-span continuous beam where the interior support is considered to be an elastic one, two extreme values of the interior support reaction could be readily found. Firstly, if the support is approximated to an unyielding one, a routine analysis can give an upper bound to the reaction. Secondly, the support could be considered to yield to such an extent that no effective reaction develops there, thus giving a lower bound value of zero. These bounds on the solution are useful as a check on the analysis, but they may also be sufficient in their own right as an approximate analysis result. Returning to the beam on an elastic support problem, experience with a similar problem may suggest that the result is only marginally affected by the yielding support so that the upper bound solution is an acceptable approximation. By definition, the true result must be less than the upper bound value and greater than the lower bound value, and both results will converge on the true result as the respective models are refined.

As a further example, two approaches to the approximate analysis of a continuous beam can be considered. If each span were to be considered as a free span, by assuming a pin connection across each interior support, then the resulting equilibrium moment diagrams would represent a limiting set of values that could be regarded as a lower bound set. Such an approach ignores the compatibility requirements of the structure and it is an approach often taken in approximate analysis. On the other hand, each support could be considered to be fully restrained against rotation, forcing compatibility but ignoring joint equilibrium. The result is well known from previous chapters as the fixed end moment solution, but here it represents an upper bound set of values for the moments. Sketching a deflected shape and assuming locations for points of inflection in a continuous beam, as discussed in section 9.1.1, is of course a refinement on the bounds approach suggested; the bounds may be easier to determine and appropriate enough in some circumstances.

While the general principles outlined can be extended to a wide range of structures, it is difficult to be more specific. In the final analysis, engineers must develop their own skills in the field of approximate analysis from experience. Continual appraisal of any known solution and comparison with previous results will enable the engineer to build up a store of information useful to approximate analysis.

9.3 PROBLEMS FOR SOLUTION

9.1 For the continuous beams of figure P9.1, sketch the deflected shape and select points of inflection. Complete an approximate analysis of the beams to give the bending moment diagrams based on equilibrium principles and the nominated points of inflection. The symmetry of the beam of figure P9.1(b) should be used to advantage.

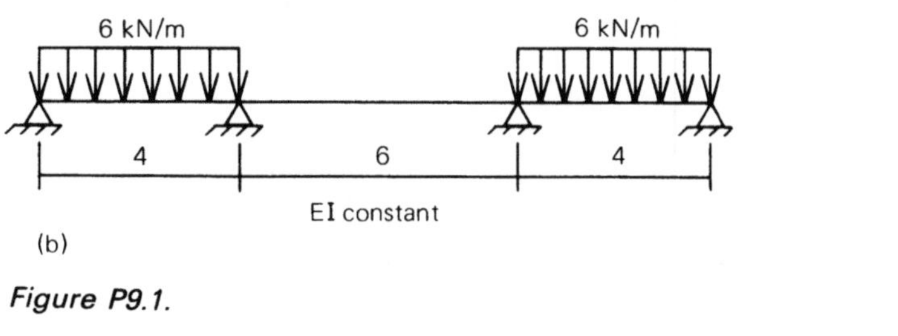

Figure P9.1.

9.2 Considering the lateral load and the vertical load as two separate load cases, analyse the plane frame of figure P9.2 by approximate methods to find the moments in the frame. Clearly show the deflected shapes for both cases and the locations of the points of inflection.

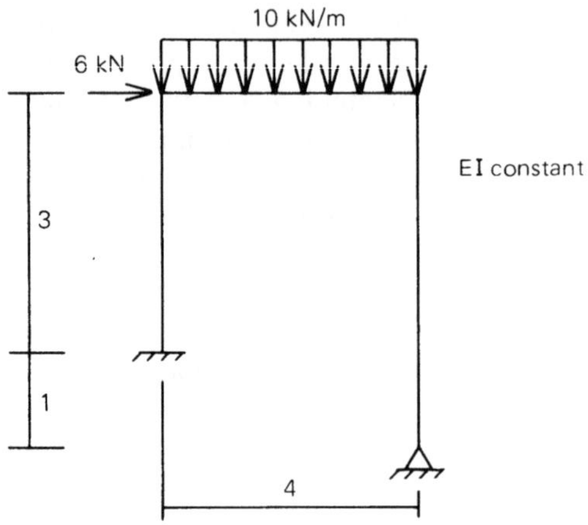

Figure P9.2.

9.3 Analyse the frame of figure P9.3 to determine the bending moment diagram and reactions by both the cantilever method and the portal method of approximate analysis. Revise the analysis for the cantilever method for the case when the interior column has twice the area of the exterior columns.

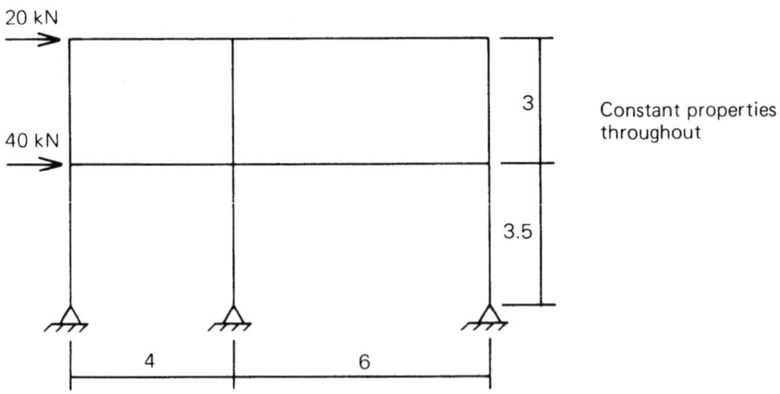

Figure P9.3.

9.4 Sketch a suitable deflected shape for the plane frames of figure P9.4 and nominate points of inflection in order that an approximate analysis can be carried out. Complete the analysis and draw the bending moment diagrams.

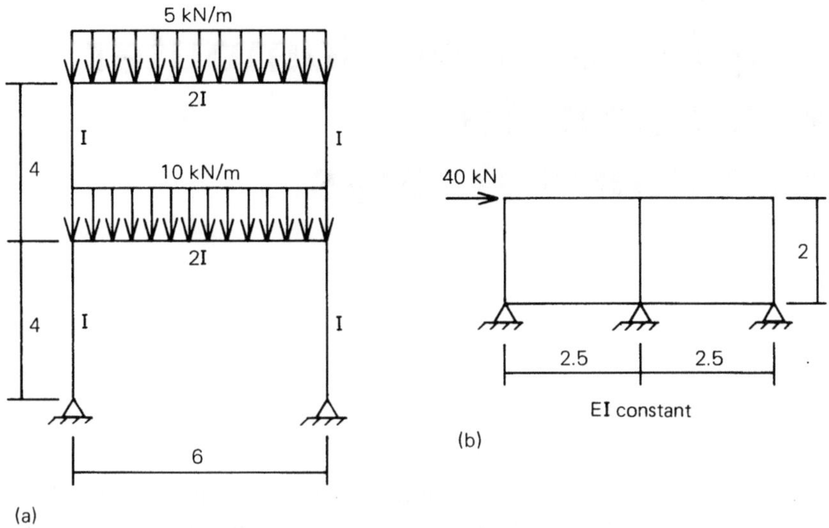

Figure P9.4.

REFERENCES

Norris, C. H. and Wilbur, J. B. (1960). *Elementary Structural Analysis*, 2nd edn, McGraw-Hill, New York.

Benjamin, J. R. (1959). *Statistically Indeterminate Structures*, McGraw-Hill, New York.

Chapter 10
Application of Computer Programs to Structural Analysis

Structural analysis, the fundamentals of which are the subject of this text, is simply a part of the overall design process in structural engineering. Since it often requires considerable computational effort though, it is not surprising to learn that the analysis was the first aspect to which digital computers were applied. What might be described as computer aided analysis is now carried out as a routine design office procedure on a wide range of computer systems.

The broader aspects of computer aided design are receiving attention also so that, in the long term, the structural engineering design process may be carried out in an integrated manner on a computer system. This will involve the integration of the various techniques of computer aided draughting, computer aided structural analysis and computer aided element design. The structural engineering design process is illustrated in the flowchart of figure 10.1, where the structural analysis component can be seen in the overall context of the design problem. This is considered to be an important point, since the analyst should not lose sight of the overall task. This text, however, is concerned primarily with the specific task of structural analysis and this chapter will only consider general aspects of computer programs related to analysis.

The emergence of the digital computer as a powerful computing machine in the 1950s led to a revised formulation of the structural analysis problem. In the following decades, matrix methods of analysis and the computer developed side by side. Matrix methods did not bring any new theory of structures to light. It was simply the fact that matrix algebra, representing a convenient way of manipulating numerical data with a computer, led to a matrix formulation of the classical methods of structural analysis. A definitive series of papers on the subject was written by Argyris (1954/55) and this work is generally regarded as the foundation for the matrix methods of analysis.

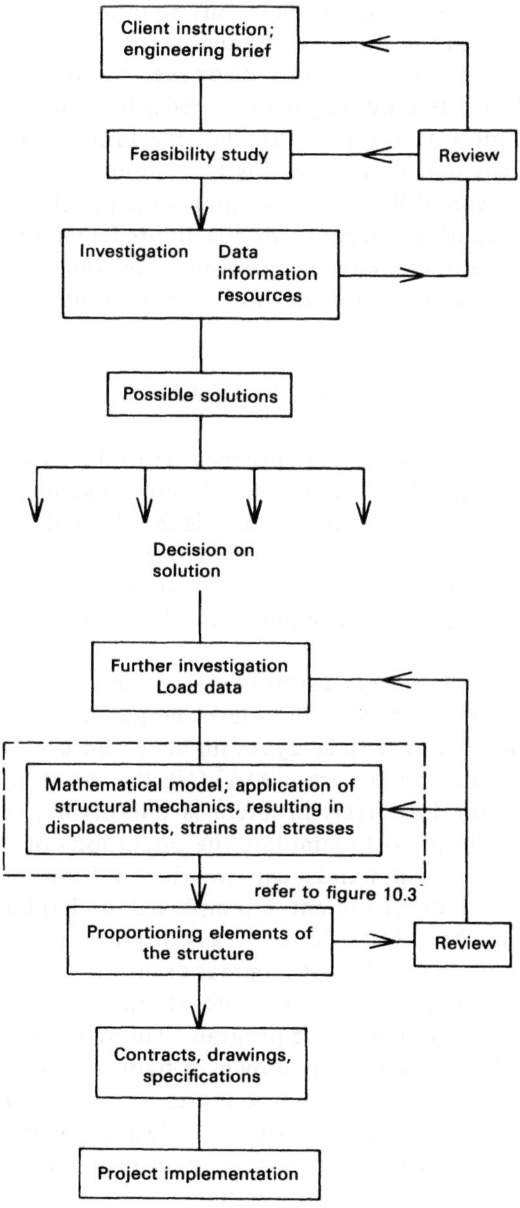

Figure 10.1 The design process.

A central feature of the solution of a structural analysis problem has always been the need to solve a set of linear simultaneous equations. This presents no particular difficulty when the number of equations is, say, less than 10; but when the number of equations is hundreds or even thousands, the task without computational aids is daunting if not impossible. Such a problem faced the analysts of aircraft frames with the development of modern jet airliners, and this provided great stimulus to computer-based structural analysis. Initially there were difficulties in coping with a problem that led to a large number of equations because of the limited memory capacity and speed of the early computers. However, advances in computer hardware have now reduced this problem to one of only incidental interest.

10.1 THE STRUCTURE OF AN ANALYSIS PROGRAM

In common with all software development, before proceeding to the structure of the program, the problem must be clearly defined and the steps to the solution must be set out. The preceding chapters of this text have done this and the mathematics involved have been demonstrated in the examples. Figure 4.5 gives a flowchart indicating the steps in the analysis of a continuous beam. The material there represents a good example to illustrate program development.

As was previously mentioned, the matrix formulation of the problem lends itself to application on a computer since high-level languages such as Fortran and BASIC can so readily cope with arrays of numbers representing matrices. As has been seen with the program MATOP, the essential basis of the solution to the structural analysis problem is one of matrix manipulation. MATOP has been designed to simulate the operations of a structural analysis program and its commands can provide a form of pseudo-code to assist in full program development. Example 6.2 of chapter 6 presents the full output file from a MATOP solution to a continuous beam problem. The essential features of the MATOP commands used in that solution are given in figure 10.2(a) and they serve as pointers to the required program modules and hence to the structure of the program. The suggested processes involved in program development are shown in figure 10.2(b). The use of MATOP to provide a pseudo-code step is not essential, but it does serve as a convenient link in the context of this text. Details of the program MATOP are given in the abstract and user manual in appendix A, where a full description of the commands is presented.

It should be immediately apparent that a major component of any structural analysis program is overlooked in both the flowchart of figure 4.5 and the pseudo-code of figure 10.2(a). In both cases the formation of the key matrices receives no attention. This is deliberate since the nature of the element stiffness matrices and the structure stiffness matrix has been

```
LOAD.K.10.10
LOAD.P.10.2
MODDG.K
SCALE. K.10000
SOLVE.K.P
PRINT.P DISPLACEMENTS
LOAD.E.4.4
SCALE.E.10000
SELECT.D.P.4.2.I.1
MULT.E.D.F
PRINT.F FORCES
repeat for each element
```

(a) Pseudo-Code

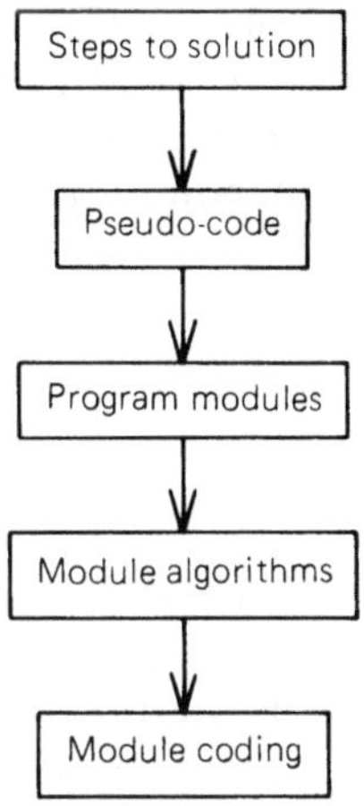

(b) Processes

Figure 10.2 Program development.

presented in earlier chapters. More specifically, MATOP has been designed as a teaching program and users are expected to formulate the required matrices based on their own understanding. However, in a fully developed structural analysis program the user need not even be aware that matrices are involved; the matrix operations are normally transparent to the user. The starting point then must be the data necessary to define the element matrices and suitable algorithms must be used to assemble the structure matrices, so that the whole process of matrix manipulation is automatic. The LOAD command, as an item of pseudo-code, implies this. Details of the techniques involved are given in a number of texts including Mosley and Spencer (1984), which includes listings of suitable analysis programs.

The structure of an analysis program has been presented against the background of the matrix stiffness method. While the majority of computer programs for the analysis of skeletal structures are based on matrix stiffness methods, the alternatives should not be overlooked. Computer programs have been developed on the basis of the matrix formulation of the flexibility method, and other structural analysis techniques may form the basis of some computer programs.

10.1.1 Data Input

Element stiffness or flexibility matrices obviously form the starting point for an analysis based on matrix methods. Such matrices have been defined in previous chapters and it can be seen that the terms are based on the geometry of the structure and the element properties. Clearly then, a major function of the input module is to provide the necessary information for the terms of the matrices to be evaluated. The data set can be described in a general way under five headings as follows:

(a) coordinate geometry;
(b) element connectivity;
(c) element properties;
(d) boundary conditions;
(e) load data.

The data is entered in a systematic manner and manipulated to form the necessary matrices.

The computer analysis starts with the mathematical model and the coordinate geometry of the model defines the nodal points with regard to a specified set of axes, known as the global axes or system axes. The element connectivity can then be expressed by specifying the node numbers that are relevant to each element. Each element specification is usually tagged with a code that can subsequently be related to the properties of the element. The element properties include properties of area and material properties such as the modulus of elasticity. Boundary conditions must also be specified by an indication of what nodes are boundary nodes and what restraints to the degrees of freedom are to be applied there. Obviously the load data forms an integral part of the input data. However, the load data is specific and unlike the rest of the data which essentially defines the characteristics of the structure. The normal specification of load data includes the loads on the nodes and a variety of load types along the elements. Other features can include multiple load cases and load case combinations with load factoring.

In a well developed program, the data entry will be carefully considered and include features such as validation of the data and editing sequences to permit a review of the data. Opportunities also exist for data generation for regular structures. For instance, with coordinate data it is frequently possible to provide only key coordinates along with details of how intermediate nodes are to be generated.

Many existing programs still operate in a batch file mode where the input data is set up to a specified format by a suitable text editor. The input file is then subsequently linked to the program during its execution. The alternative strategy is to offer the user data entry on an interactive basis with screeen prompts for the data. This is valuable for the inexperienced or infrequent user but sometimes tiresome for the experienced users. Programs which offer both modes of data input are therefore desirable. In a final sequence of interactive data input, options for storing the data on a disk file are usually presented.

10.1.2 Data Output

While it is fairly clear as to what information should be presented as output to a structural analysis program there are some features worthy of emphasis. The output can be categorised in the following manner:

(a) echo output of all input data;
(b) displacements of the structure;
(c) element actions;
(d) reactive forces and equilibrium checks;
(e) graphics options.

It is the engineer's responsibility to check the validity of the output of all structural analysis programs. The information must be carefully considered and engineering judgement must be applied to ensure that the results are sensible. The echo output of all input data is most important in this regard, to ensure that there is no conjecture about the data the program started with. The major component of the output data then involves the displacements of the structure and the actions in the elements.

In general-purpose packages capable of handling large structures, options to select items of an output file are sometimes offered. Other features often include load case combinations with optional load factors. Typically the dead loads, wind loads and live loads may have been considered in three separate load cases in the analysis. It is useful to be able to combine the results of these load cases within the scope of the program, without having to do the task manually.

The results for the reactions enable an overall equilibrium check to be applied to the structure. Sometimes this is included as part of the program operations but in any event, it is an exercise which the engineer can readily carry out with the loads and reactions known.

There is a general expectation that structural analysis programs should include options for the graphical presentation of data. Clearly the computer facilities for doing this exist and it makes little sense to have an engineer return to the desk to plot bending moment diagrams from a printout of numerical data. The graphics option is also relevant to the data input since the defined geometric model can be displayed to check its validity before the problem is actually solved. This is particularly important for complex three-dimensional models of structures like space frames. Frequently a general-purpose graphics module is incorporated into the program with standard graphics features such as zooming, windowing, rotating and screen dump commands. For output data, it is possible to display scaled deflected shapes against the undeflected model, and selected actions such as axial force, shear force and moment against all or part of the structure. The use of peripheral units such as printers and plotters to reproduce the screen graphics completes the task with hard copy output.

It is quickly appreciated that a major component of the programming effort in a structural analysis program relates to the input/output features. As much as 75 per cent of the code may in fact be devoted to the input and output of data.

10.2 MODELLING OF STRUCTURES

The central box of figure 10.1 may be expanded, as shown in figure 10.3, to illustrate the relationship between the model and the analysis process. In the theory presented in the previous chapters, the modelling has already taken place and each of the line diagrams, representing the structure for analysis, are idealisations of the concept of the finished structure. In many cases the relationship between the structural concepts and the model is self-evident, but this is not always true and it is particularly less so as the problem becomes more complex.

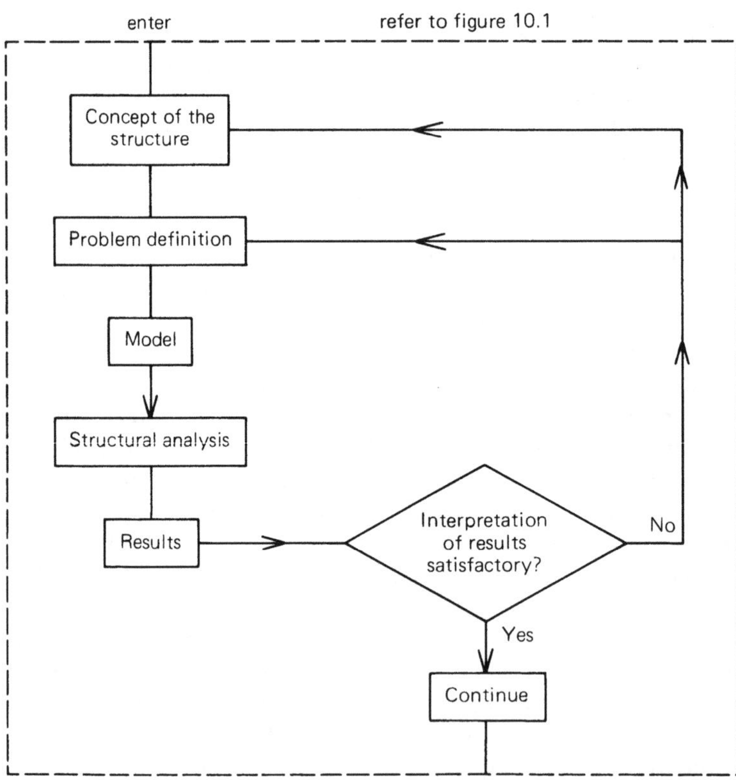

Figure 10.3 A model analysis process.

The analysis of a structure may generally be broken down into component tasks, each with its own problem definition and perhaps with a different model as a function of the task. Recognising the multiple nature of the tasks, figure 10.3 shows an important procedure to be used in the analysis of structures. While it might appear to be obvious, the importance of applying engineering judgement to the results of the analysis cannot be overstressed. There are always underlying assumptions in the model and they may not be appropriate to the task or problem definition. The fundamental principle is that the model should predict the behaviour of the structure as conceived as accurately as possible. If the results are considered to be unsatisfactory then the model, or perhaps even the problem definition, should be reviewed. Once satisfactory results have been obtained, the next task can be considered or the design process can proceed to the next phase.

Before elaborating on the nature of the model, the idea of component tasks and multiple models can be illustrated by reference to a simple example in statics. The tower crane is a structure which strongly expresses its structural form and the model in its realisation. There are several aspects to its analysis perhaps commencing with the question of overall stability, then the analysis of the boom followed by the analysis of the tower. Even considering the analysis of the boom alone, two tasks are immediately apparent: one is to calculate the reactions on the boom (or conversely the actions on the tower from the boom), and the other is to determine the forces in the boom. These two tasks have been set as an exercise in chapter 2 in problems **2.6** and **2.7**. While the details are unimportant here it can be appreciated that by taking advantage of symmetry, a two-dimensional model is adequate for calculating the reactions, while a three-dimensional model is necessary to find the forces in the boom since it is in fact a space truss.

10.2.1 Element Connections

The very nature of a skeletal structure suggests which model should be used in the analysis and, in most cases, modelling is not a difficult task. The major idealisation concerns the connection of the elements. Truss elements are assumed to be pin-connected, but they are rarely constructed in that way, while the dominant assumption in a frame is that the elements are rigidly connected at the joints. For a truss, the triangulation ensures that the dominant action is an axial one and the assumption of pin-connected elements is justified. This can be tested by analysing the same truss, firstly using a pin-connected model and secondly using a rigid connected model. For a frame, the modelling of the element connections is more important since the stability of the structure usually derives from moment transfer through the elements. A number of elements in a frame may be conceived as being pin-connected and they must both be modelled and constructed as such.

The modelling is expressed mathematically with element end moment releases as described earlier. If a connection in a frame is modelled as a pin connection, then it is anticipated that the connection will be through web cleats, permitting little moment transfer. On the other hand, if the connection is modelled as a rigid one, then steps must be taken in the subsequent connection design to ensure that moment transfer does occur. The behaviour of a connection may be intermediate between the two extremes of a pin connection and a fully rigid one. It is possible to model such flexible connections but the details will not be pursued here.

10.2.2 Boundary Conditions

The principles applied to modelling the element connections also apply to the boundary conditions. At the foundations, the connection may be modelled as a pin connection or regarded as fully fixed or with a range of other releases of action. Generally there is more scope for variation in the model at the restraints since a variety of conditions can be applied there without too much difficulty. For example, in the simplest case the supports of a structure are regarded as unyielding. On an elastic sub-grade it might be more realistic to consider the restraints as spring restraints. In many cases a computer program will allow for this but, if not, additional elements can be introduced to accommodate that behaviour if it is expected.

10.2.3 The Modelling of Non-skeletal Structures by One-dimensional Elements

It was stated earlier that skeletal structures express the nature of the mathematical model through their very form. This is certainly true during construction or once the cladding is removed. There are however clear cases of structures where the skeletal form is not immediately obvious. The first example is that of the superstructure of a bridge or simply a bridge deck. A common form of bridge deck consists of a series of longitudinal beams with an *in situ* concrete slab cast across the top of the beams. In some cases the beams are eliminated and the superstructure is simply a concrete slab. A wide range of such structures can be analysed in the skeletal form of a plane grid. This is really a specialised form of modelling but it is a useful example to illustrate the concepts of structural modelling. A significant amount of work was done in this field and it resulted in a series of recommendations concerning the modelling of bridge decks as grid structures (West, 1973). Briefly, in many cases the longitudinal elements of the grid are obvious, but what is not so obvious is the modelling of the effect of the deck slab by the transverse elements of the grid.

A second example arises with the use of shear walls in framed buildings. Beams may well be rigidly connected to a shear wall, as part of a lift shaft for example. The problem then is to model the continuous wall with a one-dimensional element of appropriate position and properties, to obtain the influence that the shear wall may have on the rest of the frame. Details will not be presented here; the idea is simply suggested to broaden the concepts of structural modelling. In any event, a note of caution should be struck. In such cases the solution often leads to assigning a very large second moment of area to the element representing the shear wall, while adjacent elements have standard values. If this is extended too far there may be computational difficulties due to ill-conditioning of the equations involved in the solution. It is difficult to be precise on this point since it is a function of the accuracy to which a particular program works. Programs operating in double precision on modern computers have a high tolerance in this regard, but the user should always be aware of the possibility of ill-conditioning.

10.3 INFLUENCE OF THE COMPUTER PROGRAM ON MODELLING

The analysis process is not confined to computer analysis and the principles are the same irrespective of how the analysis is carried out. However, if it is thought to be appropriate to use a computer program for the analysis, then the nature of the available program may well influence the model. There is a wide range of computer programs available for structural analysis, both from commercial sources and published texts and papers. In considering the use of any program, the capabilities and limitations of that program must first be understood and the notation and sign convention must be appreciated. Details are generally given in program abstracts and user manuals available with the program. To illustrate the point here, consider the limitations of a plane frame analysis program based on the matrix stiffness method and written assuming only rigid joint connections between the elements. In this case end moment releases in an element are not provided for, except at the supports where a full range of releases would probably be available. At first glance the program might be considered as unsuitable for use with a frame where one or more elements are considered to be pin-connected. However the pin connection can be modelled by including a short flexible element in the model. The element flexibility is controlled by the flexural rigidity, EI, and a value several orders of magnitude less than the prevailing values for the rest of the structure would be appropriate. The precise nature of this must be studied by testing with the program to ensure that suitable results are being obtained and that ill-conditioning does not occur. If the flexible nature of the element is likely to introduce excessive deflections, then an additional support condition may have to be imposed on the structure.

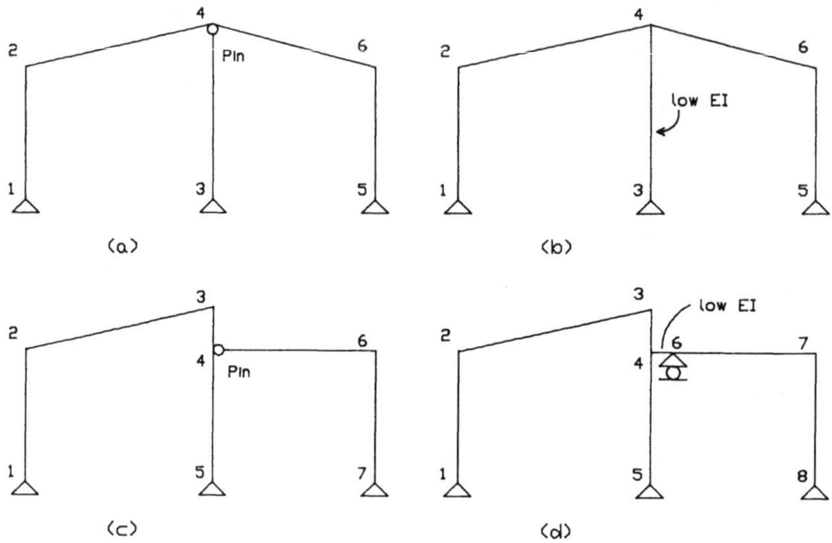

Figure 10.4 *Rigid-jointed frame modelling of pin connections.*

This feature of structural modelling is illustrated in figure 10.4 with two common cases that occur in industrial building frames. In figure 10.4(a), the end bay of a portal frame has been propped with the column element 34 pin-connected to the ridge joint. The column element can be assigned a low flexural rigidity, while the axial rigidity, EA, is maintained, and modelled as the rigid jointed frame of figure 10.4(b).

The frame of figure 10.4(c) was extensively studied in chapter 8. For a computer analysis under the circumstances outlined, the pin connection between element 46 and the rest of the frame at node 4 can be modelled by the arrangement shown in figure 10.4(d). An additional flexible element has been introduced and the element 46 could be taken as, say, 0.1 of the original length of the side bay beam. However, introducing a flexible element alone would allow node 6 to deflect excessively, when in fact the beam end must remain essentially at the same level as node 4. This is achieved by the introduction of a vertical restraint at node 6. If the element 46 is given a high axial rigidity, then the sway response of the structure will meet the requirements of the original model. The vertical reaction at node 5 of figure 10.4(c) is given approximately as the sum of the vertical reactions of nodes 5 and 6 of figure 10.4(d).

10.3.1 Additional Effects on Structures

The load types recognised by a given program may be limited in scope although any program will almost certainly accept all of the nodal loads possible. If a required load type, such as a partial uniformly distributed load, or a triangularly distributed load, is not available, then the preliminary calculation of fixed end moments can always be carried out external to the program and the nodal loads only entered as input. The final results must then be obtained by combining the results of the computer analysis, representing results for nodal loads only, with the fixed end moment solution of each element.

This approach can be extended readily to incorporate temperature loads on structures and the influence of initial strains or lack of fit, assuming that the program does not cater for such actions specifically. Fixed end actions can be readily deduced for both temperature loads and lack of fit on a variety of elements. The negating actions can then be applied as nodal loads in the analysis program with the final results again being given as a combination of the results from the program and the fixed end actions.

Two examples are given to illustrate the point. As a first example, suppose the truss of figure 10.5(a) was fabricated with the element 34 too long by $0.01L$. The restraining compressive force, F, is readily calculated in the fixed end action situation appropriate to a truss element as shown in figure 10.5(b). The restraint action determines the necessary nodal loads shown on the truss in figure 10.5(a), which can be applied in a truss analysis program in the usual way. Temperature loads on a truss can be treated in a similar manner by effectively considering each element to be initially free to expand or contract in length according to the temperature variation. The restraining forces then follow as a function of the axial rigidity of the elements.

The flexural action caused by temperature load on the portal frame of figure 8.5(a) was considered in chapter 8 as an illustration of the use of the flexibility method. The same problem can be studied here as the second example, this time in the context of the matrix stiffness method. The frame is shown again in figure 10.6(a). The fixed end moments for an element under a temperature gradient are shown in figure 10.6(b) and the values

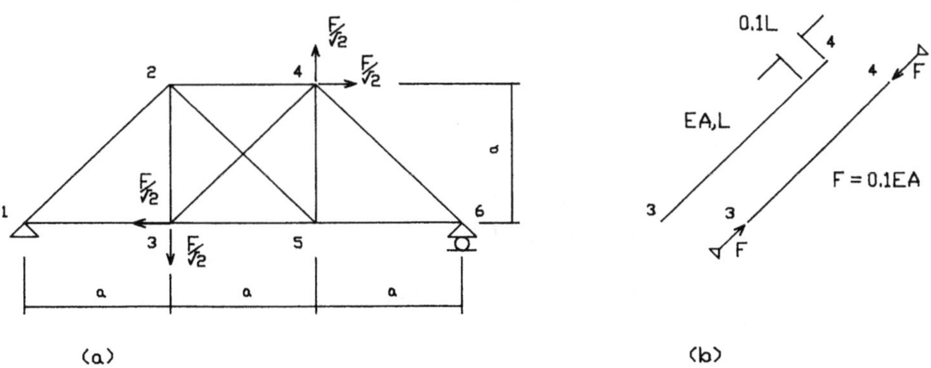

Figure 10.5 *Truss loads due to fabrication error.*

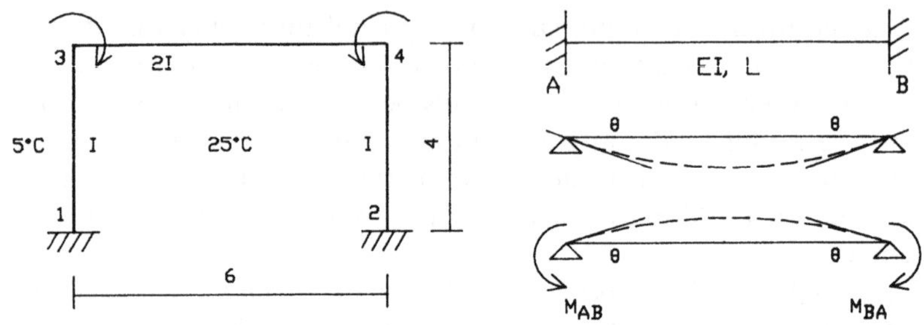

(a) Nodal Movements (b) Fixed End Moments

Figure 10.6 Frame loads due to temperature load.

have been calculated using a flexibility approach and the end moment rotation relationships developed in chapter 3. For the beam with the end moments released the rotation at either end of the beam can be calculated using virtual work. From figure 7.7 in section 7.3.3 of chapter 7, since

$$d\phi = \frac{2\alpha\Delta T}{d}\,dx$$

then the angle θ of figure 10.6(b) is

$$\theta = \frac{L}{2} \times \frac{2\alpha\Delta T}{d}$$

$$= \frac{L\alpha\Delta T}{d}$$

The fixed end moments then follow since

$$M_{AB} = \frac{2EI}{L}\theta \quad \text{and} \quad M_{BA} = -\frac{2EI}{L}\theta$$

that is

$$M_{AB} = \frac{2EI\alpha\Delta T}{d} \quad \text{and} \quad M_{BA} = -\frac{2EI\alpha\Delta T}{d}$$

It should be noted that the fixed end moments are necessarily expressed in terms of the flexural rigidity, *EI*, of the element. With the fixed end moment solution known, the necessary nodal loads for the frame of figure 10.6(a) can be determined as shown and these can be applied in a computer analysis. For this simple example, the results can also be easily found using the moment distribution method. It is interesting to note that if the frame had the same element properties throughout, then the fixed end moments due to temperature load would have balanced immediately, giving a constant moment diagram with no deflection of the structure.

10.3.2 The Use of Symmetry

In a final comment, attention is drawn to the advantages that the symmetry of the structure may offer the analyst. Although the remarks are made in the context of computer modelling, the principles can be applied to any analytical procedure.

For a symmetrical structure under symmetrical loads, the behaviour of the structure along the plane or line of symmetry is usually obvious. For example, by virtue of symmetry certain displacements and internal actions must be zero. Restraints which provide these necessary conditions can be placed along the line of symmetry and the resulting model can be analysed as representative of the behaviour of the full structure. Generally one line of symmetry can be identified in a plane frame, leading to a model of one-half of the frame for analysis. The principle can be extended through

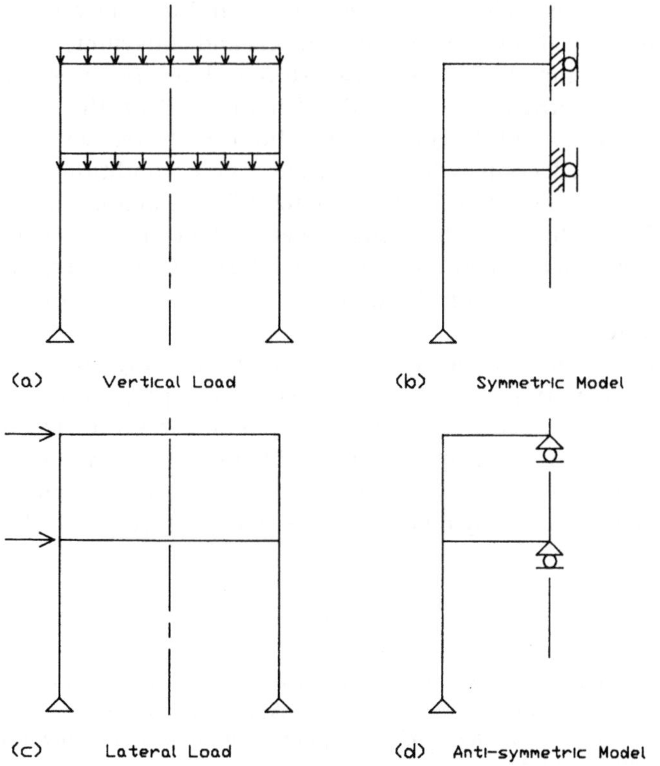

Figure 10.7 *Use of symmetry in structural modelling.*

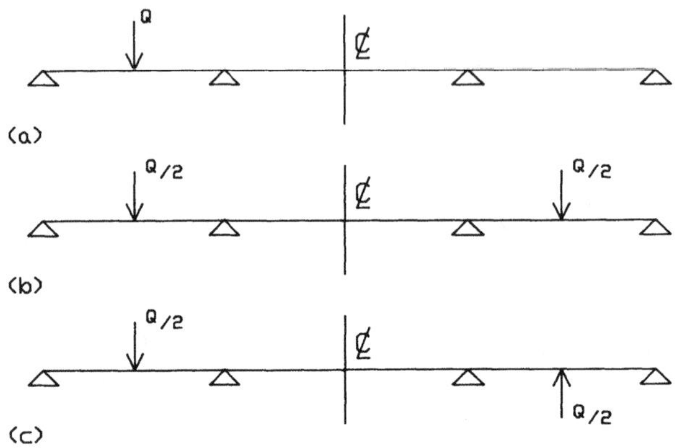

Figure 10.8 *Combining symmetrical and antisymmetrical loads.*

to more lines or planes of symmetry, reducing some models to a quarter or even one-eighth of the full model. Often the behaviour of a structure is obviously antisymmetric about a geometric line of symmetry as is the case with a symmetrical structure subjected to lateral load, and this introduces a different model. Models to accommodate symmetrical and antisymmetrical behaviour are shown in figure 10.7 using a two-storey plane frame as the example. Under the symmetrical gravity loads of figure 10.7(a), the symmetrical frame will not sway and the mid point of both the beams will not rotate although vertical deflection will occur there. These conditions are met by the model shown in figure 10.7(b). The lateral forces acting on the frame in figure 10.7(c) will cause the antisymmetric deflections shown, dictating points of inflection at the mid point of each beam with no vertical deflection. The restraints imposed on the model of figure 10.7(d) provide these necessary conditions.

For symmetrical structures subjected to asymmetric loading, analysis taking advantage of symmetry may still be used with the superposition of symmetric and antisymmetric load cases. The analysis of the symmetrical three-span continuous beam of figure 10.8(a) could proceed in this way, using the superposition of the results of the structures shown in figures 10.8(b) and 10.8(c) analysed using symmetry and antisymmetry respectively.

REFERENCES

Argyris, J. H. (1954/55). 'Energy theorems and structural analysis', *Aircraft Engineering*, Vol. 26, Oct.-Nov. 1954 and Vol. 27, Feb., Mar., Apr. and May 1955.

Mosley, W. H. and Spencer, W. J. (1984). *Microcomputer Applications in Structural Engineering*, Macmillan, London.

West, R. (1973). *Recommendations on the Use of Grillage Analysis for Slab and Pseudo-slab Bridge Decks*, Cement and Concrete Assoc. CIRIA, London.

Appendix A
MATOP (Matrix Operations Program—Version 1.2 Feb 1986)—User Manual

A.1 INTRODUCTION

Throughout the book, reference has been made to the computer program, MATOP, details of which are now presented. While the program is considered to be most useful for a student using the book, it is neither essential nor unique and any other matrix manipulation program that is available could be substituted. The program has been implemented on an IBM PC and compatible microcomputers.

A.2 FORM OF THE PROGRAM

MATOP is a program written in Fortran 77 and offers the user a series of high-level commands to carry out certain operations with matrices. The commands are a limited set but they have been selected to meet the requirements for matrix manipulations encountered in the analysis of statical structural systems. The program has been designed for an extension of its scope by the addition of other subroutines linked to additional commands of similar format.

The program may be used either interactively or in a batch mode. Used interactively, a HELP command is available and error messages are provided to guide the user, although complete data validity checks after the matrix names on a command line are not provided in this version.

A.3 OPERATION OF THE PROGRAM

The program initially requests the name of a datafile (maximum of 8 characters without any filename extension) and the mode of operation. If the batch mode is selected, the program will search for a datafile with the given filename with the extension '.INM'. For batch operation such a file should first be set up using a suitable editor. In the interactive mode, an inputfile is neither required nor created. In both modes, the series of commands, and any matrices printed by the PRINT command, are subsequently written to a file with the filename of the given datafile name with the extension '.OTM'. The output file can then be printed out at the end of the session or on a subsequent occasion. The operations are shown schematically in figure A.1.

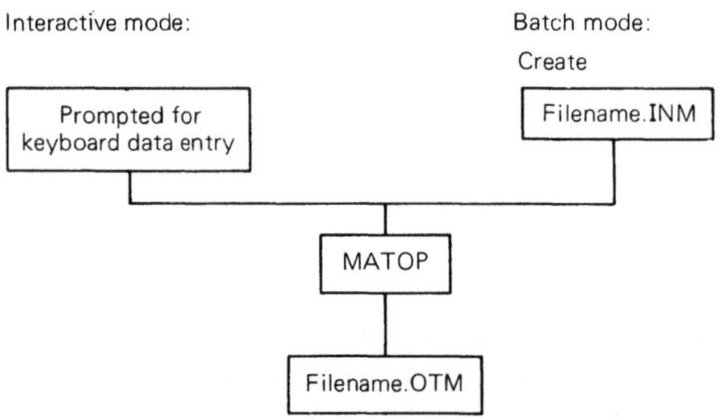

Figure A.1 Schematic presentation of operations.

It is important to note that matrices are only written to the output file by the PRINT command.

The program is limited by the total number of matrices that can be used, a maximum size of any one matrix and the total number of elements of those matrices. The maximum number of matrices permitted is 20 and a given matrix cannot have more than 20 rows or columns. Further, the total number of elements of all matrices cannot exceed 2000. However, since the DELETE command re-allocates all data storage, and resets both the number of matrices and the total number of elements, the limits can be avoided.

A.4 THE COMMAND FORMATS

The general format of a command is

Operation.[(matrix name1).(matrix name2)..,].[(n1).(n2)..,]

where each part of the command line is separated by a period (or full stop). Matrix names may use up to six alphanumeric characters and cannot be repeated as a new matrix unless the old matrix is deleted. The details of the matrix names and the numbers n1, n2 . . . , are given as required in a description of each command.

The following commands are available:

LOAD.M1.N1.N2 Allows a matrix of the name M1, with N1 rows and N2 columns, to be input. In the interactive mode, each element is entered through the keyboard, row by row, pressing the return key each time. If an error is made on data entry, the operation can be terminated by entering a non-numeric character. In batch mode, the elements of the matrix follow immediately after the LOAD command in the file, in free format with either a comma or a space separating each element.

MULT.M1.M2.M3 Multiplies the matrix M1 by M2 and stores the result in M3. The matrices must be compatible for multiplication.

SCALE.M1.N1 Multiplies the elements of the matrix M1 by the scalar quantity N1. The resulting matrix retains the name M1. N1 may be specified either as an integer or a decimal number, but it cannot be specified in E-format.

TRANS.M1.M2 Creates a new matrix M2 as the transpose of the matrix M1.

SOLVE.M1.M2 Solves the set of simultaneous linear equations expressed in the matrix form of the equation

$$A \cdot X = B$$

where M1 is the matrix, A, and M2 is the matrix, B. The solution is stored in the matrix M2 and the matrix M1 is unaltered by the routine. The equations do not have to be symmetrical.

MODDG.M1 Allows the user to modify the diagonal terms of a square matrix M1. For each diagonal element, the user is prompted either to modify the element or to pass to the next one. In the batch mode, the line following the MODDG command must give the number of elements to be modified. Successive lines then give the row number and the modified diagonal element in free format. The matrix retains the name M1.

SELECT.M1.M2.N1.N2.N3.N4 Creates a new matrix, M1, of N1 rows and N2 columns from the elements of the matrix, M2, starting at row number N3 and column number N4.

NULL.M1.N1.N2 Creates a null matrix, M1, of N1 rows and N2 columns. The diagonal terms can then be modified (refer MODDG), to give a diagonal matrix.

PRINT.M1 Causes the matrix M1 to be displayed and to be written to the output file.

HELP This command can be used to present a summary of the commands on the screen.

DELETE.M1 Deletes the matrix named M1, allowing that name to be used again. All array storage is re-allocated so that with effective use of the DELETE command, the size limitations are not significant.

REMARK.any string as a comment line! This command allows comments to be inserted in the output file. In addition each command may have an optional comment extension separated from the required command by at least one space. Such comments will appear in the print listing of the output file. The period character cannot be used in the comment.

QUIT Terminates the operations of the program.

A.5 AN EXAMPLE OF THE USE OF THE PROGRAM: SOLUTION OF SIMULTANEOUS EQUATIONS

Consider the equations

$$3x+4y+5z=12, 9$$

$$1x-1y+1z=1, 8$$

$$2x+2y-3z=1, 10$$

which may be expressed in matrix form as

$$\begin{bmatrix} 3 & 4 & 5 \\ 1 & -1 & 1 \\ 2 & 2 & -3 \end{bmatrix} \begin{Bmatrix} x \\ y \\ z \end{Bmatrix} = \begin{bmatrix} 12 & 9 \\ 1 & 8 \\ 1 & 10 \end{bmatrix}$$

and written in the notation of

$$\mathrm{A} \cdot \mathrm{X} = \mathrm{B}$$

The appropriate commands to solve for X using MATOP are then

```
LOAD.A.3.3
LOAD.B.3.2
SOLVE.A.B
PRINT.B  THE  RESULTS  FOR  X
QUIT
```

The elements of the matrices A and B are input following the respective LOAD commands.

For a batch mode of operation the datafile would be as follows:

```
LOAD.A.3.3
3   4   5
1  -1   1
2   2  -3
LOAD.B.3.2
12  9
1 8
1 10
SOLVE.A.B
PRINT.B  THE RESULTS FOR X
QUIT
```

and the resulting output file is

```
MATOP Version 1.2 Feb 1986
Matrix Operations Program- Output File :A500.OTM

LOAD.A.3.3
LOAD.B.3.2
SOLVE.A.B
PRINT.B  THE RESULTS FOR X
0.100000E+01 0.639535E+01
0.100000E+01 -.202326E+01
0.100000E+01 -.418605E+00
End of File
```

A.6 LISTING OF THE PROGRAM MATOP

```
      PROGRAM MATOP
      COMMON /A/Z,SIZE,NMN
      COMMON /B/NAME
      COMMON /C/LL,IPAGE
      COMMON /D/OFLN,MODE
      DIMENSION Z(2000),SIZE(40)
      CHARACTER*6 NAME(20)
      CHARACTER*60 COMM,MESAGE*30
      CHARACTER*8 DFLN,REPLY*3,IFLN*12,OFLN*12
      CHARACTER ESC
      DATA ESC/Z'1B'/
      DOUBLE PRECISION Z
      INTEGER SIZE
      INTEGER POS
      NMN=0
      LELEM=0
      IPAGE=1
      LL=0
      MODE=0
      CALL HEAD
      WRITE(*,10)ESC
10    FORMAT(1X,A1,'[6;14HMatrix Operations Program')
      WRITE(*,15)ESC
15    FORMAT(1X,A1,'[10;15HEnter Data Filename :')
      READ(*,20) DFLN
20    FORMAT(A8)
      POS=INDEX(DFLN,' ')
      IF(POS.EQ.0) POS=9
      WRITE(*,25)ESC
```

```
25     FORMAT(1X,A1,'[12;15HBatch or Interactive? (Type B/I) :')
       READ *, REPLY
       IF(REPLY(1:1).EQ.'B') THEN
        MODE=1
        IFLN=DFLN(:POS-1)//'.INM'
        OPEN(4,FILE=IFLN,STATUS='OLD')
       ENDIF
       OFLN=DFLN(:POS-1)//'.OTM'
       OPEN(3,FILE=OFLN,STATUS='NEW')
       CALL PAGER
90     CALL HEAD
       IF(MODE.EQ.0) THEN
        PRINT*,'Enter Command :'
        READ(*,100) COMM
100     FORMAT(A60)
       ELSE
        READ(4,100) COMM
        PRINT*,'Command read  :',COMM
        CALL CONT
       ENDIF
       IF ( COMM(1:4).EQ.'LOAD' ) THEN
       CALL LOAD(COMM,LELEM,MODE)
       ELSE IF ( COMM(1:5).EQ.'TRANS' ) THEN
       CALL TRANS(COMM,LELEM)
       ELSE IF ( COMM(1:5).EQ.'SCALE' ) THEN
       CALL SCALE(COMM)
       ELSE IF ( COMM(1:5).EQ.'PRINT' ) THEN
       CALL PRINT(COMM)
       ELSE IF ( COMM(1:4).EQ.'MULT' ) THEN
       CALL MULT(COMM,LELEM)
       ELSE IF ( COMM(1:5).EQ.'SOLVE' ) THEN
       CALL SOLVE(COMM)
       ELSE IF ( COMM(1:6).EQ.'DELETE' ) THEN
       CALL DELETE(COMM,LELEM)
       ELSE IF ( COMM(1:5).EQ.'MODDG' ) THEN
       CALL MODDG(COMM,MODE)
       ELSE IF ( COMM(1:6).EQ.'SELECT' ) THEN
       CALL SELECT(COMM,LELEM)
       ELSE IF ( COMM(1:4).EQ.'NULL' ) THEN
       CALL NULL(COMM,LELEM)
       ELSE IF ( COMM(1:7).EQ.'REMARK.') THEN
       CALL PAGER
       WRITE(3,110) COMM
       LL=LL-1
110    FORMAT(A60)
       ELSE IF ( COMM(1:4).EQ.'HELP' ) THEN
       CALL HELP
       ELSE IF ( COMM(1:4).EQ.'QUIT' ) THEN
       CALL QUIT
       GO TO 210
       ELSE
       MESAGE='Invalid Format -Try HELP'
       CALL TRAP(MESAGE)
       CALL CONT
       ENDIF
       GO TO 90
210    CALL PAGER
       WRITE(3,120)
120    FORMAT('End of File')
       ENDFILE (3)
       CLOSE (3)
       IF(MODE.EQ.1) CLOSE (4)
       END
C***
       SUBROUTINE LOAD(COMM,K1,MODE)
       COMMON /A/Z,SIZE,NMN
       COMMON /B/NAME
       COMMON /C/LL,IPAGE
       DIMENSION Z(2000),SIZE(40)
       DIMENSION TEMP(400)
       CHARACTER*6 NAME(20)
       CHARACTER*60 COMM,MATNAM*6,MESAGE*30
```

```
      CHARACTER TEST(9),TOST(11)*2
      DOUBLE PRECISION Z
      INTEGER SIZE
      INTEGER C(6)
      INTEGER FLAG
      INTEGER LL
      INTEGER IPAGE
      DATA TEST/'1','2','3','4','5','6','7','8','9'/
      DATA TOST/'10','11','12','13','14','15','16','17','18','19','20'/
      CALL LOCATE(COMM,C,IEND)
      MATNAM=COMM(C(1)+1:C(2)-1)
      CALL CHKNAM(MATNAM,FLAG)
      IF (FLAG.EQ.1) THEN
       MESAGE='Name Already used '
       CALL TRAP(MESAGE)
       CALL CONT
      ELSE
       DO 20 I=1,9
       IF (COMM(C(2)+1:C(3)-1).EQ.TEST(I)) THEN
       IR = I
       ENDIF
       IF (COMM(C(3)+1:IEND-1).EQ.TEST(I)) THEN
       IC = I
       ENDIF
20     CONTINUE
       DO 30 I=1,11
       IF (COMM(C(2)+1:C(3)-1).EQ.TOST(I)) THEN
       IR = I+9
       ENDIF
       IF (COMM(C(3)+1:IEND-1).EQ.TOST(I)) THEN
       IC = I+9
       ENDIF
30     CONTINUE
        IF(MODE.EQ.0) THEN
        K=0
         DO 50 I=1,IR
         CALL HEAD
         PRINT*,'Loading Matrix '//MATNAM
         PRINT*,'      Enter row number ',I
        PRINT*,' '
         DO 50 J=1,IC
         K=K+1
         WRITE(*,60)
50       READ(*,70,ERR=80) TEMP(K)
60       FORMAT(1H+,1X,':')
70       FORMAT(D14.0)
        GO TO 90
80       PRINT*,'Terminating Input - Matrix Deleted'
         CALL CONT
         GO TO 95
90       DO 110 I=1,IR*IC
         K1=K1+1
110      Z(K1)=TEMP(I)
        ELSE
         DO 55 I=1,IR
         READ(4,FMT=*) (Z(K1+J),J=1,IC)
         K1=K1+IC
55       CONTINUE
         CALL HEAD
         PRINT*,'Matrix '//MATNAM//' loaded'
         CALL CONT
        ENDIF
        NMN=NMN+1
        CALL WRTNAM(MATNAM,IR,IC)
        CALL PAGER
        WRITE(3,100) COMM
100     FORMAT(A60)
        LL=LL-1
95      CONTINUE
      ENDIF
      RETURN
      END
```

```
C***
      SUBROUTINE CHKNAM(MATNAM,FLAG)
      COMMON /B/NAME
      CHARACTER*6 NAME(20)
      CHARACTER*6 MATNAM
      INTEGER FLAG
      FLAG=0
      DO 100 I=1,5
      IF (MATNAM.EQ.NAME(I)) THEN
      FLAG=1
      ENDIF
100   CONTINUE
      RETURN
      END
C***
      SUBROUTINE WRTNAM(MATNAM,I,J)
      COMMON /A/Z,SIZE,NMN
      COMMON /B/NAME
      DIMENSION Z(2000),SIZE(40)
      CHARACTER*6 NAME(20)
      CHARACTER*6 MATNAM
      DOUBLE PRECISION Z
      INTEGER SIZE
      NAME(NMN)=MATNAM
      SIZE(NMN*2-1)=I
      SIZE(NMN*2)=J
      RETURN
      END
C***
      SUBROUTINE HEAD
      CHARACTER ESC
      DATA ESC /Z'1B'/
      WRITE(*,10)ESC
10    FORMAT(1X,A1,'[2J')
      PRINT*,'___________________________________________________________'
      PRINT*,' MATOP                                 Version 1.2 Feb. 1986'
      PRINT*,'-----------------------------------------------------------'
      RETURN
      END
C***
      SUBROUTINE EXTNAM(MATNAM,K,FLAG)
      COMMON /B/NAME
      CHARACTER*6 NAME(20)
      CHARACTER*6 MATNAM
      INTEGER FLAG
      FLAG=0
      DO 100 I=1,20
      IF (MATNAM.EQ.NAME(I)) THEN
      FLAG=1
      K=I
      ENDIF
100   CONTINUE
      RETURN
      END
C***
      SUBROUTINE CONT
      COMMON /D/OFLN,MODE
      CHARACTER*12 OFLN
      CHARACTER ESC
      DATA ESC/Z'1B'/
      IF(MODE.EQ.0) THEN
       PRINT 10,ESC
10     FORMAT(1X,A1,'[20;10H')
       PAUSE 'Type any key to Continue -> '
      ELSE
       PRINT 15,ESC
15     FORMAT(1X,A1,'[20;10HContinuing ....')
       DO 20 I=1,30000
20     CONTINUE
      ENDIF
      RETURN
      END
```

```
C***
      SUBROUTINE TRAP(MESAGE)
      CHARACTER*30 MESAGE
      CHARACTER ESC,BEL
      DATA ESC/Z'1B'/
      DATA BEL/Z'07'/
      PRINT*,BEL,BEL
      PRINT 10,ESC
10    FORMAT(1X,A1,'[18;10H')
      PRINT*,MESAGE
      RETURN
      END
C***
      SUBROUTINE PRINT(COMM)
      COMMON /A/Z,SIZE,NMN
      COMMON /B/NAME
      COMMON /C/LL,IPAGE
      DIMENSION Z(2000),SIZE(40)
      CHARACTER*6 NAME(20)
      CHARACTER*60 COMM,MATNAM*6,MESAGE*30
      DOUBLE PRECISION Z
      INTEGER SIZE
      INTEGER FLAG
      INTEGER C(6)
      INTEGER LL
      INTEGER IPAGE
      CALL LOCATE(COMM,C,IEND)
       MATNAM=COMM(C(1)+1:IEND-1)
       CALL EXTNAM(MATNAM,INDEX,FLAG)
       IF (FLAG.EQ.0) THEN
        MESAGE='Matrix does not Exist'
        CALL TRAP(MESAGE)
        CALL CONT
       ELSE
        CALL POINT(INDEX,IR,IC,KOUNT)
        K1=KOUNT
        CALL HEAD
        DO 40 I=1,IR
        PRINT*,'At Print: No. of Rows and Columns is: ',IR,IC
        PRINT*,'Row number :',I
        DO 45 J=1,IC
        KOUNT=KOUNT+1
        WRITE(*,50) Z(KOUNT)
50      FORMAT(1X,E12.6)
45      CONTINUE
        CALL CONT
        IF (I.NE.IR) CALL HEAD
40      CONTINUE
        CALL PAGER
        WRITE(3,100) COMM
100     FORMAT(A60)
        LL=LL-1
        DO 60 I=1,IR
        CALL PAGER
        WRITE(3,70) (Z(K1+J),J=1,IC)
        IF (IC.GT.6.AND.IC.LT.13) LL=LL-1
        IF (IC.GT.12) LL=LL-2
        LL=LL-1
        K1=K1+IC
60      CONTINUE
70      FORMAT(6(E12.6,1X))
       ENDIF
      RETURN
      END
C***
      SUBROUTINE LOCATE(COMM,C,IEND)
      CHARACTER*60 COMM
      INTEGER POS,C(6)
      K=0
      IEND=0
      DO 5 I=1,6
5     C(I)=0
```

```
C***  READ POSITION OF PERIODS AND FIRST BLANK
      DO 10 POS=1,80
      IF (COMM(POS:POS).EQ.'.') THEN
      K=K+1
      C(K)=POS
      ELSE IF (COMM(POS:POS).EQ.' ') THEN
       IF (IEND.EQ.0) THEN
       IEND=POS
       ENDIF
      ENDIF
10    CONTINUE
      RETURN
      END
C***
      SUBROUTINE POINT(INDEX,IR,IC,KOUNT)
      COMMON /A/Z,SIZE,NMN
      DIMENSION Z(2000),SIZE(40)
      DOUBLE PRECISION Z
      INTEGER SIZE
      IR=SIZE(INDEX*2-1)
      IC=SIZE(INDEX*2)
      KOUNT=0
      IF (INDEX.GT.1) THEN
      DO 30 I=1,(2*INDEX-2),2
30    KOUNT=KOUNT+SIZE(I)*SIZE(I+1)
      ENDIF
      RETURN
      END
C***
      SUBROUTINE PAGER
      COMMON /C/LL,IPAGE
      COMMON /D/OFLN,MODE
      CHARACTER*12 OFLN
      INTEGER LL
      INTEGER IPAGE
      K=0
      IF (LL.GT.0) RETURN
      IF (LL.EQ.0.AND.IPAGE.GT.1) K=7
      IF (LL.EQ.-1.AND.IPAGE.GT.1) K=6
      IF (K.NE.0) THEN
      DO 90 I=1,K
90    WRITE(3,95)
95    FORMAT(' ')
      ENDIF
      WRITE(3,110) IPAGE
110   FORMAT(/'MATOP Version 1.2 Feb 1986',44X,'Page ',I2)
      WRITE(3,112) OFLN
112   FORMAT('Matrix Operations Program- Output File :',A12/)
      IPAGE=IPAGE+1
      LL=55
      RETURN
      END
C***
      SUBROUTINE QUIT
      CALL HEAD
      PRINT*,'At Quit'
      CALL CONT
      RETURN
      END
C***
      SUBROUTINE TRANS(COMM,KOUNT)
      COMMON /A/Z,SIZE,NMN
      COMMON /B/NAME
      COMMON /C/LL,IPAGE
      DIMENSION Z(2000),SIZE(40)
      CHARACTER*6 NAME(20)
      CHARACTER*60 COMM,MATNAM*6,MESAGE*30
      CHARACTER*6 NAM1
      DOUBLE PRECISION Z
      INTEGER SIZE
      INTEGER FLAG
      INTEGER C(6)
```

```
      INTEGER LL
      INTEGER IPAGE
      CALL HEAD
      PRINT,'At Trans'
      CALL LOCATE(COMM,C,IEND)
      MATNAM=COMM(C(1)+1:C(2)-1)
      CALL EXTNAM(MATNAM,INDEX,FLAG)
      IF (FLAG.EQ.0) THEN
       MESAGE='Matrix '//MATNAM//' does not Exist'
       CALL TRAP(MESAGE)
       CALL CONT
      ELSE
       NAM1=MATNAM
       IND1=INDEX
       MATNAM=COMM(C(2)+1:IEND-1)
       CALL CHKNAM(MATNAM,FLAG)
       IF (FLAG.EQ.1) THEN
        MESAGE='Name '//MATNAM//' Already used'
        CALL TRAP(MESAGE)
        CALL CONT
       ELSE
        CALL POINT(IND1,IR1,IC1,KOUNT1)
        NMN=NMN+1
        CALL WRTNAM(MATNAM,IC1,IR1)
C***    TRANSPOSITION ROUTINE
        DO 15 I=1,IC1
        KOUNT1=KOUNT1+1
        K2=0
        DO 15 J=1,IR1
        KOUNT=KOUNT+1
        Z(KOUNT)=Z(KOUNT1+K2*IC1)
15      K2=K2+1
        CALL PAGER
        WRITE(3,100) COMM
100     FORMAT(A60)
        LL=LL-1
        PRINT*,'Matrix '//NAM1//' transposed to '//MATNAM
        CALL CONT
       ENDIF
      ENDIF
      RETURN
      END
C***
      SUBROUTINE SCALE(COMM)
      COMMON /A/Z,SIZE,NMN
      COMMON /B/NAME
      COMMON /C/LL,IPAGE
      DIMENSION Z(2000),SIZE(40)
      CHARACTER*6 NAME(20),NAM1,NAM2*20
      CHARACTER*60 COMM,MATNAM*6,MESAGE*30
      CHARACTER TEST(10)
      DOUBLE PRECISION NUMBER
      DOUBLE PRECISION Z
      INTEGER SIZE
      INTEGER FLAG
      INTEGER C(6)
      INTEGER POS
      INTEGER LL
      INTEGER IPAGE
      DATA TEST/'0','1','2','3','4','5','6','7','8','9'/
      NUMBER=0
      CALL HEAD
      PRINT,'At Scale'
      CALL LOCATE(COMM,C,IEND)
      MATNAM=COMM(C(1)+1:C(2)-1)
      CALL EXTNAM(MATNAM,INDEX,FLAG)
      IF (FLAG.EQ.0) THEN
       MESAGE='Matrix does not Exist'
       CALL TRAP(MESAGE)
       CALL CONT
      ELSE
       NAM1=MATNAM
```

```
        CALL POINT(INDEX,IR,IC,KOUNT)
        MATNAM=COMM(C(2)+1:IEND-1)
C***    THIS IS THE SCALE FACTOR
        POS=1
        IF (C(3).EQ.0) C(3)=IEND
        DO 15 I=1,C(3)-C(2)-1
        NAM2=MATNAM(POS:POS)
        DO 20 J=1,10
20      IF (NAM2.EQ.TEST(J)) N=J-1
        NUMBER=NUMBER+N*10**(C(3)-(C(2)+1+I))
        POS=POS+1
15      CONTINUE
        IF (C(3).NE.IEND) THEN
        POS=POS+1
        DO 25 I=1,IEND-C(3)-1
        NAM2=MATNAM(POS:POS)
        DO 30 J=1,10
30      IF (NAM2.EQ.TEST(J)) N=J-1
        NUMBER=NUMBER+N*10**(-I)
        POS=POS+1
25      CONTINUE
        ENDIF
        DO 35 I=1,IR
        DO 35 J=1,IC
        KOUNT=KOUNT+1
35      Z(KOUNT)=Z(KOUNT)*NUMBER
        PRINT*,'Matrix '//NAM1//' Scaled by ',NUMBER
         CALL PAGER
         WRITE(3,100) COMM
100      FORMAT(A60)
         LL=LL-1
        CALL CONT
       ENDIF
       RETURN
       END
C***
       SUBROUTINE MULT(COMM,KOUNT)
       COMMON /A/Z,SIZE,NMN
       COMMON /B/NAME
       COMMON /C/LL,IPAGE
       DIMENSION Z(2000),SIZE(40)
       CHARACTER*6 NAME(20)
       CHARACTER*60 COMM,MATNAM*6,MESAGE*30
       CHARACTER*6 NAM1,NAM2
       DOUBLE PRECISION Z
       INTEGER SIZE
       INTEGER FLAG
       INTEGER C(6)
       INTEGER LL
       INTEGER IPAGE
       CALL HEAD
       PRINT,'At Mult'
       CALL LOCATE(COMM,C,IEND)
       MATNAM=COMM(C(1)+1:C(2)-1)
       CALL EXTNAM(MATNAM,INDEX,FLAG)
       IF (FLAG.EQ.0) THEN
        MESAGE='Matrix '//MATNAM//' does not Exist'
        CALL TRAP(MESAGE)
        CALL CONT
       ELSE
        NAM1=MATNAM
        IND1=INDEX
        MATNAM=COMM(C(2)+1:C(3)-1)
        CALL EXTNAM(MATNAM,INDEX,FLAG)
         IF (FLAG.EQ.0) THEN
          MESAGE='Matrix '//MATNAM//' does not Exist'
          CALL TRAP(MESAGE)
          CALL CONT
         ELSE
          NAM2=MATNAM
          IND2=INDEX
          MATNAM=COMM(C(3)+1:IEND-1)
```

```
          CALL CHKNAM(MATNAM,FLAG)
           IF (FLAG.EQ.1) THEN
            MESAGE='Name '//MATNAM//' Already used'
            CALL TRAP(MESAGE)
            CALL CONT
           ELSE
            CALL POINT(IND1,IR1,IC1,KOUNT1)
            CALL POINT(IND2,IR2,IC2,KOUNT2)
            IF (IC1.NE.IR2) THEN
             MESAGE='Matrices Incompatible'
             CALL TRAP(MESAGE)
             CALL CONT
            ELSE
             NMN=NMN+1
             CALL WRTNAM(MATNAM,IR1,IC2)
C***        MULTIPLICATION ROUTINE
             KOUNT1=KOUNT1+1
             DO 20 I=1,IR1
             DO 20 J=1,IC2
             TEMP=0
             K2=KOUNT1+(I-1)*IC1
             K3=KOUNT2+J
             DO 25 K=1,IC1
             TEMP=Z(K2)*Z(K3)+TEMP
             K2=K2+1
25           K3=K3+IC2
C***        END OF Z ARRAY CURRENTLY AT KOUNT(LELEM)
             KOUNT=KOUNT+1
20           Z(KOUNT)=TEMP
             CALL PAGER
             WRITE(3,100) COMM
100          FORMAT(A60)
             LL=LL-1
             PRINT*,'Matrices '//NAM1//'.'//NAM2//' multiplied to '//MATN
     CAM
             CALL CONT
            ENDIF
           ENDIF
         ENDIF
       ENDIF
       RETURN
       END
C***
       SUBROUTINE SOLVE(COMM)
       COMMON /A/Z,SIZE,NMN
       COMMON /B/NAME
       COMMON /C/LL,IPAGE
       DIMENSION Z(2000),SIZE(40)
       CHARACTER*6 NAME(20)
       CHARACTER*60 COMM,MATNAM*6,MESAGE*30
       CHARACTER*6 NAM2
       DOUBLE PRECISION A(20,20),X(20,20),P(20,20)
       DOUBLE PRECISION Z
       DOUBLE PRECISION TEMP
       INTEGER SIZE
       INTEGER FLAG
       INTEGER C(6)
       INTEGER LL
       INTEGER IPAGE
       CALL HEAD
       PRINT,'At Solve'
       CALL LOCATE(COMM,C,IEND)
       MATNAM=COMM(C(1)+1:C(2)-1)
       CALL EXTNAM(MATNAM,INDEX,FLAG)
       IF (FLAG.EQ.0) THEN
        MESAGE='Matrix '//MATNAM//' does not Exist'
        CALL TRAP(MESAGE)
        CALL CONT
       ELSE
        IND1=INDEX
        MATNAM=COMM(C(2)+1:IEND-1)
```

```
      CALL EXTNAM(MATNAM,INDEX,FLAG)
       IF (FLAG.EQ.0) THEN
        MESAGE='Matrix '//MATNAM//' does not Exist'
        CALL TRAP(MESAGE)
        CALL CONT
       ELSE
        NAM2=MATNAM
        IND2=INDEX
        CALL POINT(IND1,IR1,IC1,KOUNT1)
        CALL POINT(IND2,IR2,IC2,KOUNT2)
         IF (IR1.NE.IC1.OR.IR2.NE.IC1) THEN
          MESAGE='Matrices Incompatible'
          CALL TRAP(MESAGE)
          CALL CONT
         ELSE
C***      SOLUTION ROUTINE
          K1=KOUNT2
C***      CONVERT TO TWO DIMENSIONAL ARRAYS
          DO 20 I=1,IR1
          DO 20 J=1,IR1
          KOUNT1=KOUNT1+1
20        A(I,J)=Z(KOUNT1)
          DO 25 I=1,IR1
          DO 25 J=1,IC2
          KOUNT2=KOUNT2+1
25        P(I,J)=Z(KOUNT2)
C***      FORWARD ELIMINATION
          FLAG=0
          DO 30 K=1,IR1-1
          K2=K+1
          IF ((ABS(A(K,K))-.000001).LE.0) THEN
C***      TRY ROW INTERCHANGE
            FLAG=1
            DO 32 J=K2,IR1
            IF ((ABS(A(J,K))-.000001).GT.0) THEN
             DO 34 L=K,IR1
             TEMP=A(K,L)
             A(K,L)=A(J,L)
34           A(J,L)=TEMP
             DO 36 L=1,IC2
             TEMP=P(K,L)
             P(K,L)=P(J,L)
36           P(J,L)=TEMP
             FLAG=0
             GO TO 42
            ENDIF
32           CONTINUE
42        ENDIF
          DO 30 I=K+1,IR1
            TEMP=A(I,K)/A(K,K)
            DO 35 J=K,IR1
35          A(I,J)=A(I,J)-TEMP*A(K,J)
            DO 40 L=1,IC2
40          P(I,L)=P(I,L)-TEMP*P(K,L)
30          CONTINUE
            IF ((ABS(A(IR1,IR1))-.000001).LE.0) FLAG=1
            IF (FLAG.EQ.1) THEN
             MESAGE='Matrix Singular'
             CALL TRAP(MESAGE)
             CALL CONT
            ELSE
C***      BACK SUBSTITUTION
             DO 45 L=1,IC2
45           X(IR1,L)=P(IR1,L)/A(IR1,IR1)
             DO 50 K=IR1-1,1,-1
             DO 50 L=1,IC2
             DO 60 J=IR1,K+1,-1
60           P(K,L)=P(K,L)-A(K,J)*X(J,L)
50           X(K,L)=P(K,L)/A(K,K)
C***      DISPLAY AND STORE RESULT
             CALL HEAD
             PRINT*,'Solution Matrix '//NAM2
```

```
               DO 55 I=1,IR1
55             WRITE(*,65) (X(I,J),J=1,IC2)
65             FORMAT("b",6(D12.6,1X))
               DO 70 I=1,IR1
               DO 70 J=1,IC2
               K1=K1+1
70             Z(K1)=X(I,J)
               CALL PAGER
               WRITE(3,100) COMM
100            FORMAT(A60)
               LL=LL-1
               CALL CONT
              ENDIF
           ENDIF
         ENDIF
      ENDIF
      RETURN
      END
C***
      SUBROUTINE DELETE(COMM,LELEM)
      COMMON /A/Z,SIZE,NMN
      COMMON /B/NAME
      COMMON /C/LL,IPAGE
      DIMENSION Z(2000),SIZE(40)
      DIMENSION ZTEMP(2000),SIZET(40)
      CHARACTER*6 NAME(20)
      CHARACTER*60 COMM,MATNAM*6,MESAGE*30
      CHARACTER*6 NTEMP(20)
      DOUBLE PRECISION Z
      DOUBLE PRECISION ZTEMP
      INTEGER SIZE
      INTEGER FLAG
      INTEGER C(6)
      INTEGER SIZET
      INTEGER LL
      INTEGER IPAGE
      CALL HEAD
      PRINT,'At Delete'
      CALL LOCATE(COMM,C,IEND)
       MATNAM=COMM(C(1)+1:IEND-1)
       CALL EXTNAM(MATNAM,INDEX,FLAG)
       IF (FLAG.EQ.0) THEN
        MESAGE='Matrix does not Exist'
        CALL TRAP(MESAGE)
        CALL CONT
       ELSE
        CALL POINT(INDEX,IR,IC,KOUNT)
C***    REMOVE ELEMENTS FROM Z(2000) DATABASE
        K1=0
        DO 40 I=1,LELEM
        IF (I.LT.KOUNT+1.OR.I.GT.KOUNT+IR*IC) THEN
        K1=K1+1
        ZTEMP(K1)=Z(I)
        ENDIF
40      CONTINUE
        LELEM=LELEM-IR*IC
        DO 45 I=1,LELEM
45      Z(I)=ZTEMP(I)
C***    REMOVE DIMENSIONS FROM SIZE(40) DATABASE
        K1=0
        DO 50 I=1,NMN
        IF (I.NE.INDEX) THEN
        K1=K1+1
        SIZET(2*K1-1)=SIZE(2*I-1)
        SIZET(2*K1)  =SIZE(2*I)
        ENDIF
50      CONTINUE
        NMN=NMN-1
        DO 55 I=1,NMN
        SIZE(2*I-1)=SIZET(2*I-1)
55      SIZE(2*I)  =SIZET(2*I)
C***    REMOVE NAME FROM NAME(20) DATABASE
```

```
          NMN=NMN+1
          K1=0
          DO 60 I=1,NMN
          IF (I.NE.INDEX) THEN
          K1=K1+1
          NTEMP(K1)=NAME(I)
          ENDIF
60        CONTINUE
C***      CLEAR NAME(20)
          DO 70 I=1,20
70        NAME(I)='
          NMN=NMN-1
          DO 65 I=1,NMN
65        NAME(I)=NTEMP(I)
          PRINT*,'Matrix '//MATNAM//' Deleted'
          CALL PAGER
          WRITE(3,100) COMM
100       FORMAT(A60)
          LL=LL-1
          CALL CONT
         ENDIF
        RETURN
        END
C***
        SUBROUTINE MODDG(COMM,MODE)
        COMMON /A/Z,SIZE,NMN
        COMMON /B/NAME
        COMMON /C/LL,IPAGE
        DIMENSION Z(2000),SIZE(40)
        CHARACTER*6 NAME(20)
        CHARACTER*60 COMM,MATNAM*6,MESAGE*30
        CHARACTER*3 REPLY
        DOUBLE PRECISION Z
        INTEGER SIZE
        INTEGER FLAG
        INTEGER C(6)
        INTEGER LL
        INTEGER IPAGE
        CALL HEAD
        PRINT*,'At Moddg'
        PRINT*,' '
        CALL LOCATE(COMM,C,IEND)
         MATNAM=COMM(C(1)+1:IEND-1)
         CALL EXTNAM(MATNAM,INDEX,FLAG)
         IF (FLAG.EQ.0) THEN
          MESAGE='Matrix does not Exist'
          CALL TRAP(MESAGE)
          CALL CONT
         ELSE
          CALL POINT(INDEX,IR,IC,KOUNT)
          IF(MODE.EQ.0) THEN
          DO 35 I=1,IR
           DO 35 J=1,IC
           KOUNT=KOUNT+1
           IF(I.EQ.J) THEN
          WRITE(*,45) I,J
45         FORMAT(1H+,1X,'Element ',2I3,' Modify? (Type Y/N) ')
           READ(*,50) REPLY
50         FORMAT(A3)
           IF (REPLY(1:1).EQ.'Y') THEN
           WRITE(*,55) Z(KOUNT)
55         FORMAT(1H+,1X,D12.6,' Modify to: ')
           READ(*,40) Z(KOUNT)
40         FORMAT(D14.8)
          ENDIF
          ENDIF
35         CONTINUE
          ELSE
           READ(4,FMT=*) K
           DO 70 I=1,K
           READ(4,FMT=*) IND,VALUE
           J=KOUNT+(IND-1)*IC+IND
```

```
          Z(J)=VALUE
70        CONTINUE
          PRINT*,'Matrix '//MATNAM//' modified'
         ENDIF
         CALL PAGER
         WRITE(3,100) COMM
100      FORMAT(A60)
         LL=LL-1
         CALL CONT
        ENDIF
       RETURN
       END
C***
       SUBROUTINE SELECT(COMM,LELEM)
       COMMON /A/Z,SIZE,NMN
       COMMON /B/NAME
       COMMON /C/LL,IPAGE
       DIMENSION Z(2000),SIZE(40)
       CHARACTER*6 NAME(20)
       CHARACTER*60 COMM,MATNAM*6,MESAGE*30
       CHARACTER*6 NAM1,NAM2
       CHARACTER TEST(9),TOST(11)*2
       DOUBLE PRECISION Z
       INTEGER SIZE
       INTEGER C(6)
       INTEGER FLAG
       INTEGER LL
       INTEGER IPAGE
       DATA TEST/'1','2','3','4','5','6','7','8','9'/
       DATA TOST/'10','11','12','13','14','15','16','17','18','19','20'/
       CALL LOCATE(COMM,C,IEND)
       MATNAM=COMM(C(1)+1:C(2)-1)
       CALL CHKNAM(MATNAM,FLAG)
       IF (FLAG.EQ.1) THEN
        MESAGE='Name Already used '
        CALL TRAP(MESAGE)
        CALL CONT
       ELSE
        NAM1=MATNAM
        MATNAM=COMM(C(2)+1:C(3)-1)
        CALL EXTNAM(MATNAM,INDEX,FLAG)
        IF (FLAG.EQ.0) THEN
         MESAGE='Matrix '//MATNAM//' does not Exist'
         CALL TRAP(MESAGE)
         CALL CONT
        ELSE
         NAM2=MATNAM
         CALL POINT(INDEX,IR,IC,KOUNT)
         DO 10 I=1,9
         IF (COMM(C(3)+1:C(4)-1).EQ.TEST(I)) IR1=I
10       IF (COMM(C(4)+1:C(5)-1).EQ.TEST(I)) IC1=I
         DO 15 I=1,11
         IF (COMM(C(3)+1:C(4)-1).EQ.TOST(I)) IR1=I+9
15       IF (COMM(C(4)+1:C(5)-1).EQ.TOST(I)) IC1=I+9
         DO 20 I=1,9
         IF (COMM(C(5)+1:C(6)-1).EQ.TEST(I)) IR2=I
20       IF (COMM(C(6)+1:IEND-1).EQ.TEST(I)) IC2=I
         DO 25 I=1,11
         IF (COMM(C(5)+1:C(6)-1).EQ.TOST(I)) IR2=I+9
25       IF (COMM(C(6)+1:IEND-1).EQ.TOST(I)) IC2=I+9
C***      SELECT MATRIX ELEMENTS
          CALL HEAD
          PRINT*,'At Select'
          NMN=NMN+1
          CALL WRTNAM(NAM1,IR1,IC1)
          DO 40 I=IR2,IR2+IR1-1
          K1=KOUNT+(I-1)*IC+IC2-1
          DO 40 J=1,IC1
          K1=K1+1
          LELEM=LELEM+1
40        Z(LELEM)=Z(K1)
          PRINT*,'Matrix '//NAM1//' selected from '//NAM2
```

```
          CALL PAGER
          WRITE(3,100) COMM
100       FORMAT(A60)
          LL=LL-1
          CALL CONT
        ENDIF
      ENDIF
      RETURN
      END
C***
      SUBROUTINE NULL(COMM,K1)
      COMMON /A/Z,SIZE,NMN
      COMMON /B/NAME
      COMMON /C/LL,IPAGE
      DIMENSION Z(2000),SIZE(40)
      CHARACTER*6 NAME(20)
      CHARACTER*60 COMM,MATNAM*6,MESAGE*30
      CHARACTER TEST(9),TOST(11)*2
      DOUBLE PRECISION Z
      INTEGER SIZE
      INTEGER C(6)
      INTEGER FLAG
      INTEGER LL
      INTEGER IPAGE
      DATA TEST/'1','2','3','4','5','6','7','8','9'/
      DATA TOST/'10','11','12','13','14','15','16','17','18','19','20'/
      CALL LOCATE(COMM,C,IEND)
      MATNAM=COMM(C(1)+1:C(2)-1)
      CALL CHKNAM(MATNAM,FLAG)
      IF (FLAG.EQ.1) THEN
       MESAGE='Name Already used '
       CALL TRAP(MESAGE)
       CALL CONT
      ELSE
       DO 20 I=1,9
       IF (COMM(C(2)+1:C(3)-1).EQ.TEST(I)) THEN
       IR = I
       ENDIF
       IF (COMM(C(3)+1:IEND-1).EQ.TEST(I)) THEN
       IC = I
       ENDIF
20     CONTINUE
       DO 30 I=1,11
       IF (COMM(C(2)+1:C(3)-1).EQ.TOST(I)) THEN
       IR = I+9
       ENDIF
       IF (COMM(C(3)+1:IEND-1).EQ.TOST(I)) THEN
       IC = I+9
       ENDIF
30     CONTINUE
        CALL HEAD
        PRINT*,'At Null'
        DO 110 I=1,IR*IC
        K1=K1+1
110     Z(K1)=0.0
        NMN=NMN+1
        CALL WRTNAM(MATNAM,IR,IC)
        CALL PAGER
        WRITE(3,100) COMM
100     FORMAT(A60)
        LL=LL-1
        PRINT*,'Matrix '//MATNAM//' created as null matrix'
        CALL CONT
      ENDIF
      RETURN
      END
C***
      SUBROUTINE HELP
      CALL HEAD
      PRINT*,'Matrix names may be any alphanumeric string to a maximum'
      PRINT*,'            of six characters'
      PRINT*,'For any specified matrix the'
```

```
PRINT*,'         maximum number of rows is 20'
PRINT*,'         maximum number of columns is 20'
PRINT*,'Command Formats are as follows:'
PRINT*,' '
PRINT*,'  HELP - will cause this display'
PRINT*,' '
PRINT*,'  QUIT - will terminate the session'
PRINT*,'  LOAD.matrixname.no rows.no columns'
PRINT*,'                Example: LOAD.KMAT.6.6'
PRINT*,'  PRINT.matrixname'
PRINT*,' '
PRINT*,'  MULT.matrixname1.matrixname2.newmatrixname'
PRINT*,'                Example: MULT.A.B.C  (result in C)'
CALL CONT
CALL HEAD
PRINT*,'  SOLVE.matrixname1.matrixname2'
PRINT*,'                Example: SOLVE.KMAT.P (result in P)'
PRINT*,'  TRANS.matrixname.newmatrixname'
PRINT*,'                Example: TRANS.A.B'
PRINT*,'  SCALE.matrixname.scalefactor'
PRINT*,'                Example: SCALE.KMAT.1500 (not E format)'
PRINT*,'  MODDG.matrixname'
PRINT*,'                Permits modification of diagonal elements'
PRINT*,'  DELETE.matrixname'
PRINT*,'                Deletes matrix from database'
PRINT*,'  SELECT.newmatrixname.matrixname.rn.cn.rs.cs'
PRINT*,'                Example: SELECT.D.P.4.1.6.2'
PRINT*,'  NULL.matrixname1.no rows.no columns'
PRINT*,'                Example: NULL.S.4.6'
PRINT*,'  REMARK.character string'
PRINT*,'                Example: REMARK. THE STIFFNESS MATRIX'
CALL CONT
RETURN
END
```

Appendix B
Structural Mechanics Students' Handbook—A Manual of Useful Data and Information

PART 1

B1.1 Introduction—Convention

In introducing students to beam behaviour and bending moment diagrams, it is usual to introduce a sign convention based on a rigorous mathematical approach using the first quadrant right-hand set of cartesian axes. This leads logically to the notion of positive bending moment being associated with positive curvature or, more simply, a sagging beam. Such moments are then often plotted with positive ordinates above a datum line in the conventional manner of any graph.

However it is a widely held convention that the ordinates of a bending moment diagram should be plotted off the tension face of a line diagram of the structure. Where bending moment diagrams are shown in this manual, that is the convention that has been followed. Beyond this, it is not necessary to indicate whether the bending moment is positive or negative. Although the concept of positive and negative bending can still be applied to beams, it becomes rather meaningless for frames. Only simple beam deflections are quoted in this manual and they are taken as positive downward.

In tables B1.3 and B1.3A, where end moments are given, the end moments are quoted as positive or negative according to the direction of the reactive moment developed. The convention followed here is that the anticlockwise moment on the beam and the corresponding clockwise moment on the joint are POSITIVE moments, while the clockwise moment on the beam and the corresponding anticlockwise moment on the joint are NEGATIVE. It should be appreciated that this information, of itself, is insufficient when it comes to drawing a bending moment diagram. It is the position of an applied moment with respect to the beam, together with its sense, which determines the tension face and hence the way in which the bending moment diagram is drawn. The classic illustration of this is the case of a simply supported beam with an anticlockwise moment (positive) applied firstly at the left-hand end and secondly at the right-hand end.

Table B1.1 shows some standard information relating to simply supported beams under transverse load. In table B1.2 some useful properties of

Table B1.1 Simple beam moments and deflections (uniform EI)

BEAM	Max. Moment	Max. Deflection
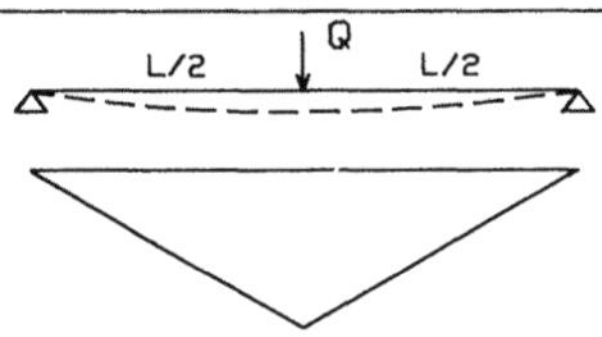	$QL/4$	$QL^3/48EI$
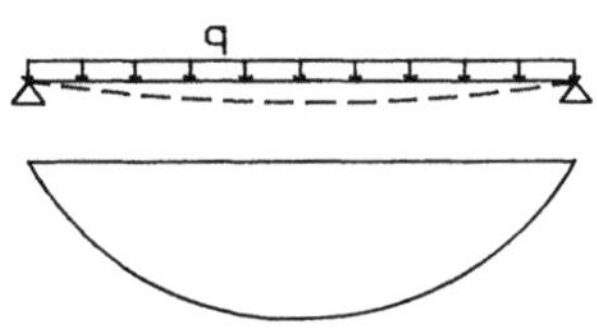	$qL^2/8$	$5qL^4/384EI$
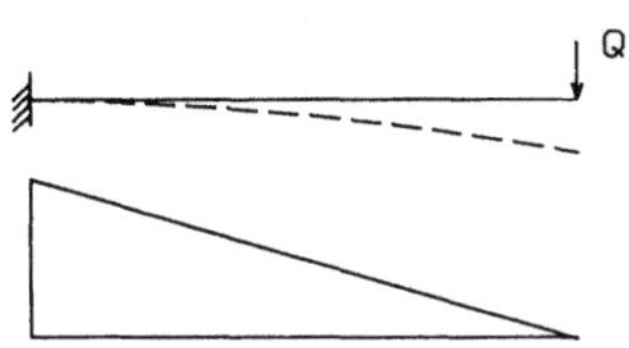	QL	$QL^3/3EI$
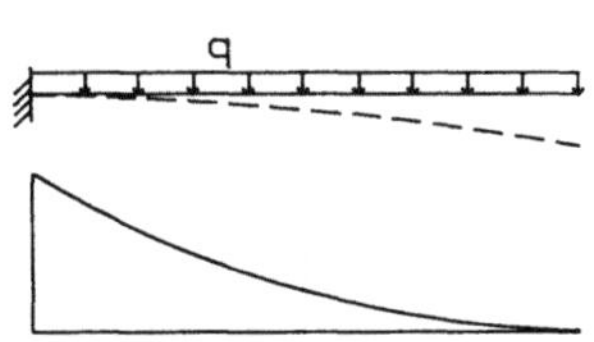	$qL^2/2$	$qL^4/8EI$
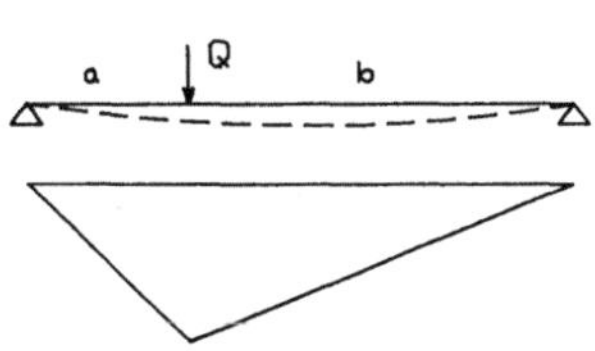	Qab/L	$\frac{Qa(b^3(L+a)^3/243)^{0.5}}{EIL}$
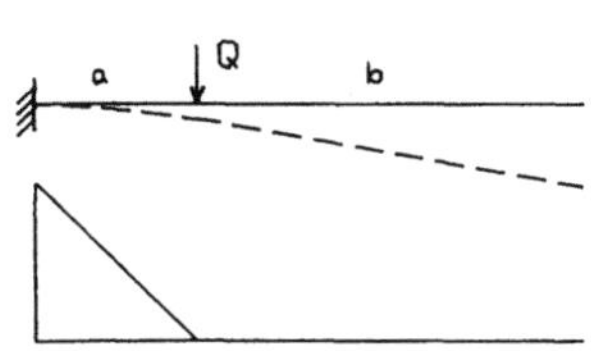	Qa	$Qa^2(2L+b)/6EI$

Table B1.2 Some properties of area

SHAPE	Centroid Distance $\bar{x}$	Area
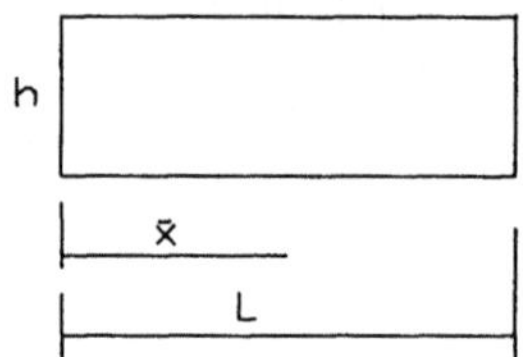	L/2	Lh
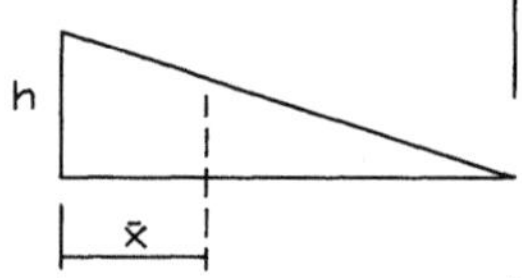	L/3	Lh/2
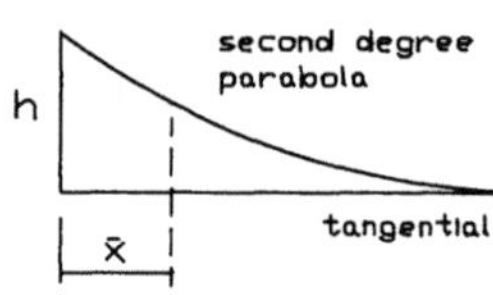	L/4	Lh/3
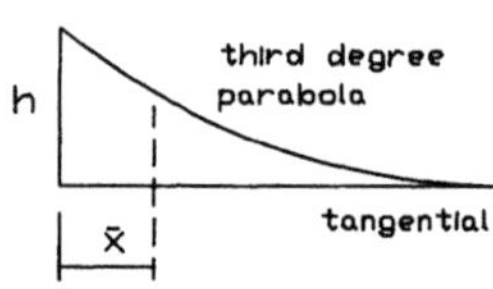	L/5	Lh/4
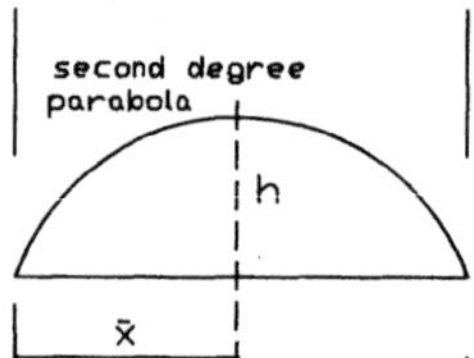	L/2	2Lh/3
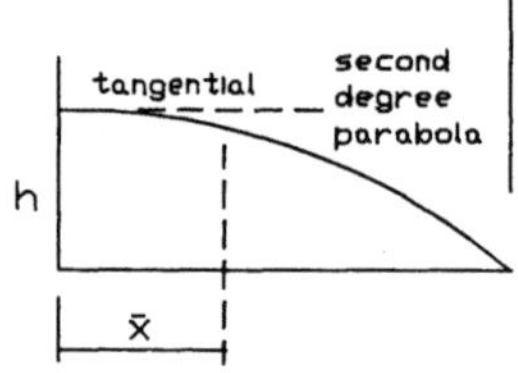	3L/8	2Lh/3

Table B1.3 Indeterminate beam end moments—transverse loads (uniform *EI*)

END A	A BEAM B	END B
$+QL/8$	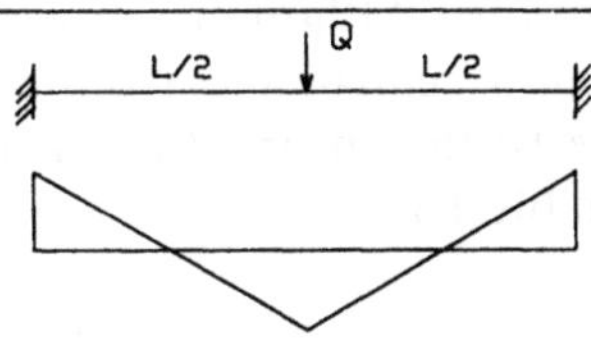	$-QL/8$
$+qL^2/12$	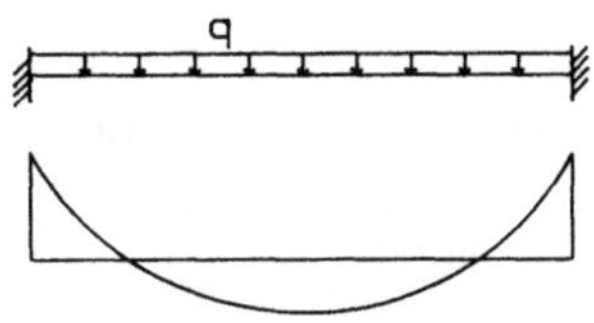	$-qL^2/12$
$+3QL/16$	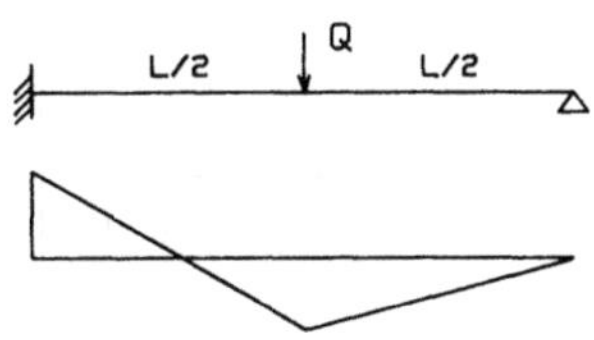	
$+qL^2/8$	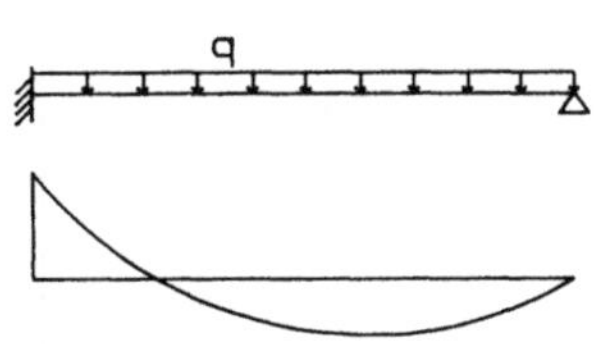	
$+Qab^2/L^2$	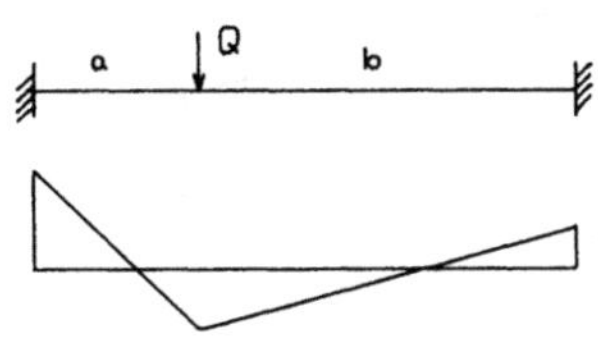	$-Qa^2b/L^2$
$+\dfrac{Qa(L-a)(2L-a)}{2L^2}$	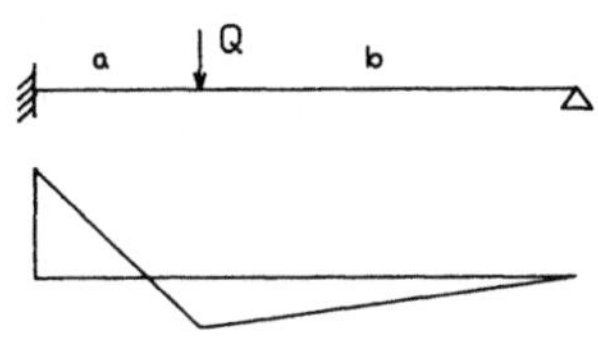	

area are given. This information is frequently used in the application of the moment area theorems in the calculation of slopes and deflections. Table B1.5 is also relevant to deflection calculations since the standard integrals arise in the application of the principle of virtual forces to determine deflections. Tables B1.3 and B1.3A present standard fixed end moments for a beam under various conditions. Although the vertical reactions are not shown there, they can be readily calculated from equilibrium. Table B1.6 completes the set with some further properties of area. See sections B1.2 and B1.3 for discussion of table B1.4.

Table B1.3A Indeterminate beam end moments—translation only (uniform EI)

END A	A BEAM B	END B
$+6EI\Delta/L^2$	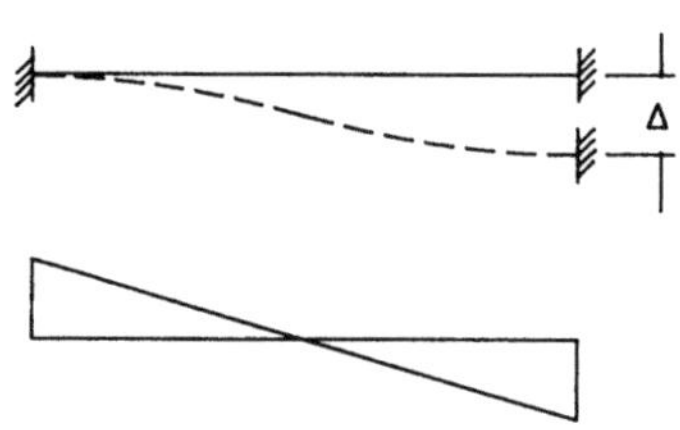	$+6EI\Delta/L^2$
$+3EI\Delta/L^2$	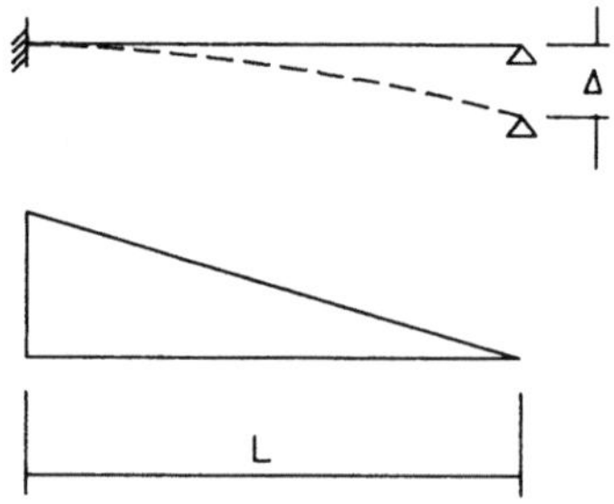	

Table B1.4 Beam end rotations under transverse load

Definitions	
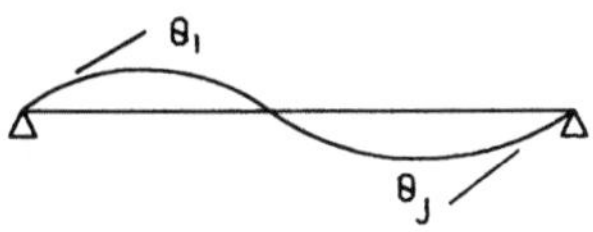	Flexural Properties EI Length L Total Load = Q $\alpha = a/L$; $\beta = b/L$; $\gamma = c/L$
Load Type	**End Rotations**
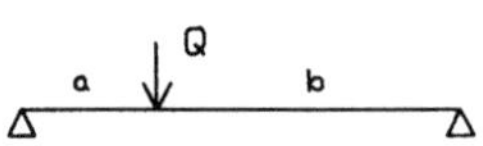	$EI\ \theta_I = -Qb(L^2 - b^2)/6L$ $EI\ \theta_J = Qa(L^2 - a^2)/6L$
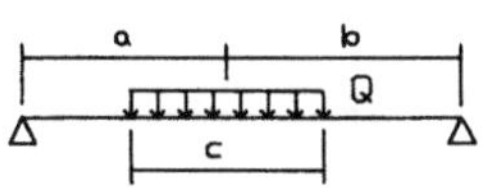	$EI\ \theta_I = -Qb[4a(b+L)-c^2]/24L$ $EI\ \theta_J = Qa[4b(a+L)-c^2]/24L$
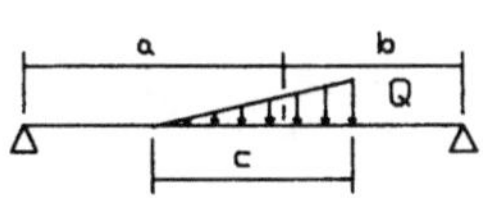	$EI\ \theta_I = -QL^2[270(\beta-\beta^3)-\gamma^2(45\beta+2\gamma)]/1620$ $EI\ \theta_J = +QL^2[270(\alpha-\alpha^3)-\gamma^2(45\alpha-2\gamma)]/1620$
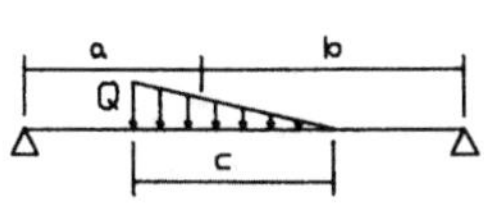	$EI\ \theta_I = -QL^2[270(\beta-\beta^3)-\gamma^2(45\beta-2\gamma)]/1620$ $EI\ \theta_J = +QL^2[270(\alpha-\alpha^3)-\gamma^2(45\alpha+2\gamma)]/1620$

Table B1.5 Standard integrals relating to moment diagrams

The Integral: $\int_0^L g_1(x).g_2(x)\,dx$			
$g_1(x)$	$g_2(x)$		
	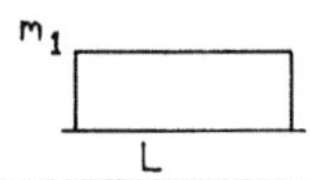	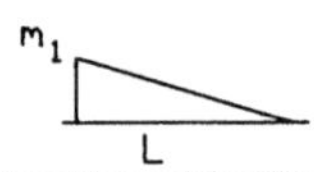	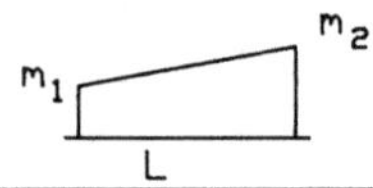
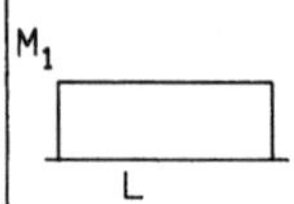	$L\, M_1 m_1$	$\frac{L}{2} M_1 m_1$	$\frac{L}{2} [M_1 (m_1 + m_2)]$
	$\frac{L}{2} M_1 m_1$	$\frac{L}{3} M_1 m_1$	$\frac{L}{6} [M_1 (2m_1 + m_2)]$
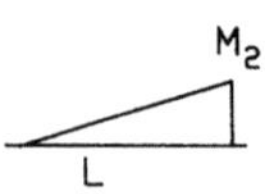	$\frac{L}{2} M_2 m_1$	$\frac{L}{6} M_2 m_1$	$\frac{L}{6} [M_2 (m_1 + 2m_2)]$
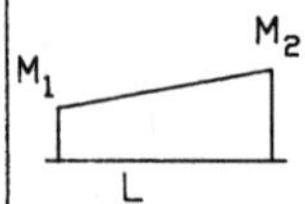	$\frac{L}{2} [m_1 (M_1 + M_2)]$	$\frac{L}{6} [m_1 (2M_1 + M_2)]$	$\frac{L}{6} [M_1 (2m_1 + m_2) + M_2 (m_1 + 2m_2)]$
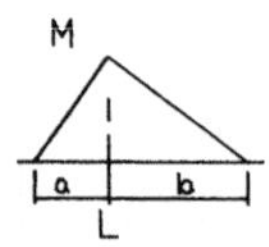	$\frac{L}{2} M\, m_1$	$\frac{L}{6} M\, m_1 (1 + \frac{b}{L})$	$\frac{L}{6} M[m_1 (1 + \frac{b}{L}) + m_2 (1 + \frac{a}{L})]$

Table B1.5
(continued)

The Integral: $\int_0^L g_1(x).g_2(x)\,dx$			
$g_1(x)$	$g_2(x)$ 2		
(all second degree parabolas)	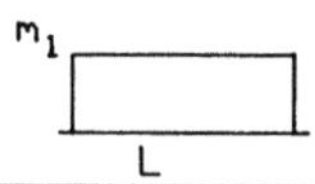	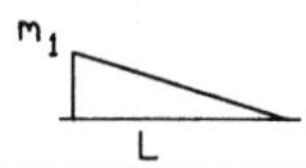	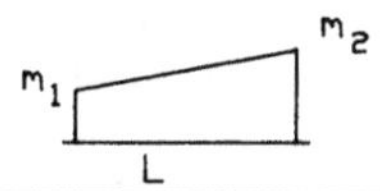
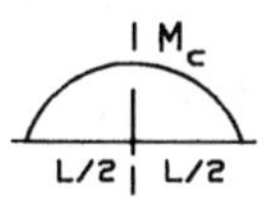	$\frac{2L}{3} M_c m_1$	$\frac{L}{3} M_c m_1$	$\frac{L}{3} [M_c(m_1+m_2)]$
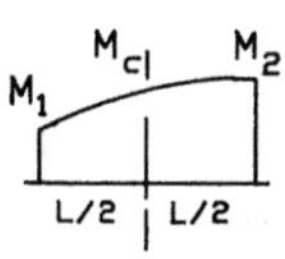	$\frac{L}{6} [(M_1+4M_c+M_2)m_1]$	$\frac{L}{6} [(M_1+2M_c)m_1]$	$\frac{L}{6} [(M_1+2M_c)m_1 + (2M_c+M_2)m_2]$
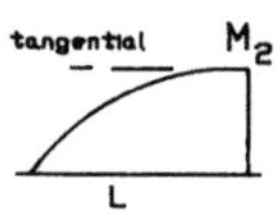	$\frac{2L}{3} M_2 m_1$	$\frac{L}{4} M_2 m_1$	$\frac{L}{12} [M_2(3m_1+5m_2)]$
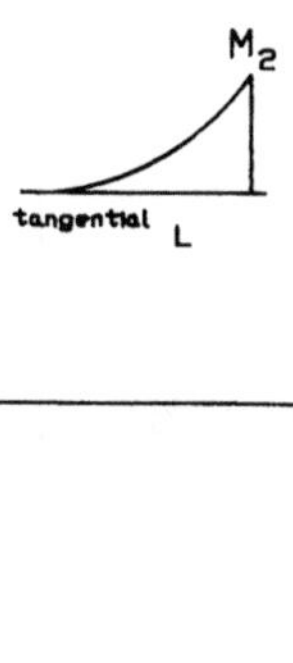	$\frac{L}{3} M_2 m_1$	$\frac{L}{12} M_2 m_1$	$\frac{L}{12} [M_2(m_1+3m_2)]$

Table B1.6 Second moments of area

SHAPE	Ixx	Iyy
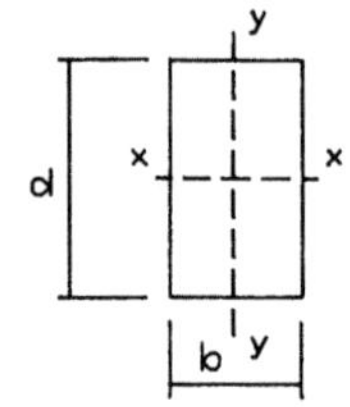	$bd^3/12$	$db^3/12$
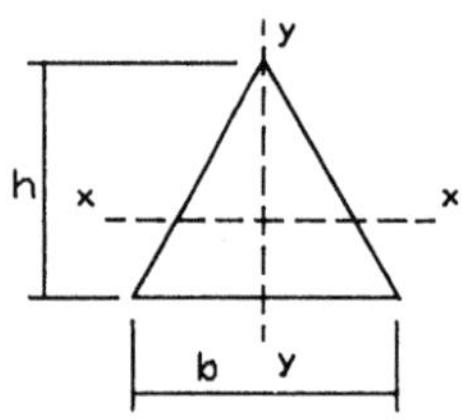	$bh^3/36$	$hb^3/48$
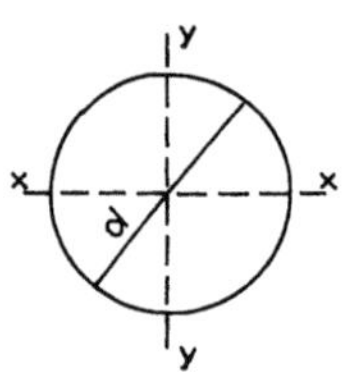	$\pi d^4/64$	$\pi d^4/64$
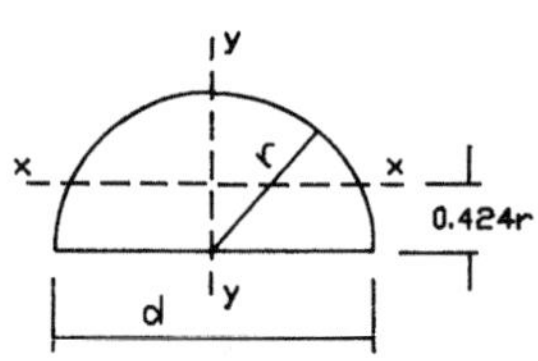	$0.1098r^4$	$\pi d^4/128$
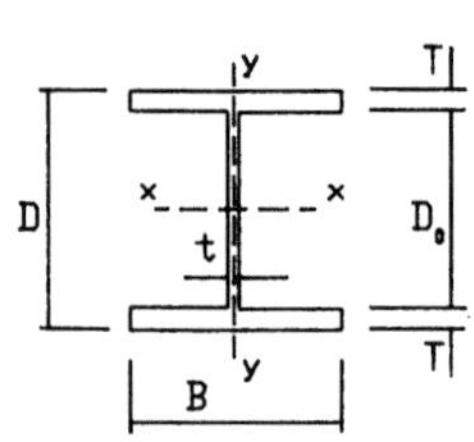	$B(D^3-D_0^3)/12+tD_0^3/12$	$2TB^3/12+D_0t^3/12$

B1.2 Summary of the Slope–Deflection Equations

In the linear elastic analysis of beams and frames, the slope–deflection equations represent a powerful technique. For a beam element subjected to end moments and shear only, the equations may be written as

$$m_{12}=\frac{4EI}{L}\theta_1+\frac{2EI}{L}\theta_2-\frac{6EI}{L^2}\Delta$$

$$m_{21}=\frac{2EI}{L}\theta_1+\frac{4EI}{L}\theta_2-\frac{6EI}{L^2}\Delta$$

Since, in general $\Delta=d_2-d_1$, the equations become

$$m_{12}=\frac{4EI}{L}\theta_1+\frac{2EI}{L}\theta_2+\frac{6EI}{L^2}d_1-\frac{6EI}{L^2}d_2$$

$$m_{21}=\frac{2EI}{L}\theta_1+\frac{4EI}{L}\theta_2+\frac{6EI}{L^2}d_1-\frac{6EI}{L^2}d_2$$

For beams subject to end rotations only, obviously the equations become

$$m_{12}=\frac{4EI}{L}\theta_1+\frac{2EI}{L}\theta_2$$

$$m_{21}=\frac{2EI}{L}\theta_1+\frac{4EI}{L}\theta_2$$

The sign convention is again defined by reference to the following diagram where all the terms are shown in the POSITIVE sense.

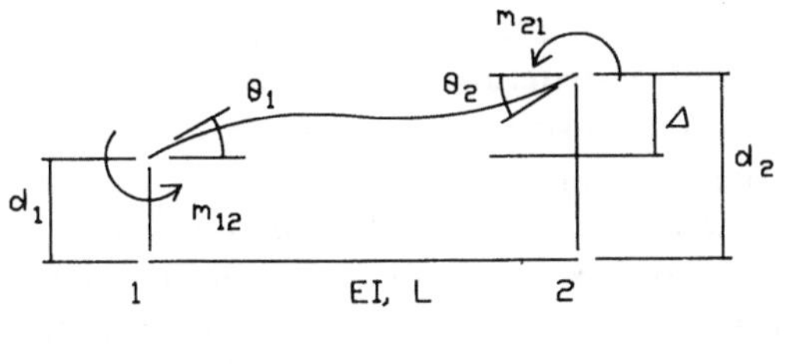

B1.3 Use of Table B1.4—Fixed End Moment Calculation

In conjunction with the slope–deflection equations, table B1.4 may be used to calculate some more general fixed end moments for beams. If the free span end rotations under transverse loads are known, then the fixed end moment must be that moment that would counteract the end rotation. This can be readily calculated by substituting for the given end rotation, with a change in sign, in the slope–deflection equations.

For example, consider the case of a beam under a general point load. From table B1.4:

$$\theta_1 = -\frac{Qb(L^2-b^2)}{6EIL}$$

$$\theta_2 = +\frac{Qa(L^2-a^2)}{6EIL}$$

Hence

$$m_{12} = \frac{4EI}{L}\left(\frac{Qb(L^2-b^2)}{6EIL}\right) + \frac{2EI}{L}\left(-\frac{Qa(L^2-a^2)}{6EIL}\right)$$

$$= \frac{Q}{L^2}\left(\frac{2b(L^2-b^2)-a(L^2-a^2)}{3}\right)$$

$$= \frac{Q}{L^2}\left(\frac{a^3+2bL^2-aL^2-2b^3}{3}\right)$$

which reduces to

$$m_{12} = +\frac{Qab^2}{L^2}$$

as required.

In general, the information is best handled numerically and it is not expected that algebraic expressions would be evaluated.

PART 2

B2.1 Introduction—Convention

Element stiffness matrices are used to describe the relationship between the actions on the nodes of an element of a structure and the corresponding displacements at the nodes in response to those actions. In this context the element stiffness matrix provides the link between the element actions and element displacements such that when the element displacement vector is pre-multiplied by the element stiffness matrix, the element actions are given.

The derivation of the element stiffness matrix is a function of both the element type, the governing stress–strain law and equilibrium. Significantly, the form of the resulting element stiffness matrix for a given type is also dependent on the sign convention adopted in defining the problem and the way in which the displacement and action vectors are defined. For this reason, in each case where an element stiffness matrix is presented, it is given in the matrix relationship between the action and the displacement vector. In addition, the specific terms are also defined in an accompanying diagram where each action and displacement is shown in the POSITIVE sense. On this basis the diagrams also serve to define the sign convention.

The beam element stiffness matrix is presented in three forms. Firstly for the standard continuous beam, then for the same element with a moment release at the left hand end, and finally with a moment release at the right hand end. The element stiffness matrix for a column element, ignoring axial deformation, is then presented. This may be used effectively in the analysis of simple rectangular frames which are considered to be axially rigid. A second form of the matrix allows for a moment release at the base.

The general plane frame element stiffness matrix is then given as a six by six matrix. This matrix includes axial deformation and may be used in a general two dimensional analysis of a rigid jointed plane frame. The appropriate plane truss element stiffness matrix may be deduced from the matrix given and a similar matrix is given for use in the analysis of plane grids. The plane frame coordinate transformation matrix accompanies the presentation of the plane frame element stiffness matrix, and the coordinate transformation matrix for a plane truss analysis can be seen as a subset of the given matrix. All of the preceding matrices can be seen as a subset of the twelve by twelve general space frame element stiffness matrix, which is given to conclude the Appendix. Attention is also drawn to the general form of the transformation matrix from which preceding transformation matrices can be found.

B2.2 Continuous Beam Element

The element stiffness matrix for a beam element is given by the following relationship:

$$\begin{Bmatrix} v_{12} \\ m_{12} \\ \hline v_{21} \\ m_{21} \end{Bmatrix} = \left[\begin{array}{cc|cc} \dfrac{12EI}{L^3} & \dfrac{6EI}{L^2} & -\dfrac{12EI}{L^3} & \dfrac{6EI}{L^2} \\ \dfrac{6EI}{L^2} & \dfrac{4EI}{L} & -\dfrac{6EI}{L^2} & \dfrac{2EI}{L} \\ \hline -\dfrac{12EI}{L^3} & -\dfrac{6EI}{L^2} & \dfrac{12EI}{L^3} & -\dfrac{6EI}{L^2} \\ \dfrac{6EI}{L^2} & \dfrac{2EI}{L} & -\dfrac{6EI}{L^2} & \dfrac{4EI}{L} \end{array}\right] \begin{Bmatrix} d_1 \\ \theta_1 \\ \hline d_2 \\ \theta_2 \end{Bmatrix}$$

where the terms are defined by

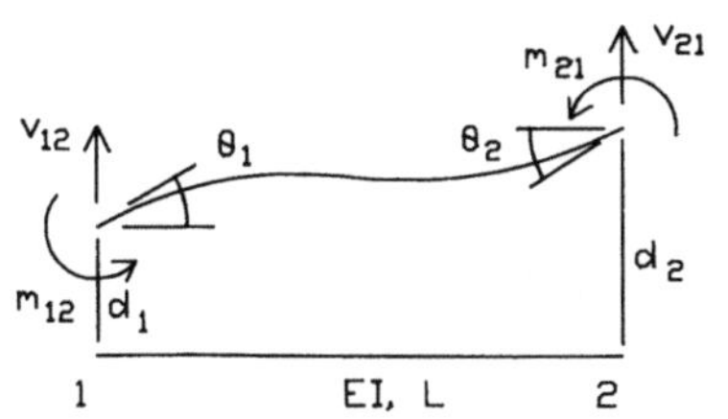

B2.3 Continuous Beam Element—LHE-pinned Moment Release

The element stiffness matrix is given by the following relationship:

$$\begin{Bmatrix} v_{12} \\ m_{12} \\ \hline v_{21} \\ m_{21} \end{Bmatrix} = \left[\begin{array}{cc|cc} \frac{3EI}{L^3} & 0 & -\frac{3EI}{L^3} & \frac{3EI}{L^2} \\ 0 & 0 & 0 & 0 \\ \hline -\frac{3EI}{L^3} & 0 & \frac{3EI}{L^3} & -\frac{3EI}{L^2} \\ \frac{3EI}{L^2} & 0 & -\frac{3EI}{L^2} & \frac{3EI}{L} \end{array}\right] \begin{Bmatrix} d_1 \\ \theta_1 \\ \hline d_2 \\ \theta_2 \end{Bmatrix}$$

where the terms are defined by

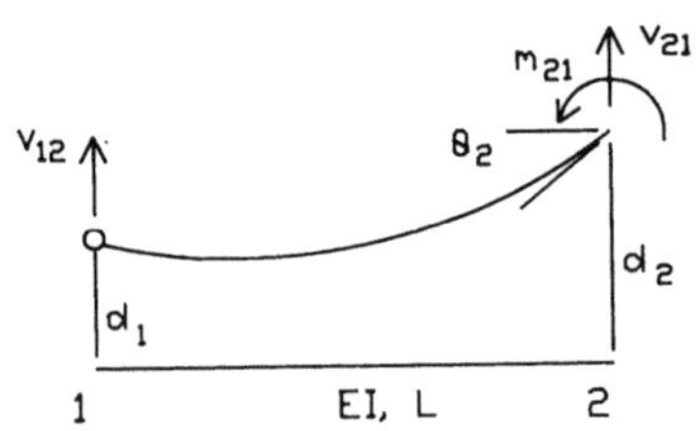

B2.4 Continuous Beam Element—RHE-pinned Moment Release

The element stiffness matrix is given by the following relationship:

$$\begin{Bmatrix} v_{12} \\ m_{12} \\ \hline v_{21} \\ m_{21} \end{Bmatrix} = \left[\begin{array}{cc|cc} \frac{3EI}{L^3} & \frac{3EI}{L^2} & -\frac{3EI}{L^3} & 0 \\ \frac{3EI}{L^2} & \frac{3EI}{L} & -\frac{3EI}{L^2} & 0 \\ \hline -\frac{3EI}{L^3} & -\frac{3EI}{L^2} & \frac{3EI}{L^3} & 0 \\ 0 & 0 & 0 & 0 \end{array}\right] \begin{Bmatrix} d_1 \\ \theta_1 \\ \hline d_2 \\ \theta_2 \end{Bmatrix}$$

where the terms are defined by

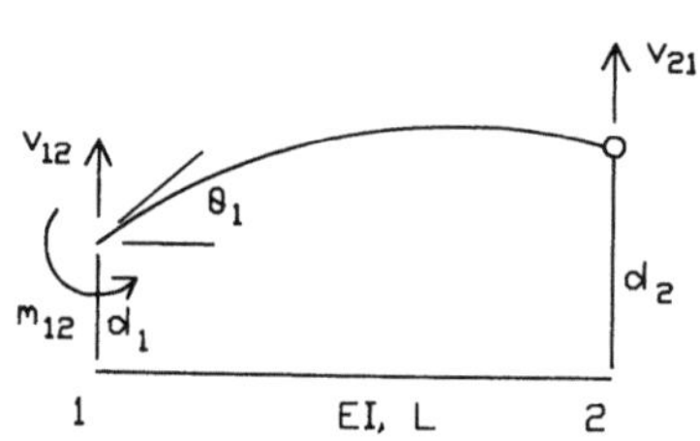

B2.5 Column Element

The element stiffness matrix for a column element, ignoring axial deformation, is given by the following relationship:

$$\begin{Bmatrix} v_{12} \\ m_{12} \\ \hline v_{21} \\ m_{21} \end{Bmatrix} = \left[\begin{array}{cc|cc} \frac{12EI}{L^3} & -\frac{6EI}{L^2} & -\frac{12EI}{L^3} & -\frac{6EI}{L^2} \\ -\frac{6EI}{L^2} & \frac{4EI}{L} & \frac{6EI}{L^2} & \frac{2EI}{L} \\ \hline -\frac{12EI}{L^3} & \frac{6EI}{L^2} & \frac{12EI}{L^3} & \frac{6EI}{L^2} \\ -\frac{6EI}{L^2} & \frac{2EI}{L} & \frac{6EI}{L^2} & \frac{4EI}{L} \end{array}\right] \begin{Bmatrix} d_1 \\ \theta_1 \\ \hline d_2 \\ \theta_2 \end{Bmatrix}$$

where the terms are defined by

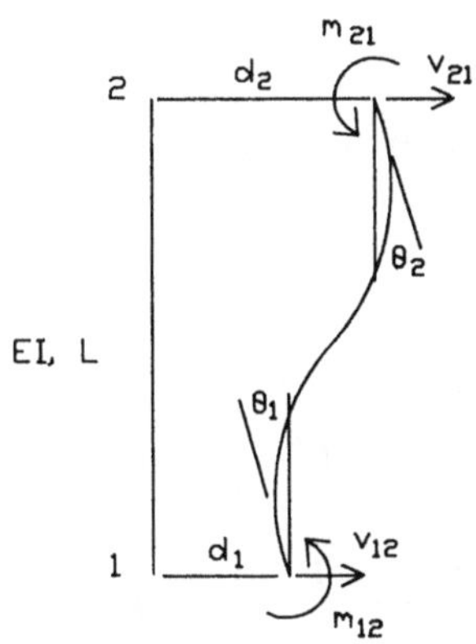

B2.6 Column Element—Base-pinned

The element stiffness matrix for a column element, ignoring axial deformation and having a pinned base, is given by the following relationship:

$$\begin{Bmatrix} v_{12} \\ m_{12} \\ \hline v_{21} \\ m_{21} \end{Bmatrix} = \left[\begin{array}{cc|cc} \frac{3EI}{L^3} & 0 & -\frac{3EI}{L^3} & -\frac{3EI}{L^2} \\ 0 & 0 & 0 & 0 \\ \hline -\frac{3EI}{L^3} & 0 & \frac{3EI}{L^3} & \frac{3EI}{L^2} \\ -\frac{3EI}{L^2} & 0 & \frac{3EI}{L^2} & \frac{3EI}{L} \end{array}\right] \begin{Bmatrix} d_1 \\ \theta_1 \\ \hline d_2 \\ \theta_2 \end{Bmatrix}$$

where the terms are defined by

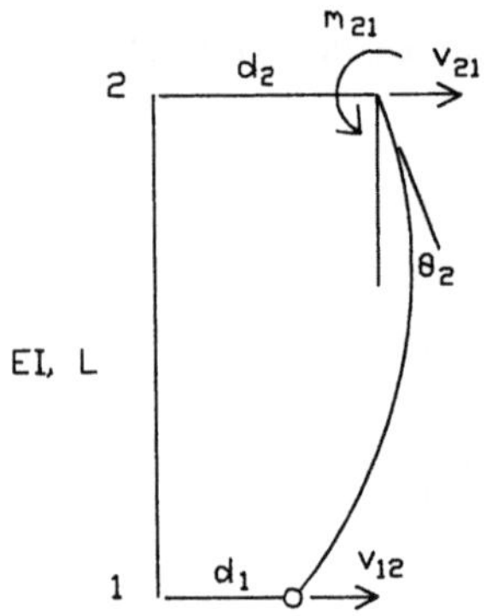

B2.7 General Plane Frame Element

The element stiffness matrix for a plane frame element is given by the following relationship:

$$\begin{Bmatrix} p_{12} \\ v_{12} \\ m_{12} \\ \hline p_{21} \\ v_{21} \\ m_{21} \end{Bmatrix} = \left[\begin{array}{ccc|ccc} \frac{EA}{L} & 0 & 0 & -\frac{EA}{L} & 0 & 0 \\ 0 & \frac{12EI}{L^3} & \frac{6EI}{L^2} & 0 & -\frac{12EI}{L^3} & \frac{6EI}{L^2} \\ 0 & \frac{6EI}{L^2} & \frac{4EI}{L} & 0 & -\frac{6EI}{L^2} & \frac{2EI}{L} \\ \hline -\frac{EA}{L} & 0 & 0 & \frac{EA}{L} & 0 & 0 \\ 0 & -\frac{12EI}{L^3} & -\frac{6EI}{L^2} & 0 & \frac{12EI}{L^3} & -\frac{6EI}{L^2} \\ 0 & \frac{6EI}{L^2} & \frac{2EI}{L} & 0 & -\frac{6EI}{L^2} & \frac{4EI}{L} \end{array}\right] \begin{Bmatrix} s_1 \\ d_1 \\ \theta_1 \\ \hline s_2 \\ d_2 \\ \theta_2 \end{Bmatrix}$$

where the terms are defined by

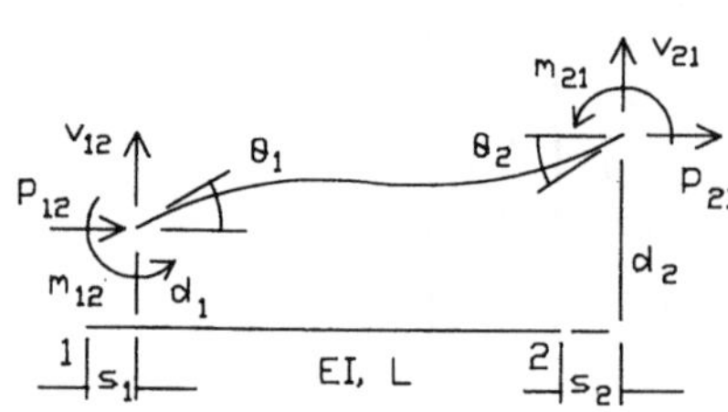

Plane Frame Transformation Matrix

The coordinate transformation matrix allowing for transformation between element actions or displacements expressed in global coordinates, and the same actions or displacements expressed in local coordinates is given by:

$$\left[\begin{array}{ccc|ccc} \cos\alpha & \sin\alpha & 0 & 0 & 0 & 0 \\ -\sin\alpha & \cos\alpha & 0 & 0 & 0 & 0 \\ 0 & 0 & 1 & 0 & 0 & 0 \\ \hline 0 & 0 & 0 & \cos\alpha & \sin\alpha & 0 \\ 0 & 0 & 0 & -\sin\alpha & \cos\alpha & 0 \\ 0 & 0 & 0 & 0 & 0 & 1 \end{array}\right]$$

where the global and local axes are defined by

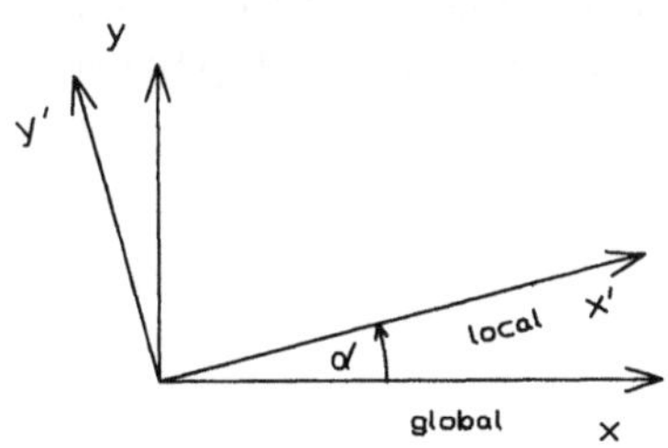

B2.8 Plane Grid Element

The element stiffness matrix for a plane grid element is given by the following relationship:

$$\left\{\begin{array}{c} t_{12} \\ m_{12} \\ v_{12} \\ \hline t_{21} \\ m_{21} \\ v_{21} \end{array}\right\} = \left[\begin{array}{ccc|ccc} \frac{GJ}{L} & 0 & 0 & -\frac{GJ}{L} & 0 & 0 \\ 0 & \frac{4EI}{L} & -\frac{6EI}{L^2} & 0 & \frac{2EI}{L} & \frac{6EI}{L^2} \\ 0 & -\frac{6EI}{L^2} & \frac{12EI}{L^3} & 0 & -\frac{6EI}{L^2} & -\frac{12EI}{L^3} \\ \hline -\frac{GJ}{L} & 0 & 0 & \frac{GJ}{L} & 0 & 0 \\ 0 & \frac{2EI}{L} & -\frac{6EI}{L^2} & 0 & \frac{4EI}{L} & \frac{6EI}{L^2} \\ 0 & \frac{6EI}{L^2} & -\frac{12EI}{L^3} & 0 & \frac{6EI}{L^2} & \frac{12EI}{L^3} \end{array}\right] \left\{\begin{array}{c} \theta_{x_1} \\ \theta_{y_1} \\ d_1 \\ \hline \theta_{x_2} \\ \theta_{y_2} \\ d_2 \end{array}\right\}$$

where the terms are defined by:

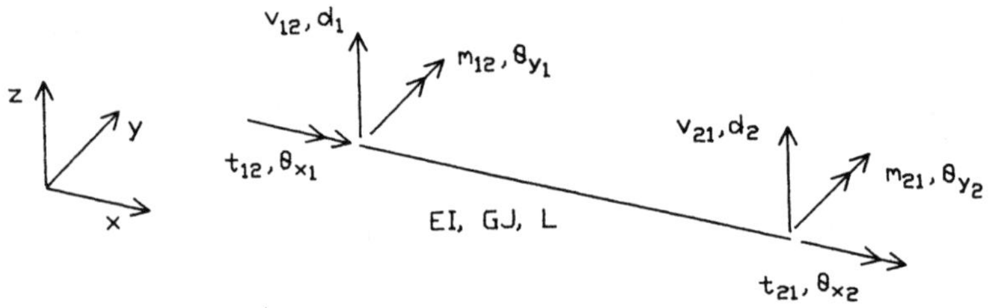

Plane Grid Transformation Matrix

For the plane grid element stiffness matrix expressed in the above manner, the coordinate transformation matrix is identical to that given for the plane frame element. The angle α is defined in the following relationship between the local and global axes:

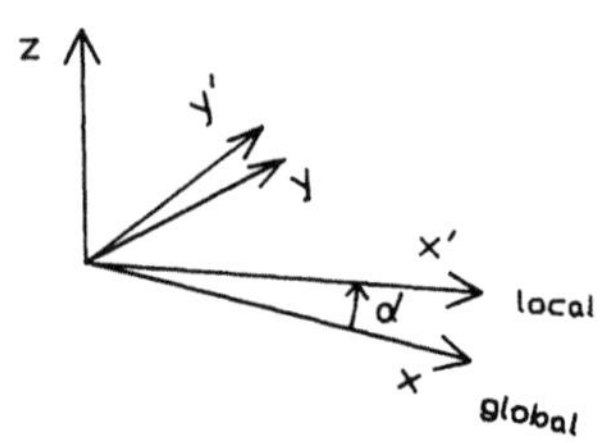

B2.9 Space Frame Element

The element stiffness matrix for a space frame element is given by the following relationship:

$$\begin{Bmatrix} p_{x_1} \\ p_{y_1} \\ p_{z_1} \\ m_{x_1} \\ m_{y_1} \\ m_{z_1} \\ \hline p_{x_2} \\ p_{y_2} \\ p_{z_2} \\ m_{x_2} \\ m_{y_2} \\ m_{z_2} \end{Bmatrix} = \left[\begin{array}{c|c} k_{11} & k_{12} \\ \hline k_{21} & k_{22} \end{array} \right] \begin{Bmatrix} d_{x_1} \\ d_{y_1} \\ d_{z_1} \\ \theta_{x_1} \\ \theta_{y_1} \\ \theta_{z_1} \\ \hline d_{x_2} \\ d_{y_2} \\ d_{z_2} \\ \theta_{x_2} \\ \theta_{y_2} \\ \theta_{z_2} \end{Bmatrix}$$

where the terms are defined by

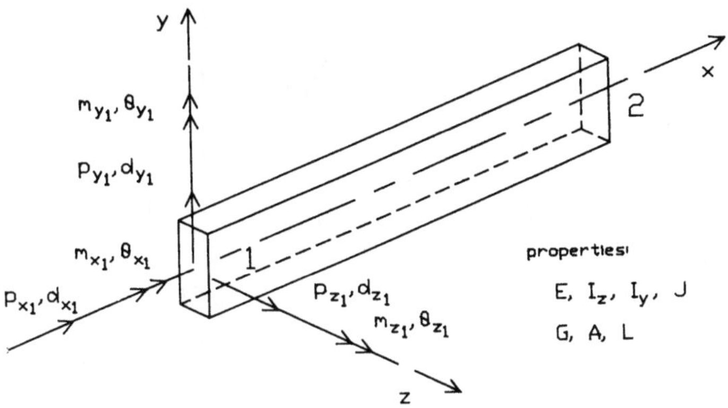

and a general notation for element actions and displacements has been introduced, similar to that given in reference [B.1].

Further:

$$k_{11} = \begin{bmatrix} \frac{EA}{L} & 0 & 0 & 0 & 0 & 0 \\ 0 & \frac{12EI_z}{L^3} & 0 & 0 & 0 & \frac{6EI_z}{L^2} \\ 0 & 0 & \frac{12EI_y}{L^3} & 0 & -\frac{6EI_y}{L^2} & 0 \\ 0 & 0 & 0 & \frac{GJ}{L} & 0 & 0 \\ 0 & 0 & -\frac{6EI_y}{L^2} & 0 & \frac{4EI_y}{L} & 0 \\ 0 & \frac{6EI_z}{L^2} & 0 & 0 & 0 & \frac{4EI_z}{L} \end{bmatrix}$$

$k_{22} = k_{11}$, but with the signs of the off diagonal elements reversed. And

$$k_{12} = \begin{bmatrix} -\frac{EA}{L} & 0 & 0 & 0 & 0 & 0 \\ 0 & -\frac{12EI_z}{L^3} & 0 & 0 & 0 & \frac{6EI_z}{L^2} \\ 0 & 0 & -\frac{12EI_y}{L^3} & 0 & -\frac{6EI_y}{L^2} & 0 \\ 0 & 0 & 0 & -\frac{GJ}{L} & 0 & 0 \\ 0 & 0 & \frac{6EI_y}{L^2} & 0 & \frac{2EI_y}{L} & 0 \\ 0 & -\frac{6EI_z}{L^2} & 0 & 0 & 0 & \frac{2EI_z}{L} \end{bmatrix}$$

with $k_{21} = k_{12}^{\mathrm{T}}$.

The space frame transformation matrix operating on the 12 by 12 element stiffness matrix has the form

$$T = \left[\begin{array}{c|c} R & 0 \\ \hline 0 & R \end{array}\right] (12 \times 12)$$

where

$$R = \left[\begin{array}{c|c} R_0 & 0 \\ \hline 0 & R_0 \end{array}\right] \quad \text{and} \quad R_0 = \begin{bmatrix} l_1 & m_1 & n_1 \\ l_2 & m_2 & n_2 \\ l_3 & m_3 & n_3 \end{bmatrix}$$

and l, m and n, $i = 1, 3$, are the direction cosines of the local x, y and z axes respectively with respect to the global x, y and z axes respectively.

A more convenient expression of the matrix R is given as:

$$R_0 = \begin{bmatrix} L_x/L & L_y/L & L_z/L \\ -\frac{L_xL_y \cos\gamma - LL_z \sin\gamma}{L\sqrt{(L_x^2 + L_z^2)}} & \frac{\sqrt{[(L_x^2 + L_z^2)]} \cos\gamma}{L} & \frac{(-L_yL_z \cos\gamma + LL_x \sin\gamma)}{L\sqrt{(L_x^2 + L_z^2)}} \\ \frac{L_xL_y \sin\gamma - LL_z \cos\gamma}{L\sqrt{(L_x^2 + L_z^2)}} & -\frac{\sqrt{[(L_x^2 + L_z^2)]} \sin\gamma}{L} & \frac{L_yL_z \sin\gamma + LL_x \cos\gamma}{L\sqrt{(L_x^2 + L_z^2)}} \end{bmatrix}$$

where L_x, L_y and L_z are the projections of the length of the element on to the global x, y and z axes respectively, and γ is the angle between the global axes x–y plane and the element axes x–y plane.

Users should note that this form of the matrix R_0 is indeterminate when the centroidal x-axis of an element is coincident with the global y-axis. Further details are given in reference [B.1].

Reference

B.1. Coates, R. C., Coutie, M. G. and Kong, F. K., *Structural Analysis*, 2nd edn, Van Nostrand Reinhold, London, 1980.

Index

MIX
Papier aus verantwortungsvollen Quellen
Paper from responsible sources
FSC® C105338

If you have any concerns about our products,
you can contact us on
ProductSafety@springernature.com

In case Publisher is established outside the EU,
the EU authorized representative is:
Springer Nature Customer Service Center GmbH
Europaplatz 3, 69115 Heidelberg, Germany

Printed by Libri Plureos GmbH
in Hamburg, Germany